LABORATORY MANUAL
for
HUMAN BIOLOGY

2nd Edition

**David Morton
Joy B. Perry
and James W. Perry**

BROOKS/COLE
CENGAGE Learning™

Australia • Brazil • Japan • Korea • Mexico • Singapore • Spain • United Kingdom • United States

BROOKS/COLE
CENGAGE Learning™

Laboratory Manual for Human Biology, Second Edition
David Morton, Joy B. Perry, and
James W. Perry

Publisher/Executive Editor: Yolanda Cossio

Senior Acquisitions Editor: Peggy Williams

Developmental Editor: Elizabeth Momb

Editorial Assistant: Shana Baldassari

Marketing Manager: Tom Ziolkowski

Marketing Communications Manager: Linda Yip

Content Project Management: PreMediaGlobal

Creative Director: Rob Hugel

Art Director: John Walker

Senior Print Buyer: Judy Inouye

Permissions Acquisitions Manager:
Don Schlotman

Cover Image: © LWA-Dann Tardif/Corbis

Compositor: PreMediaGlobal

For product information and technology assistance, contact us at
Cengage Learning Customer & Sales Support, 1-800-354-9706
For permission to use material from this text or product,
submit all requests online at **www.cengage.com/permissions.**
Further permissions questions can be emailed to
permissionrequest@cengage.com

Library of Congress Control Number: 2010938242

Student Edition:

ISBN-13: 978-0-8400-4943-8

ISBN-10: 0-8400-4943-9

Brooks/Cole
20 Davis Drive
Belmont, CA 94002–3098
USA

Cengage Learning is a leading provider of customized learning solutions with office locations around the globe, including Singapore, the United Kingdom, Australia, Mexico, Brazil, and Japan. Locate your local office at:
international.cengage.com/region

Cengage Learning products are represented in Canada by Nelson Education, Ltd.

For your course and learning solutions, visit **www.cengage.com.**

Purchase any of our products at your local college store or at our preferred online store **www.cengagebrain.com.**

Printed in the Canada
1 2 3 4 5 6 7 14 13 12 11 10

Dedications

To my loving granddaughters—Catherine, Gwendolyn, Emma and Sarah Morton, who inspire me with their fresh sense of wonder about our world.
—*David Morton*

Dedicated to my parents, Alfred and Catherine Bissonnette, who gave me a deep love of nature and the opportunity to explore it.
—*Joy B. Perry*

In memory of Aldo Leopold, who died the year I was born, but who changed my life as an undergraduate with his essay "Thinking Like a Mountain."
—*James W. Perry*

CONTENTS

Greetings from the authors! We're happy that you are examining our work. We predict you'll find the information below an invaluable introduction to the second edition of this laboratory manual.

You may be familiar with our well-tested *Laboratory Manual for General Biology* for Starr's texts. This lab manual continues that successful format. In the second edition, we have improved readability, streamlined the exercises, and sharpened the focus on recent developments in human biology.

This manual is designed for students at the college entry level and assumes no previous college biology or chemistry courses. Many students plan a course in human biology as their only college biology experience. Others build on this experience with further courses such as anatomy and physiology or genetics. Our exercises provide the foundation necessary for today's diverse students to be successful in human biology and to make informed decisions about biological questions in their everyday lives.

What's Important to Us

In preparing this lab manual, we paid particular attention to pedagogy, clarity of procedures and terminology, illustrations, and practicality of materials used. The equipment and supplies used in the exercises are readily available from biological and laboratory supply houses. We've attempted to keep instrumentation and materials as simple and inexpensive as possible.

We believe strongly that all students should have a functional understanding of the methods of science. Accordingly, in many exercises, we emphasize data acquisition and analysis, as well as hypothesis testing.

Pedagogy

Conceptually, we wrote the lab manual to include the knowledge necessary to seamlessly continue to higher level courses. Within each exercise information flows from the stated objectives to the laboratory activities and, finally, to the Post-Lab Questions.

We worked to make each portion of the exercise part of a continuous thought. Thus, we do not wait until the Post-Lab Questions to ask students to record pertinent information including conclusions when it is more appropriate to do so within the body of the procedure. Connections of exercise portions to each other and to other topics are clearly stated.

The exercises in this human biology lab manual are written so conscientious students can accomplish the objectives of each exercise with minimal guidance from an instructor. The Procedure sections of the exercises are more detailed and step by step than in most other manuals. Instructions follow a natural progression of thought so the instructor need not conduct every movement.

We realize there is wide variation in the amount of time each instructor has to devote to laboratory activities. To provide maximum flexibility for the instructor, the procedure portions of the exercises are divided by major headings. After the introduction has been studied, portions of the procedures can be deleted or presented as demonstrations without sacrificing the pedagogy of the exercise as a whole.

It's our experience that if discussion of the topic has occurred before the lab is begun, students find the exercise much more relevant and understandable. We create this situation in our courses, and thus no time is spent on a lecture-style introduction in the lab itself before the exercise begins.

Terms required to accomplish objectives are **boldfaced**. Scientific names, precautionary statements, and other terms to be emphasized are in *italics*.

Most illustrations of microscopic specimens are labeled to provide orientation and clarity. A few are unlabeled but are provided with leaders to which students can attach labels. In other cases, more can be gained by requesting that students do simple drawings. Space has been included in the manual for these, with boxes for drawings of macroscopic specimens and circles for microscopic specimens.

We included flexible quiz options to allow you to assess your students' preparedness and learning. Each exercise has a set of Pre-Lab Questions. We have found, through nearly 40 years of experience, that students left to their own initiative typically come to the laboratory unprepared to do the exercise. Few read the exercise beforehand. One answer to this is to incorporate a pre-exercise activity.

In our own courses, we have students take the pre-lab quiz consisting of the questions in their lab manual arranged in a different order to discourage memorization. These quizzes are available in our Instructor's Manual. Our quiz takes literally ten minutes of lab time and counts as a portion of the lab grade, thus rewarding students for preparation.

Other instructors have told us they use the pre-lab questions to assess learning after the exercise has been completed. We encourage you to be creative with the manual; do what works best for you and your students.

Text Organization

Each exercise includes:

1. **Objectives:** A list of desired outcomes

2. **Introduction:** To pique student interest, indicate relevance, and provide background

3. **Materials:** A list for each portion of the exercise so that a student can quickly gather the necessary supplies. Materials are listed "per student," "per pair," "per group," and "per lab room."

4. **Procedures:** Including safety notes, illustrations of apparatus, figures to be labeled, drawings to be made, tables for recording data, graphs to be drawn, and questions that lead to conclusions. The procedures are listed in easy-to-follow numbered steps.

5. **Pre-Lab Questions:** Ten multiple-choice questions that students should be able to answer after reading the exercise but before entering the laboratory

6. **Post-Lab Questions:** That draw upon knowledge gained from doing the exercise and that the student should be able to answer after finishing the exercise. These questions assess both recall and understanding.

   We believe the Post-Lab Questions should draw on the knowledge gained by observations. Consequently, we've incorporated illustrations into the Post-Lab Questions. These illustrations typically are similar, but not identical, to those in the procedures. Thus, they assess students' ability to apply knowledge gained during the exercise to a new situation.

New Features in This Edition

In response to reviewers' suggestions and our personal convictions, we have incorporated practice in the methods of science throughout the manual. In addition, students are often directed to search for current topical information online. We scrutinized every exercise for opportunities to make the writing clearer and more succinct while updating content to reflect current knowledge. We summarize the highlights of changes for each exercise here:

- **EXERCISE 1:** Learning About the Natural World—This exercise represents a major expansion of an appendix from the previous edition. Students are asked to perform short, noninvasive experiments and observations on group members in the context of the various elements of the scientific method. Also, they are asked to analyze scientific articles.

- **EXERCISE 2:** Observing Microscopic Details of the Natural World—The text was shortened to increase clarity and to stress more practical aspects of the use

of microscopes to examine the natural world. The number of illustrations was increased significantly.

- **EXERCISE 3:** Chemistry of Life: Biological Molecules and Diet Analysis—Dietary guidelines and recommendations have been completely updated and now culminate in a new visual comparison of students' actual versus recommended food group consumption.

- **EXERCISE 4:** Cell Structure—A streamlined focus considers the structure and function of eukaryotic cells only.

- **EXERCISE 5:** Cell Membranes and How They Work—Substantial revisions include a streamlined diffusion and osmosis demonstration. A new section applying principles of osmosis to cholera symptoms and physiology reinforces the relevance of membrane function to health.

- **EXERCISE 6:** Enzymes, the Catalysts of Life—This exercise has been completely revised using a simple, quantitative assay of catalase activity. After learning the method, students design and evaluate tests of hypotheses regarding factors affecting catalase activity.

- **EXERCISE 7:** Homeostasis—A section has been added on the taking of blood pressure and its importance as a homeostatic variable. Again, the text was edited to increase clarity. The concept of feedforward is introduced.

- **EXERCISE 8:** How Mammals Are Constructed—Some illustrations were replaced by better ones. The section on organ analysis was removed, and in its place, material stressing greater emphasis on organ identification was added.

- **EXERCISE 9:** Support and Movement: Human Skeletal and Muscular Systems—The text was shortened to increase clarity. Many images of the human skeleton were added to help students' recognition and identification of bone organs.

- **EXERCISE 10:** External Anatomy and Organs of the Digestive and Respiratory Systems—The external anatomy was added along with many images of fetal pig dissections. Students and their instructors were given choices other than actual dissection of a pig. Additional alternatives were to use instructor dissections or human torso models.

- **EXERCISE 11:** Organs of the Circulatory, Urinary, and Reproductive Systems—The circulatory system was moved from the previous exercise to this one to accommodate the addition of a section on external features. The dissection of the nervous system was dropped to make room and because most of its concepts were covered elsewhere in the manual.

- **EXERCISE 12:** Human Blood and Circulation—The text was extensively edited for brevity and clarity. Several images were replaced.

- **EXERCISE 13:** Human Respiration—The text was extensively edited for brevity and clarity and the modernization of concepts. Several images were replaced.

- **EXERCISE 14: Human Sensations, Reflexes, and Reactions**—The text was extensively edited for brevity and clarity. Several images were replaced.

- **EXERCISE 15: Structure and Function of the Sensory Organs**—The text was extensively edited for brevity and clarity. Several images were replaced, especially in the section on the eye and vision.

- **EXERCISE 16: Reproduction and Development**—The text was extensively edited for brevity and clarity. Several images were replaced and added.

- **EXERCISE 17: Cell Reproduction**—The focus of this exercise is sharpened by removal of discussion of cancer. Expanded terminology of chromosome structure is introduced, preparing students for exercises in heredity and molecular genetics. Discussion of gametogenesis is reduced to avoid duplication with Exercise 16.

- **EXERCISE 18: Human Heredity**—Exploration and practice in understanding patterns of Mendelian inheritance, especially in humans and other animals, are followed by application of this knowledge to two hereditary human diseases, cystic fibrosis and hemophilia.

- **EXERCISE 19: DNA, Genes, Cancer, and Biotechnology**—Simple, rapid methods to extract and identify DNA from strawberries replace more complicated procedures using bacterial DNA. Consideration of cancers as disorders resulting from faulty gene control is linked to preparation of stained slides of HeLa cells, demonstrating their atypical chromosome numbers. Finally, new biotechnology modules demonstrate the utility of molecular biology techniques to the field of forensic science, replacing a less reliable and more time-consuming bacterial transformation activity.

- **EXERCISE 20: Evidences of Evolution**—This exercise is substantially revised. An experiment on the effects of natural selection on a dandelion population required materials that were difficult to obtain. It has been replaced with an activity demonstrating variability in the effects of natural selection based on characteristics of predators and prey modeled from common food and household items. A discussion of human evolution

has been updated and simplified, focusing on more easily obtained specimens.

- **EXERCISE 21: Human Impact on the Environment**—Begins with a new activity analyzing environmental consequences of converting natural areas to residential and commercial land uses. Methods for assessing stream water quality have been simplified.

Instructor's Manual

There is no need to worry about "Where am I going to get that?" or "How do I prepare this?" Our Instructor's Manual includes:

- Procedures for preparing reagents, materials, and equipment
- Scheduling information for materials needing advance preparation
- Approximate quantities of material needed
- Answers for Pre-Lab Questions
- Answers for Post-Lab Questions
- Answers to Mendelian genetics problems (Exercise 18)
- Tear-out sheets of Pre-Lab Questions in an order different from those in the lab manual for instructors who wish to duplicate and use them as quizzes
- Vendors and item numbers for supplies
- Text and preparation guide for optional experiments and activities

And in the End . . .

We would like to express our special thanks to Elizabeth Momb and other individuals at Cengage Learning involved in this project, including Peggy Williams and Shana Baldassari.

There are very few things in life that are perfect. We don't suppose that this lab manual is one of them. We think your students will enjoy the exercises. Perhaps you and they will find places where rephrasing will make the activity better. Please, contact us to express your opinions and any ideas you wish to share; encourage your students to do likewise.

David Morton
dmorton@frostburg.edu

Joy B. Perry
joy.perry@uwc.edu

James W. Perry
james.perry@uwc.edu

To The Student

Welcome! In this course, you will learn things about yourself and your surroundings that will broaden and enrich your life. You will have the opportunity to marvel at the microscopic world, to be fascinated by the events occurring in your body at this very moment, and to gain an appreciation for the world around you.

We can offer a number of suggestions to make your collegiate experience in biology a pleasant one. The first step toward that goal has been taken by us; we have written a laboratory manual that is "user friendly." You will be able to hear the authors speaking with you as though we were there to share your experience. All of us share a personal belief that the more we make you feel comfortable with us, the more likely you are to share our enthusiasm for biology. It would be naïve for us to suppose that each and every one of you will be science majors at graduation. But one thing we all must realize is that we are citizens of "spaceship Earth." The fate of our spaceship is largely in your hands because you are the decision makers of the future. As has been so aptly stated, "We inherited the earth from our parents and grandparents, but we are only its caretakers for our children and grandchildren."

As caretakers, we need to be informed about the world around us. That's why we enroll in colleges and universities with the hope of gaining a liberal education. In doing so, we establish a basis on which to make educated decisions about the future of the planet. Each exercise in this manual contains a lesson in life that is of a more global nature than the surroundings of your human biology laboratory.

To enhance your biology education, take the initiative to put yourself at the best possible advantage: Don't miss class; read your text assignment routinely; and read the laboratory exercise *before* you come to the lab.

Each exercise in the manual is organized in the same way:

1. *Objectives* tell exactly what you should learn from the exercise. If you wish to know what will be on the exam, consult the objectives for each exercise.

2. The *Introduction* provides you with background information for the exercise and is intended to stimulate your interest.

3. The *Materials List* for each portion of the exercise allows you to determine at a glance whether you have all the necessary supplies needed to do the activity. Check with your instructor if anything is missing.

4. The *Procedure* for each section, in easy-to-follow, step-by-step fashion, describes the activity. Within the procedure, spaces are provided to make required drawings, answer in-text questions, describe observations, and record data. Some in-text questions are posed asking you to draw conclusions about an activity you are engaged in. You'll find a lot of illustrations, most of which are labeled, and others that are not but have leaders for you to attach labels. The terms to be used as labels are found in the procedure and in a list accompanying the illustration. Sometimes we believe it is best for you to make a simple drawing, and we have inserted boxes or circles for your sketches. When appropriate, tables and graphs are present for recording your data.

4. *Pre-Lab Questions* can be answered easily by simply reading the exercise. They're meant to "set the stage" for the lab period by emphasizing some of the more salient points.

5. *Post-Lab Questions* are intended to be done after the laboratory is completed. Some are straightforward interpretations of what you have done, but others require additional thought and perhaps some research in your textbook or on the Internet. In fact, some have no "right" or "wrong" answer at all!

In our experience, students are much too reluctant to ask questions for fear of appearing stupid. Remember, there is no such thing as a stupid question. Speak up! Think of yourselves as "basic learners" and your instructors as "advanced learners." Interact and ask questions so that you and your instructors can further your respective educations.

Laboratory Supplies and Procedures

Materials and Supplies Kept in the Lab at All Times

The materials listed below will likely always be available in the lab room. Familiarize yourself with their location prior to beginning the exercises.

- Compound light microscopes
- Dissection microscopes
- Glass microscope slides
- Coverslips
- Lens paper
- Tissue wipes
- Plastic 15-cm rulers
- Dissecting needles
- Razor blades
- Glassware of various types and sizes
- Assorted glassware-cleaning brushes
- Detergent for washing glassware
- Distilled water
- Hand soap
- Paper towels
- Safety equipment (see separate list)

Laboratory Safety

None of the exercises in this manual are inherently dangerous. Some of the chemicals are corrosive (causing burns to the skin); others are poisonous if ingested or inhaled in large amounts. Contact with your eyes by otherwise innocuous substances may result in permanent eye injury; *remember, when your sight is lost, it's probably lost forever*. Locate the safety items described below and then study the list of basic safety rules.

1. *Shower*
   Do not be put off by the lack of a drain; this isolates any spill to the room in which it occurs.

2. *Eyewash bottle or eye bath*
   If any substance is splashed in your eyes, wash them thoroughly.

3. *Fire extinguisher*
   Read the directions for use of the fire extinguisher.

4. *Fire blanket*
   If someone's clothing catches on fire, wrap the blanket around the individual and roll the person on the floor to smother the flames.

5. *First-aid kit*
   Minor injuries such as small cuts can be treated effectively in the lab. Open the first-aid kit to determine its contents.

6. *Safety goggles*
   Eye protection should be worn during some exercises as directed in your lab manual or by your instructor.

7. *Laboratory gloves*
   Hand protection should be worn as directed in your lab manual or by your instructor.

Safety Rules

1. Do not eat, drink, or smoke in the laboratory.

2. Wash your hands with soap and warm water before leaving the laboratory.

3. When heating a test tube, point the mouth of the tube away from yourself and other people.

4. Always wear shoes in the laboratory.

5. Keep extra books and clothing in designated places so your work area is as uncluttered as possible.

6. If you have long hair, tie it back when in the laboratory.

7. Read labels carefully before removing substances from a container. *Never return a substance to a container.*

8. Discard used chemicals and materials into appropriately labeled containers. Certain chemicals should *not* be washed down the sink; these will be indicated by your instructor.

CAUTION Report all accidents and spills to your instructor immediately!

Instructions for Washing Laboratory Glassware

1. Place contents to be discarded in proper waste container as described in the exercise or as directed by your instructor.

2. Rinse the glassware with tap water.

3. Add a small amount of glassware-cleaning detergent.

4. Scrub the glassware with an appropriately sized brush.

5. Rinse the glassware with tap water until detergent disappears.

6. Rinse the glassware three times with distilled water (dH_2O). When glassware is clean, dH_2O sheets off rather than remaining on the surface in droplets.

7. Allow the glassware to dry in an inverted position on a drying rack (if available).

Learning About the Natural World

OBJECTIVES

After completing this exercise, you will be able to

1. define *scientific method, mechanist, vitalist, cause and effect, induction, deduction, experimental group, control group, independent variable, dependent variable, controlled variables, correlation, theory,* and *principle.*

2. explain the nature of scientific knowledge.

3. describe the basic steps of the scientific method.

4. state the purpose of an experiment.

5. explain the difference between cause and effect and correlation.

6. describe the design of a typical research article in biology.

Introduction

To appreciate biology or, for that matter, any scientific subject, you need to understand how the **scientific method** is used to gather that knowledge. We use the scientific method to test the predictions of possible answers to questions about nature in ways that we can duplicate or verify. Answers supported by test results are added to the body of scientific knowledge and contribute to the concepts presented in your textbook and other science books. Although these concepts are as up to date as possible, they are always open to further questions and modifications.

One of the roots of the scientific method can be found in ancient Greek philosophy. The natural philosophy of Aristotle and his colleagues was mechanistic rather than vitalistic. A **mechanist** believes that only natural forces govern living things, along with the rest of the universe. A **vitalist** believes that the universe is at least partially governed by supernatural powers. Mechanists look for interrelationships between the structures and functions of living things and the processes that shape them. Their explanations of nature deal in **cause and effect**—the idea that one thing is the result of another thing. (For example, fertilization of an egg initiates the developmental process that forms an adult.) In contrast, vitalists often use purposeful explanations of natural events. (The fertilized egg strives to develop into an adult.) Although statements that ascribe purpose to things often feel comfortable to the writer, try to avoid them when writing lab reports and scientific papers.

Aristotle and his colleagues developed three rules to examine the laws of nature: first, carefully observe some aspect of nature; second, examine these observations as to their similarities and differences; and third, produce a principle or generalization about the aspect of nature being studied. An example is the principle that all mammals nourish their young with milk, which made it difficult to accept two egg–laying mammals later found In Australia.

The major defect of natural philosophy was that it accepted the idea of *absolute truth.* This belief suppressed the testing of principles after they had been formulated. Another example is Aristotle's belief in spontaneous generation, the principle that some life can arise from nonliving things (e.g., maggots from spoiled meat). This belief survived more than 2000 years of controversy before being discredited by Louis Pasteur in 1860. Rejection of the idea of absolute truth coupled with the testing of principles either by experimentation or by further pertinent observation is the essence of the modern scientific method.

1.1 Modern Scientific Method

Although there is not one universal scientific method, Figure 1-1 illustrates the general process.

MATERIALS

Per lab room:

- Blindfold
- Plastic beakers with an inside diameter of about 8 cm stuffed with cotton wool
- Four or five liquid crystal thermometers

PROCEDURE

A. Observation

As with natural philosophy, the scientific method starts with careful observation. An investigator may make observations from nature or from the words of other investigators, which are published in books or research articles in scientific journals and are available in the storehouses of human knowledge (e.g., libraries and the Internet). One subject we all have some knowledge of is the human body. The first four rows of Table 1-1 list some observations about the human body. The fifth row is blank so you

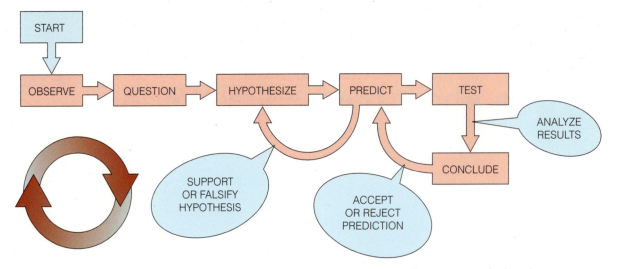

FIGURE 1-1 The scientific method. Support or falsification of the hypothesis usually necessitates further observations, adjustment to the question, and modification of the hypothesis. Once started, the scientific method cycles over and over again, each turn further refining the hypothesis.

can fill in the steps of the scientific method for another observation about the human body or about anything else you and your instructor wish to investigate.

B. Question

In the second step of the scientific method, *we ask a question* about these observations. The quality of this question depends on how carefully the observations were made and analyzed. Table 1-1 includes questions raised by the listed observations.

C. Hypothesis

Now we *construct a hypothesis*—that is, we derive by inductive reasoning a possible answer to the question. **Induction** is a logical process by which all known observations are combined and considered before producing a possible answer. Table 1-1 includes examples of hypotheses.

D. Prediction

Next we *formulate a prediction*—we assume the hypothesis is correct and predict the result of a test that reveals some aspect of it. This is deductive or "if-then" reasoning. **Deduction** is a logical process by which a prediction is produced from a possible answer to the question asked. Table 1-1 lists a prediction for each hypothesis.

E. Experiment or Pertinent Observations

Now we *perform an experiment or make pertinent observations* to test the prediction.

1. Along with the other members of your lab group, choose one prediction from Table 1-1. Coordinate your choice with the other lab groups so that each group tests a different prediction.

2. In an experiment of classical design, the individuals or items under study are divided into two groups: an **experimental group** that is treated with (or

possesses) the independent variable and a **control group** that is not (or does not). Sometimes there is more than one experimental group. Sometimes subjects participate in both groups, experimental and control, and are tested both with and without the treatment.

In any test, there are three kinds of variables. The **independent variable** is the treatment or condition under study. The **dependent variable** is the event or condition that is measured or observed when the results are gathered. The **controlled variables** are all other factors, which the investigator attempts to keep the same for all groups under study.

3. Here are some hints about each of the predictions listed in Table 1-1.

 (a) To test the prediction in row I of Table 1-1, follow these directions:

 (i) With arms dangling at their sides, identify in as many group members as possible a vein segment either on the back of the hands or on the forearms (Figure 1-2).

 (ii) Demonstrate that blood flows in vein segments: Note their collapse caused by gravity speeding up blood flow when each subject's arm is raised above the level of the heart.

 (iii) Lower the arm to a position below the heart and stop blood flow through the vein segment by permanently pressing a finger on the swelling farthest from the heart.

 (iv) Use a second finger to squeeze the blood out of the vein segment past the next swelling toward the heart.

 (v) Remove the second finger and note whether blood flows back into the vein segment.

 (vi) Remove the first finger and note whether blood flows back into the vein segment.

TABLE 1-1   Some Observations About the Human Body

Observation	Question	Hypothesis	Prediction
I. Veins containing blood are seen under the skin. Swellings present along the vein are often located where veins join together.[a]	What is the function of the swellings?	Swellings contain one-way valves that allow blood to flow only toward the heart.	If these valves are present, then blood flows only from vein segments farther from the heart to the next segments nearer the heart and never in the opposite direction.
II. People have two ears.	What is the advantage of having two ears?	Two ears allow us to locate the sources of sounds.	If the hypothesis is correct, then blocking hearing in one ear will impair our ability to determine a sound's source.[b]
III. People can hold their breath for only a short period of time.	What factor forces a person to take a breath?	The buildup of carbon dioxide derived from the body's metabolic activity stimulates us to take a breath.	If the hypothesis is correct, then people will hold their breath a shorter time just after exercise compared with when they are at rest.
IV. Normal body temperature is 98.6°F.	Is all of the body at the same 98.6°F temperature?	The skin, or at least some portion of it, is not 98.6°F.	If the hypothesis is correct, then a liquid crystal thermometer will record different temperatures on the forehead, back of the neck, and forearm.
V. _____	_____	_____	_____

[a] The portion of the vein between swellings is called a segment and is illustrated in Figure 1-2.
[b] This is especially the case for high-pitched sounds because higher frequencies travel less easily directly through tissues and bones.

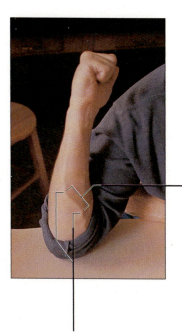

vein segment

valves

FIGURE 1-2 Veins under the skin. *(Photo by D. Morton.)*

(vii) Repeat v–vii, only reverse the order of the swellings—first stop blood flow through the swelling nearest the heart and then squeeze the blood out of the vein segment in a direction away from the heart past the next swelling.

(b) To test the prediction in row II of Table 1-1, you will need a blindfold and a beaker stuffed with cotton wool to block hearing in one ear. Quantify how well each blindfolded group member can point out the source of a high-pitched sound by estimating the angle (0–180 degrees) between the line from source to subject and the line along which the subject points to identify the source.

(c) To test the prediction in row III, each group member in turn will need to exercise in a safe way. Either run in place or *follow instructions given to you by your instructor.*

(d) To test the prediction in row IV, you will need a liquid crystal thermometer to measure skin temperature in three places for each group member.

4. In your own words, describe your procedure to test your group's prediction and identify its variables below.

_____

_____

_____

_____

_____

_____

_____

Independent variable: _____

_____

Dependent variable: _____

_____

Controlled variables: _____

_____

5. Show the above section to your instructor and get approval to perform your test.

F. Test

After your instructor's approval, perform your procedure and record your results in Table 1-2. You will need to label the column headings for your particular test. You may not need all of the rows and columns to accommodate your data. Describe or present your results in one of the bar charts in Figure 1-3.

G. Conclusion

1. The last step in one cycle of the scientific method is to _make a conclusion_. Use the results of the experiment or pertinent observations to evaluate your hypothesis. If your prediction does not occur, it is rejected, and your hypothesis or some aspect of it is falsified. If your prediction does occur, you may conditionally accept your prediction, and your hypothesis is supported. However, you can never completely accept or reject any hypothesis; all you can do is state a probability that one is correct or incorrect. To quantify this probability, scientists use a branch of mathematics called _statistical analysis_.

2. Even if the prediction is rejected, it does not necessarily mean that the treatment caused the result. A coincidence or the effect of some unforeseen and thus uncontrolled variable could be causing the result. For this reason, the results of experiments and observations must be _repeatable_ by the original investigator and others.

3. Even if the results are repeatable, it does not necessarily mean that the treatment caused the result. _Cause and effect_, especially in biology, is rarely proven in experiments. We can, however, say that the treatment and result are correlated. A **correlation** is a relationship between the independent and the dependent variables.

　　The following example should illustrate these concepts. Severe narrowing of a coronary artery branch reduces blood flow to the heart muscle downstream. This region of heart muscle gets insufficient oxygen and cannot contract and may die, resulting in a heart attack. The initial cause is the narrowing of the artery, and the final effect is the heart attack. Perhaps the heart attack victim smoked cigarettes. Smoking cigarettes is one of several factors that make a person more likely to have a heart attack. This is based on a correlation between smoking and heart attacks in the general population, but we cannot say for sure that the smoking caused the heart attack.

4. Write a likely conclusion for your experiment or pertinent observation. Statistics are not required, but if you know how, apply the correct statistical test before writing your conclusion.

_____

_____

_____

_____

_____

_____

H. Theories and Principles

When exhaustive experiments and observations consistently support an important hypothesis, it is accepted as a **theory.** A theory that stands the test of time may be elevated to the status of a **principle** (Figure 1-4). Theories and principles are always considered when new hypotheses are formulated. However, similar to hypotheses, theories and principles can be modified or even discarded in the light of new knowledge. Biology, like life itself, is not static but is constantly changing.

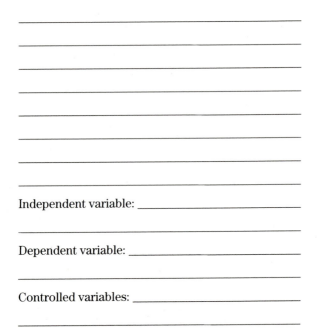

TABLE 1-2	Generalized Data Sheet		
Subject			
1			
2			
3			
4			
5			
Average			

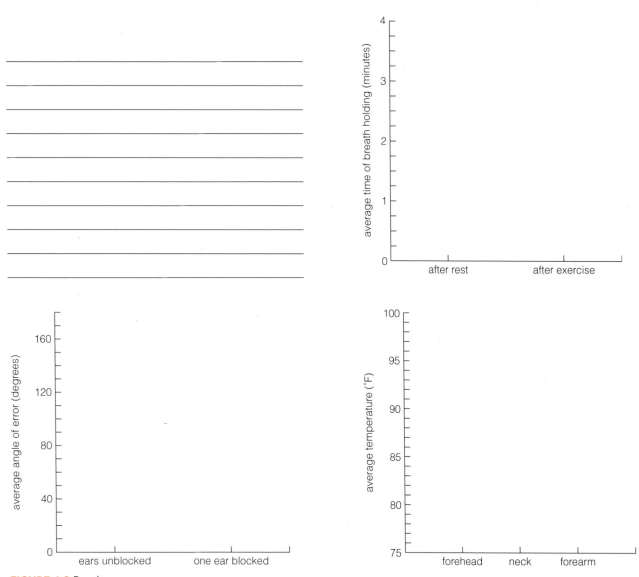

FIGURE 1-3 Results.

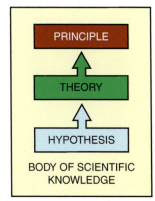

FIGURE 1-4 Theories and Principles: When exhaustive experiments and observations consistently support an important hypothesis, it is accepted as a **theory**. A theory that stands the test of time may be elevated to the status of a **principle**. Theories and principles are always considered when new hypotheses are formulated. However, like hypotheses, theories and principles can be modified or even discarded in the light of new knowledge. Biology, like life itself, is not static but is constantly changing.

1.2 Research Article

The account of one or several related cycles of the scientific method is usually initially reported in depth in a research article published in a scientific journal. The goal of the scientific community is to be cooperative as well as competitive. Writing research articles allows scientists to share knowledge. Scientists provide enough information so that other scientists can repeat the experiments or pertinent observations they describe. The journal *Science* along with several others presents its research articles in narrative form, and many of the details of the scientific method are understood rather than stated. However, adherence to the modern scientific method is expected, and the scientific community understands that it is as important to expose mistakes as it is to praise new knowledge.

MATERIALS

Per student:

- A typical research article in biology
- A research article from *Science*

PROCEDURE

1. Check the design of a typical research article in biology and list the titles of its various sections.

    (a) Example: Abstract (summary of the paper)

    (b)_____

    (c)_____

    (d)_____

    (e)_____

    (f)_____

2. Read the article and then fill in the blanks in the following statements or answer the questions.

    (a) Any changes in the dependent variable are described in the _____ section.

    (b) Which section contains the details necessary to repeat this experiment or observation?

    _____

    (c) Which section contains the questions being asked, the predictions, or the hypotheses?

    _____

    (d) The _____ section contains the conclusions.

3. Look at an article from the journal *Science*. What steps of the scientific method are included in the narrative?

    _____

    _____

    _____

_____ **1.** The natural philosophy of Aristotle and his colleagues was
 (a) mechanistic
 (b) vitalistic
 (c) a belief in absolute truth
 (d) a and c

_____ **2.** A person who believes that the universe is at least partially controlled by supernatural powers can best be described as a(n)
 (a) teleologist
 (b) vitalist
 (c) empiricist
 (d) mechanist

_____ **3.** The first step of the scientific method is to
 (a) ask a question
 (b) construct a hypothesis
 (c) observe carefully
 (d) formulate a prediction

_____ **4.** Which series of letters lists the first four steps of the scientific method (see question 3) in the correct order?
 (a) a, b, c, d
 (b) a, b, d, c
 (c) c, a, b, d
 (d) d, c, a, b

_____ **5.** In an experiment, the subjects or items being investigated are divided into the experimental group and
 (a) the non-experimental group
 (b) the control group
 (c) the statistics group
 (d) none of these choices

_____ **6.** The variables that investigators try to keep the same for both the experimental and the control groups are
 (a) independent
 (b) controlled
 (c) dependent
 (d) a and c

_____ **7.** Variables that are _always_ different between the experimental and the control groups are
 (a) independent
 (b) controlled
 (c) dependent
 (d) a and c

_____ **8.** The results of an experiment
 (a) do not have to be repeatable
 (b) should be repeatable by the investigator
 (c) should be repeatable by other investigators
 (d) must be both b and c

_____ **9.** The detailed report of an experiment is usually published in a
 (a) newspaper
 (b) book
 (c) scientific journal
 (d) magazine

_____ **10.** When two variables are correlated, it means
 (a) there is no relationship between them
 (b) there is a relationship between them
 (c) that one caused an effect in the other
 (d) a and c

EXERCISE 1 Learning About the Natural World

POST-LAB QUESTIONS

INTRODUCTION

1. How does the modern scientific method differ from the natural philosophy of the ancient Greeks?

1.1 Modern Scientific Method

2. List the six steps of one full cycle of the scientific method.

a.

b.

c.

d.

e.

f.

3. What is tested by an experiment?

4. Within the framework of an experiment, describe the

a. independent variable

b. dependent variable

c. controlled variables

5. Is the statement, "In most biology experiments, the relationship between the independent and the dependent variable can best be described as cause and effect," true or false? Explain your answer.

6. Is a scientific principle taken as absolutely true? Explain your answer.

1.2 Research Article

7. What is the function of research articles?

8. Identify three sources where research articles can be found.

Food for Thought

9. Describe how you have applied or could apply the scientific method to an everyday problem.

10. Do you think the differences between religious and scientific knowledge make it difficult to debate points of perceived conflict between them? Explain your answer.

Observing Microscopic Details of the Natural World

OBJECTIVES

After completing this exercise, you will be able to

1. define *magnification, resolving power, contrast, field of view, parfocal, parcentral, depth of field,* and *working distance.*

2. describe how to care for a compound light microscope.

3. recognize and give the function of the parts of a compound light microscope.

4. accurately align a compound light microscope.

5. correctly use a compound light microscope.

6. make a wet mount.

7. correctly use a dissecting microscope.

8. describe the usefulness of the phase-contrast, transmission electron, and scanning electron microscopes.

9. use your skills to enjoy a fascinating world unavailable to the unaided eye.

Introduction

Light microscopes contain transparent glass lenses, which focus light rays emanating from a specimen to produce an image of a specimen on the retina, the light-sensitive layer of the eye. Table 2-1 presents the three most important properties of lenses and their images.

2.1   Compound Light Microscope

In everyday life, the eye lens projects onto the retina a focused image of an object held no closer than about 10 cm. At this distance, details separated by 0.1 mm are visible. Most cells and related structures are smaller than this, and a light microscope is needed to see them. A microscope placed between the eye and a specimen (usually a section or thin object(s) mounted on a glass slide) acts to bring the specimen very close to the eye so you can focus on its details. Ultimately, the greater the proportion of the retina covered by the final image of the specimen, the greater its magnification. It does this by producing a series of magnified images.

Magnification without enough resolving power is referred to as *empty*, and with a light microscope, the maximum useful magnification is about 1000 times the diameter of the specimen (1000×). Above this value, additional details are missing. Furthermore, adequate contrast is needed to see the details preserved in an image. Dyes are usually added to sections of biological specimens to increase contrast.

Similar to automobiles, there are many models of compound light microscopes, and these instruments have numerous accessories that may or may not be present. Typical examples are shown in Figure 2-1, and one is diagrammed in Figure 2-2. If your microscope differs significantly from Figure 2-2, your instructor will give you an unlabeled diagram. If the instructor assigns you a specific microscope for your lab work, record its identification code in the second column of the first row in Table 2-2.

MATERIALS

Per student:

- Compound light microscope
- Lens paper
- Lint-free cloth (optional)

TABLE 2-1   Important Lens and Image Properties

Property	Definition
Magnification	The amount that the image of an object is enlarged (e.g., 100×)
Resolving power	The extent to which object detail in an image is preserved during the magnifying process
Contrast	The degree to which image details stand out against their background

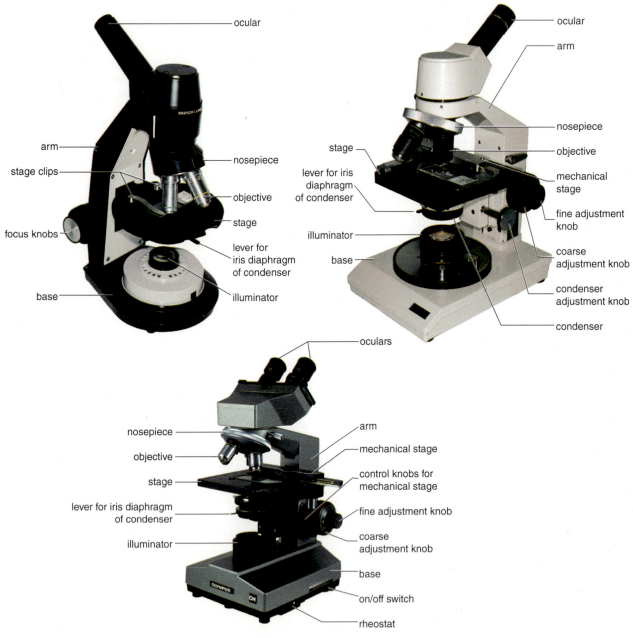

FIGURE 2-1 Compound light microscopes. *(Photos by D. Morton and J. W. Perry.)*

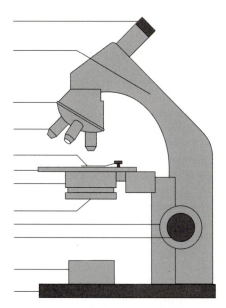

Labels: ocular, objective, arm, base, illuminator, condenser, lever (rotating disk) for iris diaphragm of condenser, stage, stage clip, coarse adjustment knob, fine adjustment knob, nosepiece

FIGURE 2-2 Compound light microscope.

TABLE 2-2	Characteristics of My Microscope	
Characteristic	**Description**	**Function**
Code		identification of my microscope
Light source		
Condenser		
Stage		
Focusing knobs		
Objectives		
Ocular(s)		

- Unlabeled diagram of the compound light microscope model used in your course (optional)
- Prepared slide with a whole mount of stained diatoms
- Prepared slide with the mounted letter *e*
- Index card
- Prepared slide with crossed colored threads coded for thread order
- Prepared slide with a section of the mammalian kidney
- Prepared slide with unstained fibers

Per student group (4):
- Bottle of lens-cleaning solution (optional)
- Dropper bottle of immersion oil (optional)

Per lab room:
- Labeled chart of a compound light microscope

PROCEDURE

A. Care of a Compound Light Microscope

1. To carry a microscope to and from your lab bench, grasp the **arm** (Figure 2-1) with your dominant hand and support the **base** (Figure 2-1) with the other hand, always keeping the microscope upright. If the arm is wide, there should be a carrying indent or something similar to help you grasp it. *Do not try to carry anything else at the same time.* Label the arm and the base on Figure 2-2 or on the diagram given to you by your instructor.

CAUTION Never wipe a glass lens with anything other than lens paper.

2. Remove the dust cover and clean the exposed parts of the optical system. Blow off any loose dust that may be on the ocular and then gently brush off any remaining dust with a piece of lens paper.

   If the lens of the ocular is still dirty, breathe on it and gently polish it with a rotary motion using a fresh piece of lens paper. If the ocular lens is still dirty, and *with your instructor's approval*, clean it with a piece of lens paper moistened with lens-cleaning solution.

3. Always remember that your microscope is a precision instrument. Never force any of its moving parts.

4. It is just as difficult to see clearly through a dirty slide as through a dirty microscope. Clean dirty slides with a lint-free cloth or with lens paper before using.

5. At the end of an exercise, make sure the last slide has been removed from the stage and *rotate the nosepiece so that the low-power objective is in the light path.* If your instrument focuses by moving the body tube, turn the coarse adjustment so that it is racked all the way down. If your microscope has an electric cord, neatly fold it up on itself and tie it with a plastic strap or rubber band. Otherwise, wind the cord around the base of the arm of the microscope.

6. Replace the dust cover before returning your microscope to the cabinet.

B. Parts of the Compound Light Microscope

Now that you know how to care for your microscope, remove the instrument assigned to you from the cabinet and place it on your lab bench. Use Figure 2-1 and the chart on the wall of your lab room to identify the various parts of your microscope. Read each step below and manipulate the parts *only where indicated.* Before you start, make sure the low-power objective is in the light path.

1. *Light source.* A compound microscope uses transmitted light to illuminate a transparent specimen usually mounted on a glass slide. Newer microscopes have a built-in illuminator (Figure 2-1). Locate the **illuminator;**

the *off/on switch*; and perhaps also a *rheostat*, which is used to vary the intensity of the light. On some models, the switch and the rheostat are combined. Turn on the light source and look through the ocular. If the illuminator has a rheostat, adjust the intensity so the light is not too bright.

(a) Label the illuminator on Figure 2-2 or on your instructor's diagram.

(b) Describe the light source in Table 2-2, and then state its function.

2. *Condenser.* For maximum resolving power, a **condenser**—with a *condenser lens* and *iris diaphragm*—focuses the light source on the specimen so that each of its points is evenly illuminated. The **lever for the iris diaphragm** of the condenser is used to open and close the condenser. Instead of a lever, there may be a rotating ring.

Establish whether there is a condenser adjustment knob (Figure 2-1) to set the height of the condenser. *Do not turn the knob; you will learn how to use it later.*

There may be a *filter holder* under the condenser with a blue or frosted glass disk. Many microscope manufacturers believe that blue light is more pleasing to the eye because when used with an incandescent bulb, it produces a color balance similar to daylight conditions. Also, theoretically at least, blue light gives better resolving power because of its shorter wavelength. The frosted glass disk scatters light and can be useful in producing even illumination at low magnifications.

(a) Label the condenser and lever for the iris diaphragm of the condenser on Figure 2-2 or on your instructor's diagram.

(b) Describe the condenser and its related parts in Table 2-2 and then state its function.

3. *Stage.* Either a pair of *stage clips* or a *mechanical stage* holds a specimen mounted on a glass slide in place suspended over a central hole (Figure 2-1).

(a) If your microscope has a mechanical stage, skip to part b. If your microscope has stage clips, place a prepared slide of stained diatoms under their free ends. Never remove the stage clips because they make it easier to move a slide in small increments. Skip b and do part c next.

(b) Position a prepared slide of stained diatoms on the stage by releasing the tension on the spring-loaded movable arm of the mechanical stage (Figure 2-3a). There are two knobs to the right or left of the stage: one to move the specimen forward and backward and the other to move it laterally (Figure 2-1). Label the stage and stage clips (or mechanical stage) on Figure 2-2 or your instructor's diagram.

(c) On most mechanical stages, each direction has a vernier scale so you can easily locate interesting fields again and again. A vernier scale consists of two scales running side by side, a long one in millimeters and a short one, 9 mm in length and divided into 10 equal subdivisions. To take a reading, note the whole number on the long scale coinciding with or just below the zero line of the short scale. If the whole number of the long scale and the zero of the short scale coincide, the first place after the decimal point is zero. Otherwise, the first place after the decimal point is the value of the line on the short scale that coincides (or nearly coincides) with one of the next nine lines after the whole number on the long scale. For example, the correct reading of the vernier scale in Figure 2-3b is 19.6 mm.

(d) Describe the stage and its related parts in Table 2-2 and then state its function.

4. *Focusing knobs* (Figure 2-1). The **coarse focus adjustment knob** is for use with the lower-power objectives, and the **fine-focus adjustment knob** is for critical focusing, especially with the higher power objectives. On most modern microscopes, you move the stage of the instrument up and down to focus the specimen. Modern microscopes usually have a *preset focus lock*, which stops the stage at a particular height. After setting this lock, you can lower the stage with the coarse focus knob to facilitate changing of the specimen and then raise it to focusing height without fear of colliding the specimen against the objective. There may also be a *focus tension adjustment knob*, usually located inside of the left-hand coarse focus knob.

(a) Turn the coarse focus knob. Do you turn the knob toward you or away from you to bring the slide and objective closer together?

_____

(b) Label the coarse and fine-focus adjustment knobs on Figure 2-2 or on your instructor's diagram.

(c) Describe the focusing knobs in Table 2-2 and then state their function. Note if a preset focus lock or focus adjustment knob is present.

5. *Objectives.* The compound light microscope has at least two magnifying lenses, the objective and the ocular (Figure 2-1). The objective scans the specimen. Most microscopes have several objectives mounted on a revolving **nosepiece**. The magnifying power of each objective is labeled on its side. Usually included are these objectives: a $4\times$ *low-power* or scanning, a $10\times$ *medium-power* (Figure 2-4a), an about $40\times$ *high-dry*, and perhaps an about $100\times$ *oil-immersion objective* (Figure 2-4b). The other number often labeled on the side of nosepiece objectives is the **numerical aperture** (NA). The larger the NA is, the greater the resolving power and useful magnification.

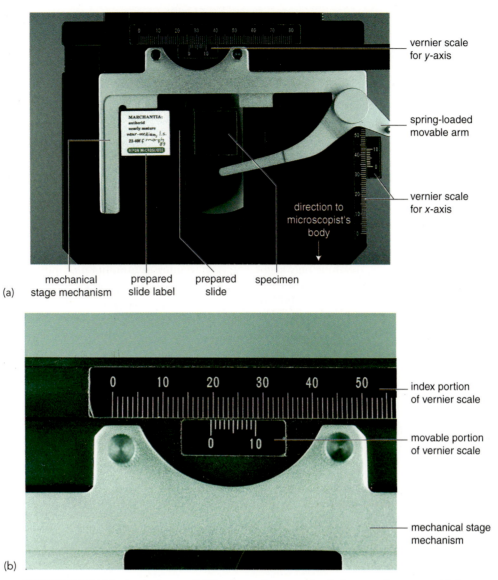

(a)

(b)

FIGURE 2-3 (a) Mechanical stage (b) Vernier scale on mechanical stage of compound light microscope. The correct reading is 19.6. *(Photos by D. Morton and J. W. Perry.)*

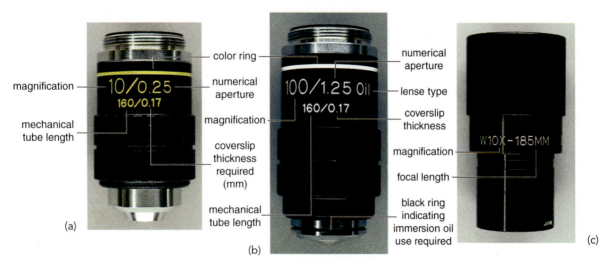

FIGURE 2-4 (a) 10× objective removed from microscope. (b) 100× oil immersion objective removed from microscope. (c) Ocular removed from microscope. *(Photos by D. Morton and J. W. Perry.)*

TABLE 2-3 — Objectives Present on My Compound Light Microscope

Objective	Objective Magnifying Power (ObMP×)	Total Magnifying Power (ObMP × OcMP = ____×)	Numerical Aperture (NA)
Low-power			
Medium-power			
High-dry			
Oil-immersion			

In a properly aligned microscope, the objectives are **parfocal.** That is, when an objective has been focused, you can rotate to another one, and the image will remain in coarse focus, requiring only slight movement of the fine-focus knob. Objectives are also **parcentral,** meaning that the center of the field of view remains about the same for each objective. The **field of view** is the circle of light you see when looking into the microscope.

Often, objectives have different lengths; the lower-power objectives are shorter than the higher-power ones. That is, the working distance of objectives decreases with magnification. **Working distance** is the space between the objective lens and the slide. Therefore, the higher the power of the objective in use, the closer the objective is to the slide—and the more careful you must be.

(a) Record the magnifying power and NA of the objectives on your microscope in Table 2-3. If your instrument does not have a particular objective, indicate that it is not present (NP).

(b) Label the nosepiece and objective on Figure 2-2 or on your instructor's diagram.

(c) Describe the objectives in Table 2-2 and then state their function.

6. *Ocular.* The magnifying lens you look into is called an **ocular** (Figure 2-4c). Oculars are generally 10×.

(a) Because each objective has a different magnifying power, the total magnification is calculated by multiplying the magnifying power of the ocular by that of the objective in use. What is the ocular magnification power (OcMP) of the ocular(s) on your microscope?

_____ ×

Calculate the total magnification for each of your microscope's ocular and objective combinations and then record them in Table 2-3.

(b) Label the ocular on Figure 2-2 or on your instructor's diagram.

(c) Describe the ocular in Table 2-2 and then state its function.

(d) Your microscope will have one or two oculars mounted on a *monocular* or *binocular head,* respectively. There may be a pointer mounted in an ocular so that you can easily show a specimen detail to your instructor or another student. For a monocular microscope, it's best to use your dominant eye to look down the ocular, keeping your other eye open. Is your microscope monocular or binocular?

_____

(e) If your microscope is monocular, determine your dominant eye.

- Look at a small object on the far wall of your room with both eyes open.

- Form your thumb and index finger of one hand into a circle and place this circle in your line of sight at arm's length so it surrounds the object.

- Close your right eye. If the object shifts out of the circle to your left, your right eye is probably dominant. If the object remains in the circle, your left eye is probably dominant.

- This time, close your left eye and go through the process again. If the object shifts to the right, your left eye is dominant. If the object remains within the circle, your right eye is dominant. The more pronounced the shift, the greater the dominance. If there is no shift, neither eye is dominant.

C. Aligning a Compound Light Microscope with In-Base Illumination and a Condenser with an Iris Diaphragm

Aligning your microscope properly will not only help you see specimen detail clearly but will also protect your eyes from strain.

1. Rotate the nosepiece until the medium-power objective is in the light path. Open the iris diaphragm.

2. If it is not already there, place the prepared slide of stained diatoms on the stage; center and carefully focus on it. *Skip steps 3 and 4 if your microscope is monocular. Skip step 5 if your microscope does not have a control to adjust the height of the condenser.*

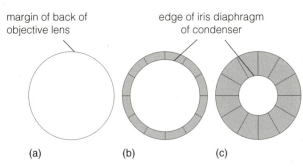

margin of back of objective lens

edge of iris diaphragm of condenser

(a)   (b)   (c)

FIGURE 2-5 Correct setting for condenser iris diaphragm. Drawing (b) is correct. In drawing (a) you cannot see the edge of the iris diaphragm. In (c), the diaphragm has been closed too much.

3. If your microscope is binocular, adjust the interpupillary distance. Hold a different ocular tube with each hand and while looking at the specimen, pull the tubes apart or push them together until you see one field of view. After making this adjustment, read and record the number off the scale.

My interpupillary distance is _____.

From now on, you can set the interpupillary distance at this number.

4. Now compensate for any difference in diopter between the lenses of each eye.

(a) *If there is one diopter adjustment ring around the left ocular tube,* cover your left eye with an index card and focus your microscope using the fine-focus knob. Now uncover the left eye and cover your right one. Use the diopter adjustment ring to bring the specimen into focus.

(b) *If both ocular tubes have a diopter adjustment ring,* set the left one to the same number as the interpupillary distance, cover your right eye with an index card, and focus on the specimen. Then uncover the right eye and cover your left one. Use the diopter adjustment ring on the right ocular tube to bring the specimen into focus.

5. Place a sharp point (pencil, dissecting needle, or some similar object) on top of the illuminator and bring the silhouette into sharp focus by adjusting the height of the condenser.

(a) *If the ocular on your microscope is removable (and with the permission of your instructor),* carefully slide it out and put the ocular open end down on a piece of lens paper in a safe place. Then, while looking down the ocular tube, adjust the iris diaphragm until the edge of the aperture lies just inside the margin of the back lens element of the objective (Figure 2-5). When done, replace the ocular.

(b) *If the ocular cannot be removed,* close the condenser diaphragm and then open it until there is no further increase in brightness. Now close it again, stopping when you see the brightness begin to diminish.

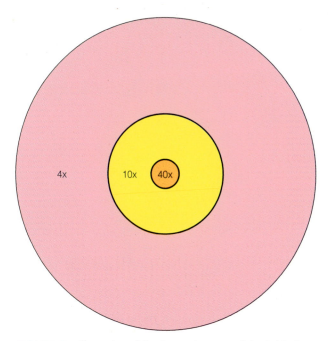

4x   10x   40x

FIGURE 2-6 Illustration of the decreasing area of the field of view when different magnification objectives are used with a 10× ocular. The actual area of each circle has been enlarged by 10×.

(c) Models that have rings that open and close the iris diaphragm of the condenser usually have numbers associated with them that match the magnification of the objective being used. Every time you switch objectives, you need to reset the ring to the correct position.

6. If your microscope has a rheostat, adjust the illumination to a level that lets you see specimen detail and that is comfortable for your eyes. To maintain the same illumination at higher magnifications, you will have to increase its intensity.

7. For best results, repeat steps 5 and 6 every time you use a different objective.

D. Using Different Magnifications

It is safest to observe a specimen on a slide first with low power and then, step by step, with higher power objectives. This way you avoid colliding the objective and slide together.

Because the magnification is in diameters, the area of the field of view decreases dramatically with increasing magnification (Figure 2-6). It follows that it is easier to use a lower power objective to locate a specific specimen detail. This is why the low-power objective is sometimes called the *scanning objective*. Also, if you lose a specimen detail at higher magnification, it is always easier to find it again if you switch to a lower power objective.

Now follow these steps to use each objective.

1. Rotate the low-power objective into the light path.

2. If it is not already there, place a prepared slide of stained diatoms on the stage, securing it with either the stage clips or the movable arm of the mechanical stage.

FIGURE 2-7 Drawing of the letter *e* as seen through the ocular (_____×).

3. Look through the ocular. Bring the diatoms into focus using the *coarse focus knob*. Adjust the illumination as described in steps 5 and 6 of Section C. At this magnification, the diatoms appear small. Center a diatom by moving the slide.

4. Rotate the nosepiece so the medium-power objective is in the light path. Adjust the illumination. Focus the diatom.

5. Rotate the nosepiece so the high-dry objective is in the light path. Adjust the illumination. Focus the diatom using the *fine-focus adjustment knob*.

E. Orientation of the Image Compared with the Specimen

1. If you have not already done so, remove the prepared slide of diatoms and replace it with a prepared slide with the letter *e*. With the medium-power objective in the light path, position the slide with the specimen (the letter *e*) right side up on the stage. Center the *e* in the field of view and carefully bring it into focus.

2. In Figure 2-7, draw the image of the *e* as you see it through the ocular. Record the total magnification used in the line at the end of the legend.

    Is the image right side up or upside down compared with the specimen?

    _____

    Compared with the specimen, is the image backward as well as upside down?

    (yes or no) _____

    In summary, the image is *inverted* with respect to the specimen.

3. Move the specimen to the right while watching it through the microscope. In which direction does the image move?

    _____

4. Move the specimen away from you. In which direction does the image move?

    _____

5. Remove the slide and put it away.

F. Depth of Field

The **depth of field** is the distance through which you can move the specimen and still have it remain in focus. Remember, the working distance—the space between the objective lens and the coverslip—decreases with increasing magnifying power. Therefore, the higher the power of the objective in use, the closer the objective is to the slide—and the more careful you must be.

1. Obtain a prepared slide of three crossed colored threads. *This exercise requires care because you are probably not yet adept at focusing on a specimen.* When you have the threads in focus (using first the low-power objective and then the medium-power objective), you need only use the fine-focus knob to focus with the high-dry objective. After switching to the high-dry objective, try rotating the fine-focus knob ½ turn away from you and then a full turn toward you. If you have not found the plane of focus, next try 1½ turns away from you and 2 full turns toward you and so on. If you work deliberately, you will find the plane of focus and will not crack the coverslip.

    (a) How many threads are in focus using the

    low-power objective?_____

    medium-power objective? _____

    high-dry objective? _____

    (b) With which objective is it easiest to focus a specimen? _____

    (c) At which magnification is it most difficult to focus a specimen? _____

2. Specimens have depth. Continue using the prepared slide of three crossed colored threads.

    (a) Use the high-dry objective to determine the order of the three threads mounted on the slide and record the results in Table 2-4. (Each slide label has a code on it. When you believe you have discovered the correct order, check with your instructor to find out if you are correct.)

    (b) Focusing carefully with the fine-focus knob, move from the bottom to the upper thread.

TABLE 2-4	Order of Threads
Location	**Color**
Closest to slide	
Middle	
Closest to coverslip	

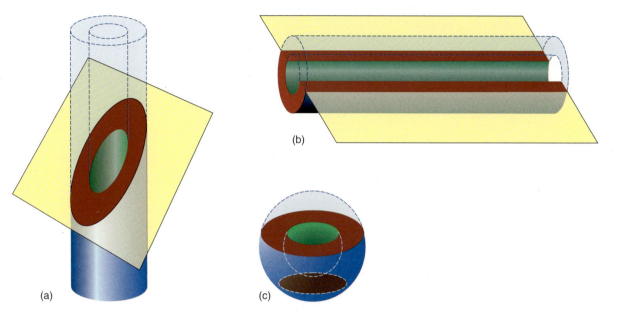

(a)   (b)   (c)

FIGURE 2-8 Difficulties encountered in interpreting the three-dimensional shape of objects from sections. Compare the transverse (cross) section of the cylinder with the distorted oblique section (a) and the longitudinal section (b). A similar shape to the transverse section of a cylinder results when a hollow ball is sectioned (c). A section through wall of the hollow ball results in a solid shape.

Did you move the knob away from or toward you? _____

3. Remove and put away the slide.

4. Viewing sections of three-dimensional structures makes interpretation of the original shape quite difficult.

    (a) Examine a slide of the cortex of the mammalian kidney with your compound microscope (Figure 2-8). Hypothesize as to the three-dimensional shape of the structures labeled *renal corpuscles* and *nephron tubules*.

The shape of renal corpuscles is

_____.

The shape of nephron tubules is

_____.

    (b) Check the textbook to see if your hypotheses are acceptable.

G. Using the Iris Diaphragm to Improve Contrast

1. Place a specimen of unstained fibers on the stage. Locate and focus on these fibers using the medium-power objective. Make sure the condenser and iris diaphragm are correctly set.

2. Close the iris diaphragm.

Does this procedure increase or decrease contrast?

_____.

Although this procedure is useful when viewing specimens with low contrast, it should be used only as a last resort because resolving power is also decreased.

3. Remove and put away the slide.

H. Units of Measurement

The basic metric unit of length at the light-microscopic level is the micrometer (μm).

$1000\ \mu\text{m} = 1\ \text{mm}$

$1\ \mu\text{m} = 0.001\ \text{mm}$

How many nanometers are there in 1 mm? _____ nm

How many millimeters are there in 1 nm? _____ mm

2.2 How to Make a Wet Mount

In the mid-seventeenth century, Robert Hooke used a microscope to discover tiny, empty compartments in thin shavings of cork. He named them *cells*. Repeating this historic observation is a good way to learn how to prepare a wet mount.

MATERIALS

Per student:

- Compound microscope, lens paper, a bottle of lens-cleaning solution (optional), a lint-free cloth (optional)
- Cork
- Razor blade
- Glass microscope slide
- Glass coverslip
- Dissecting needle

Per student group (4):

- Dropper bottle of distilled water (dH$_2$O)

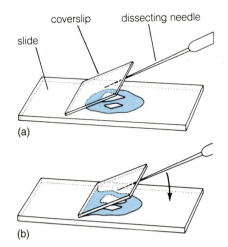

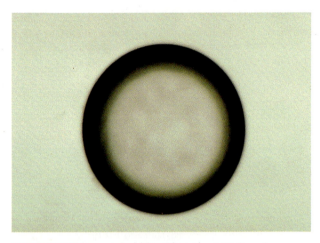

FIGURE 2-9 How to make a wet mount.

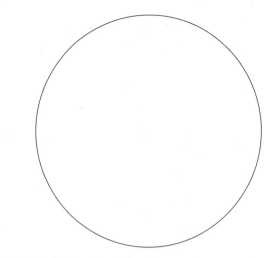

FIGURE 2-11 Drawing of the microscopic structure in a cork shaving (_____×).

FIGURE 2-10 Free air bubble. *(Photo by J.W. Perry.)*

FIGURE 2-12 Drawing of microorganisms (_____×).

PROCEDURE

1. Carefully use a razor blade to cut a number of *very thin shavings* from a cork stopper. Place them on a glass microscope slide.

2. Gently add a drop of distilled water.

3. Place one end of a glass coverslip to the right or left of the specimen so that the rest of the slip is held at a 45-degree angle over the specimen (Figure 2-9a).

4. Slowly lower the coverslip with a dissecting needle so as not to trap air bubbles (Figure 2-9b).

5. Observe the wet mount, first at low magnification and then with higher power. Air may be trapped either in the cork or as free bubbles (Figure 2-10). Trapped air will appear dark and refractive around its edges. This effect is caused by sharply bending rays of light. Draw what you see in Figure 2-11. Note the total magnification used to make the drawing.

6. Clean and replace the slide and coverslip as indicated by your instructor.

2.3 Microscopic Observations

Examining the microscopic world is both challenging and fun. Most of the macroscopic world has been explored, but the microscopic world is barely touched. Yet the microbes in it are essential to our very existence. So be an explorer and see what you can discover!

MATERIALS

Per student:

- Compound microscope, lens paper, a bottle of lens-cleaning solution (optional), a lint-free cloth (optional)
- Glass microscope slide
- Glass coverslip

Per student group (4):

- Pond water or some other mixed culture in a dropper bottle
- Dropper bottle of Protoslo®

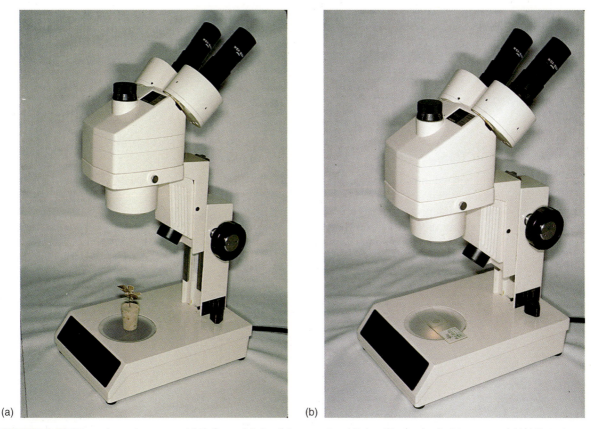

(a)   (b)

FIGURE 2-13 Dissecting microscope (a) Reflected light, (b) transmitted light. *(Photos by D. Morton and J.W. Perry.)*

Per lab room:

- Reference books for the identification of microorganisms

PROCEDURE

1. Obtain a drop of pond water or other mixed culture from the bottom of the bottle.

2. Add a drop of Protoslo®. This methyl cellulose solution slows down any swimming microorganisms.

3. Make a wet mount (Figure 2-9).

4. Observe the wet mount with your compound microscope. Start at the upper left corner of the coverslip and scan the wet mount with the low-power objective. When you find something interesting, focus on it and switch to the medium-power objective and then, if necessary, the high-dry objective.

5. Draw what you find on Figure 2-12 and note the total magnification.

6. Attempt to identify what you found using the resource books provided by your instructor. If successful, write its name under your drawing.

7. Clean and replace the slide and coverslip as indicated by your instructor.

8. Put away your compound microscope as described on page 11.

2.4 Dissecting Microscope

Dissecting microscopes (Figure 2-13) have a large working distance between the specimen and the objective lens. They are especially useful in viewing larger specimens (including thicker slide-mounted specimens) and in manipulating the specimen (e.g., when dissection of a small structure or organism is required).

The large working distance also allows for illumination of the specimen from above (reflected light) as well as from below (transmitted light). Light reflected from the specimen shows surface features better than transmitted light.

MATERIALS

Per student group:

- Dissecting microscope
- Specimens appropriate for viewing with the dissecting microscope (e.g., a prepared slide with a whole mount of a small organism, bread mold, an insect mounted on a pin stuck in a cork, a small flower)

PROCEDURE

1. Under a dissecting microscope, view one or more of the specimens provided by your instructor. What is the magnification range of this microscope?

    _____ × to _____ ×

2. Is the image of the specimen inverted as in the compound light microscope? (yes or no) _____

3. Describe the type of illumination used by your dissecting microscope. Is there a choice?

_____

_____

2.5 Other Microscopes

In future exercises, you will examine pictures taken with other types of microscopes. Some will be of living cells taken with a phase-contrast microscope (Figure 2-14a) or similar instrument, including those using the Nomarski process (Figure 2-14b). Others will be of very thin-sectioned, heavy metal–stained specimens taken with a transmission electron microscope or TEM (Figure 2-14c). Still others will be of precious metal–coated surfaces produced by signals from the scanning electron microscope or SEM (Figure 2-14d). Table 2-5 summarizes the technology and use of these microscopes.

MATERIALS

Per student group:

- Photographs of TEM micrographs (negatives) or digital images
- Photographs of SEM negatives or digital images

PROCEDURE

1. Examine some photographs of TEM micrographs (negatives) or digital images. The darker areas are more electron dense in the specimen than the lighter areas.

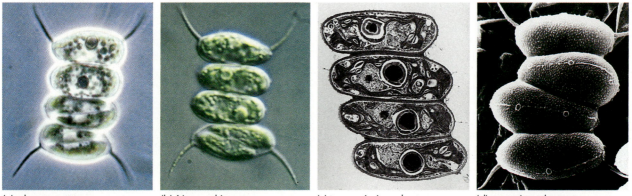

(a) phase contrast   (b) Nomarski process   (c) transmission electron   (d) scanning electron

FIGURE 2-14 How different types of microscopes reveal detail in cells of the green alga *Scenedesmus. (Photos courtesy J. Pickett-Heaps.)*

TABLE 2-5	Other Microscopes	
Microscope	**Technology**	**Use**
Phase-contrast and Nomarski process	Converts phase differences in light to differences in contrast	Observation of low-contrast specimens (often living)
Transmission electron	Increases resolving power by using electrons in a vacuum and magnetic lenses instead of light and glass lenses, respectively	Preservation of greater specimen detail allows for magnifications up to 1,000,000× or more (usually dead materials)
Scanning electron	Forms a TV-like picture from a secondary electron signal, which is emitted from surface points excited by a thin beam of electrons drawn across	Investigation of the fine structure of surfaces (usually dead specimens); the surface in a raster pattern

2. Now look at some photographs of SEM negatives or digital images. The lighter areas correspond to the emission of greater numbers of secondary electrons from that part of the specimen; the darker areas emit fewer secondary electrons.

3. What type of microscope (compound light, dissecting, phase-contrast, TEM, or SEM) would you use to examine the specimens listed in Table 2-6?

TABLE 2-6	Microscope Use
Specimen	**Microscope**
Living surface of the finger	
Dye-stained slide of a section of the finger	
Gold-coated bacteria on a single cell of the finger	
Unstained section of a biopsy from the finger	
Heavy metal–stained, very thin section of the finger	

_____ **1.** Magnification

(a) is the amount that an object's image is enlarged

(b) is the extent to which detail in an image is preserved during the magnifying process

(c) is the degree to which image details stand out against their background

(d) focuses light rays emanating from an object to produce an image.

_____ **2.** Resolving power

(a) is the amount that an object's image is enlarged

(b) is the extent to which detail in an image is preserved during the magnifying process

(c) is the degree to which image details stand out against their background

(d) focuses light rays emanating from an object to produce an image

_____ **3.** A lens

(a) is the amount that an object's image is enlarged

(b) is the extent to which detail in an image is preserved during the magnifying process

(c) is the degree to which image details stand out against their background

(d) focuses light rays emanating from an object to produce an image

_____ **4.** Contrast

(a) is the amount that an object's image of a object is enlarged

(b) is the extent to which detail in an image is preserved during the magnifying process

(c) is the degree to which image details stand out against their background

(d) focuses light rays emanating from an object to produce an image

_____ **5.** The maximum useful magnification for a light microscope is about

(a) $100\times$

(b) $1000\times$

(c) $10,000\times$

(d) $100,000\times$

_____ **6.** The two image-forming lenses of a compound light microscope are

(a) the condenser and objective

(b) the condenser and ocular

(c) the objective and ocular

(d) none of these choices

_____ **7.** Dyes are usually added to sections of biological specimens to increase

(a) resolving power

(b) magnification

(c) contrast

(d) all of the above

_____ **8.** If the magnification of the two image-forming lenses are both $10\times$, the total magnification of the image will be

(a) $1\times$

(b) $10\times$

(c) $100\times$

(d) $1000\times$

_____ **9.** The distance through which a microscopic specimen can be moved and still have it remain in focus is called the

(a) field of view

(b) working distance

(c) depth of field

(d) magnification

_____ **10.** Electron microscopes differ from light microscopes in that

(a) electrons are used instead of light

(b) magnetic lenses replace glass lenses

(c) the electron path has to be maintained in a high vacuum

(d) a, b, and c are all true

EXERCISE **2** **Observing Microscopic Details of the Natural World**

POST-LAB QUESTIONS

2.1 Compound Light Microscope

1. What is the function of the following parts of a compound light microscope?

   a. condenser lens

   b. iris diaphragm

   c. objective

   d. ocular

2. In order, list the lenses in the light path between a specimen viewed with the compound light microscope and its image on the retina of the eye.

3. What happens to contrast and resolving power when the aperture of the condenser (i.e., the size of the hole through which light passes before it reaches the specimen) of a compound light microscope is decreased?

4. What happens to the field of view in a compound light microscope when the total magnification is increased?

5. Describe the importance of the following concepts to microscopy.

   a. magnification

   b. resolving power

   c. contrast

6. Which photomicrograph of unstained cotton fibers was taken with the iris diaphragm closed?_____

(a)

(b)

(Photos by J.W. Perry.)

7. Describe how you would care for and put away your compound light microscope at the end of the lab.

2.2 How to Make a Wet Mount

8. Describe how to make a wet mount.

2.4 Dissecting Microscope

2.5 Other Microscopes

9. A camera mounted on a _____ microscope took this photo of a cut piece of cork.

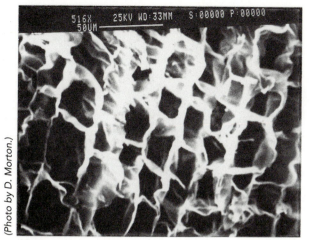

(Photo by D. Morton.)

(321X)

Food for Thought

10. Why were humans unaware of microorganisms for most of their history?

Chemistry of Life: Biological Molecules and Diet Analysis

OBJECTIVES

After completing this exercise, you will be able to

1. define *vitamin, mineral, carbohydrate, monosaccharide, disaccharide, polysaccharide, lipid, protein, amino acid,* and *calorie*.

2. describe the basic structures of carbohydrates, lipids, triglycerides, and proteins.

3. recognize positive and negative tests for carbohydrates, lipids, and proteins.

4. identify the roles that carbohydrates, lipids, proteins, minerals, and vitamins play in the body's construction and metabolism.

5. test food substances to determine the presence of some biological molecules.

6. identify common dietary sources of nutrients.

7. compare your intake of nutrients and calories with the components and relative proportions of a recommended diet.

Introduction

Why do you eat? Because food tastes good? Because you get hungry? Like all animals, our bodies are programmed to ensure that we provide it with an adequate supply of food. These food items provide us with the energy stored in the chemical bonds of the food molecules plus the raw materials from which cellular and tissue components are built. Hunger evolved to stimulate us to seek foods that provide enough nutrients for survival and health.

Food nutrients include minerals and vitamins plus the larger biological (organic, i.e., carbon-based) molecules known as carbohydrates, lipids, and proteins. As it happens, these last nutrients are three of the four major groups of large organic molecules of which all cells are made. The fourth group is the nucleic acids that store and control the genetic instructions within a cell. These crucial molecules are not usually used by the body as nutrients, so they will be studied in another exercise.

Vitamins are necessary organic molecules that our bodies cannot construct internally; we require vitamins from our diets, although in relatively small amounts. **Minerals** are required inorganic (noncarbon-containing) nutrients such as calcium and potassium. Good health, then, depends on eating and drinking the proper balance and quantities of these nutrients.

In this exercise, you will establish tests to identify the presence of carbohydrates, lipids, and proteins and then determine the presence or absence of these molecules in some common foods. You will also analyze your own diet and compare it with a diet recommended for maintaining good health.

3.1 Identification of Large Biological Molecules

You will learn some simple tests for carbohydrates, lipids, and proteins in a variety of substances, including food products. *Most of the reagents used are not harmful; however, observe all precautions listed and perform the experiments only in the proper location as identified by your instructor. Clean out test tubes between tests in the designated location.*

MATERIALS

Per student group (4):

- China marker
- 10 test tubes in test tube rack
- Dropper bottles (*or* bottles and plastic pipettes) of:
  - Distilled water (dH$_2$O)
  - Onion juice
  - Potato juice

- Hamburger juice
- Cream
- Colorless nondiet soft drink
- Colorless diet soft drink
- Glucose solution
- Fructose solution
- Lemon juice
- Starch solution
- Vegetable oil
- Egg albumin
- Benedict's reagent
- Biuret reagent
- Lugol's solution
- Sudan IV dye

- Piece of uncoated paper (grocery bag)
- Test tube clamp
- Hot plate *or* ring stand with wire gauze support and Bunsen burner
- 250- *or* 400-mL beaker with boiling beads or stones
- Vortex mixer (optional)

A. Carbohydrates

Carbohydrates are composed of carbon, hydrogen, and oxygen. A **carbohydrate** is a simple sugar or a larger molecule composed of multiple sugar units. Those composed of a single sugar molecule are called **monosaccharides.** Two examples are fructose, sometimes called fruit sugar, and glucose (Figure 3-1), a sugar that commonly provides the most immediate source of energy to cells. Monosaccharides are easily used within cells as energy sources.

Two monosaccharides can be bonded together to form a **disaccharide.** Examples of disaccharides are sucrose (common table sugar), maltose (found in many seeds), and lactose (milk sugar). If many monosaccharides are bonded together, the resulting long carbohydrate molecule is called a **polysaccharide** (Figure 3-2).

Animals, including humans, store glucose energy in the form of glycogen, a highly branched polysaccharide chain of glucose molecules. Starch, a key energy-storing polysaccharide in plants, is an important food molecule for humans. Plant cells are surrounded by a tough cell wall mostly made of another chain of glucose molecules, cellulose. Our digestive system does not break cellulose apart very well. Plant materials thus provide bulky cellulose fiber, which is necessary for good health.

Many monosaccharide molecules react upon heating with a copper-containing compound called Benedict's reagent, changing the reagent color from blue to orange or red. A disaccharide may or may not

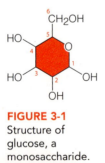

FIGURE 3-1
Structure of glucose, a monosaccharide.

react with Benedict's solution, depending on how the bonding of the component monosaccharides took place.

Polysaccharides do not react with Benedict's reagent. Other tests are available for some polysaccharides, the most common of which is Lugol's test for starch. In this test, a dilute solution of potassium iodide reacts with the starch molecules to form a dark blue (nearly black) colored product.

1. TEST FOR SUGARS USING BENEDICT'S SOLUTION

PROCEDURE

a. Half-fill the beaker with tap water and apply heat with a hot plate or burner to bring the water to a gentle boil.

b. Using the china marker, number the test tubes 1 to 10.

c. Add 10 drops of each test solution into the numbered test tubes as listed in Table 3-1.

d. Add 10 drops of Benedict's solution to each test tube and then agitate the mixture by shaking the tubes from side to side or with a vortex mixer, if available. Record the color of the mixture in Table 3-1 in the column "Initial Color."

e. Heat the tubes in the boiling water bath for about 3 minutes. Remove the tubes with the test tube clamp and place in the rack. Record any color changes that have taken place in the column "Color After Heating."

A cloudy precipitate will form that varies from green or yellow (+) to orange or red (++) to brown (+++) in color, indicating increasing concentrations of sugars. Record your conclusions regarding the presence or absence and concentration of simple sugars.

What is the purpose of the tube containing distilled water?

_____

Rank the test substances in apparent order of sugar concentration, from none (0) to most (+++):

_____

_____

_____

2. TEST FOR STARCH USING LUGOL'S SOLUTION

PROCEDURE

a. Using the china marker, number the test tubes 1 to 8.

b. Add 10 drops of the correct test solutions into the test tubes as described in Table 3-2.

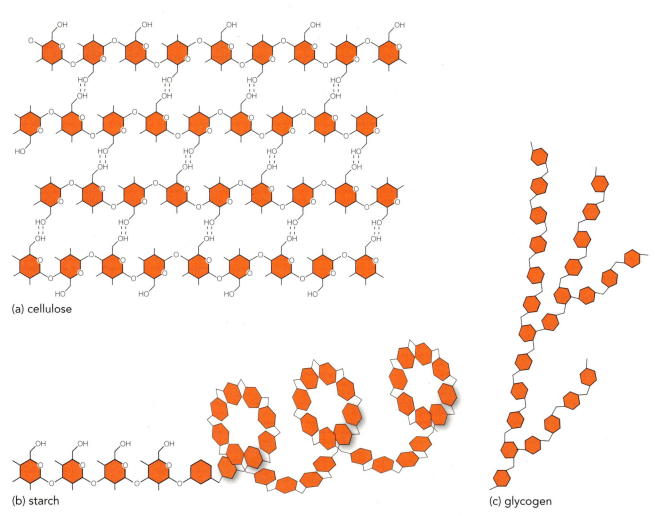

(a) cellulose

(b) starch

(c) glycogen

FIGURE 3-2 Polysaccharides are made of one or more chains of many sugar units. This diagram shows the structure of (a) cellulose, (b) starch, and (c) glycogen. Glucose is the basic building block of all of these carbohydrates.

TABLE 3-1	Benedict's Test for Simple Sugars			
Tube No.	Contents	Initial Color	Color After Heating	Sugar Content (0, +, ++, + + +)
1	Distilled water			
2	Onion juice			
3	Potato juice			
4	Cream			
5	Colorless nondiet soft drink			
6	Colorless diet soft drink			
7	Glucose			
8	Fructose			
9	Lemon juice			
10	Starch			

TABLE 3-2 Lugol's Test for Starch

Tube No.	Contents	Initial Color	Color After Adding Lugol's Solution	Starch Present? (Yes or No)
1	Distilled water			
2	Onion juice			
3	Potato juice			
4	Cream			
5	Colorless nondiet soft drink			
6	Glucose			
7	Lemon juice			
8	Starch			

c. Record the color of each solution in the column "Initial Color" in Table 3-2.

d. Add 3 drops of Lugol's solution to each test tube and then agitate the mixture by shaking the tubes from side to side or with a vortex mixer, if available. Record the color of the mixture in the column "Color After Adding Lugol's Solution." Also record your conclusions regarding the presence or absence of starch in each test solution.

What is the purpose of the tube with distilled water?

_____

What is the purpose of the tube with starch solution?

_____

B. Lipids

Lipids are oily or greasy compounds that do not dissolve in water but do dissolve in organic solvents such as ether or chloroform. Lipids provide long-term energy storage in cells, among other functions, and are very diverse. Substances we think of as fats and oils are examples of lipids.

Triglycerides are lipids having up to three fatty acids (hydrocarbon chains with an acid [—COOH] group at one end) attached to a glycerol molecule; fats are rich sources of stored energy (Figure 3-3).

Fats are triglycerides that tend to be solid at room temperature because of the saturation of their fatty acids with single covalent bonds. Some common examples of fats include the animal fats in butter and lard.

Oils are triglycerides that tend to be liquid at room temperature. Similar to fats, they are composed of fatty acids and glycerol, but their fatty acids have one or more double covalent bonds, making them unsaturated.

One of the simplest tests for lipids is to determine whether they leave a grease spot on a piece of uncoated paper, such as a paper grocery bag. Another test uses Sudan IV, a chemical commonly used to identify fats and oils in microscopic preparations. Sudan IV is a reddish liquid that dissolves in lipids and can be used to indicate their presence in a test tube.

1. UNCOATED PAPER TEST

PROCEDURE

a. Place a drop of each test substance listed in Step 2 on a *labeled* spot on the grocery bag piece. Set it aside to dry for 10 minutes.

b. After 10 minutes, describe the appearance of each spot on the paper.

Distilled water _____

Onion juice _____

Potato juice _____

Hamburger juice _____

Cream _____

Vegetable oil _____

Which substances appear to contain lipids?

_____

_____

What was the purpose of testing vegetable oil?

_____

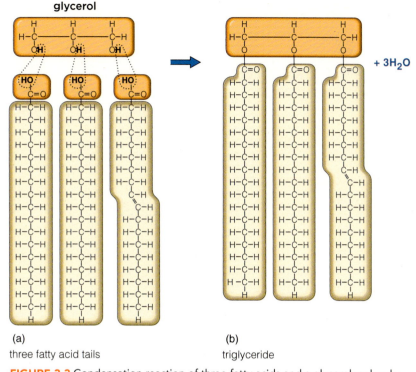

glycerol

+ 3H₂O

(a)
three fatty acid tails

(b)
triglyceride

FIGURE 3-3 Condensation reaction of three fatty acids and a glycerol molecule to form a triglyceride.

2. SUDAN IV TEST

PROCEDURE

a. With a china marker, number the test tubes 1 to 6.

b. Add 20 drops of dH_2O to each test tube and then 20 drops of each substance indicated in Table 3-3 to the appropriate test tube.

c. Add 9 drops of Sudan IV to each tube, agitate by shaking the tube side to side, and then add 10 more drops of dH_2O to each tube.

d. Record your results in Table 3-3.

Which test, the uncoated paper test or the Sudan IV test, do you think is most sensitive to small quantities of lipids? Why?

_____

_____

C. Proteins

Proteins are a diverse group of biological molecules with a wide range of functions in the human body. Many are structural components of muscle, bone, hair, and nails, among other tissues. Others are enzymes that speed up cellular reactions that would otherwise take years to occur. Movement of structures within cells (e.g., during cell division) and of sperm cells also depends on proteins. There are still further functions for specific proteins. For example, egg albumin (egg white) protects and nourishes a developing embryo.

All proteins are complex chains of **amino acids,** whose general structure is illustrated in Figure 3-4. Humans cannot synthesize about half of the amino acids; these *essential amino acids* must be obtained through our foods.

Although there are only about 20 amino acids, innumerable kinds of proteins result from different amino acid sequences. A protein begins to form when two or more amino acids are linked together by peptide bonds (bonds formed by condensation reactions between the amino group of one amino acid and the acid group of another). Multiple amino acids joined together form a polypeptide chain (Figure 3-5). Then the chain folds and twists as links form between adjacent parts of the chain.

Several tests are used for proteins, including Biuret reagent, which indicates the presence of peptide bonds. The greater the number of peptide bonds, the more intense the bluish color reaction with Biuret.

1. TEST FOR PROTEINS WITH BIURET REAGENT

PROCEDURE

a. Using the china marker, number the test tubes 1 to 6.

b. Add 20 drops of the correct test solutions into the test tubes as described in Table 3-4.

c. Add 10 drops of Biuret reagent to each test tube and agitate the mixture by shaking the tubes from side to side or with a vortex mixer, if available.

TABLE 3-3 Sudan IV Test for Lipids

Tube No.	Contents	Observations After Addition of Sudan IV	Conclusions
1	Distilled water		
2	Onion juice		
3	Potato juice		
4	Hamburger juice		
5	Cream		
6	Vegetable oil		

TABLE 3-4 Biuret Test for Protein

Tube No.	Contents	Color Reaction	Conclusions
1	Distilled water		
2	Onion juice		
3	Potato juice		
4	Hamburger juice		
5	Cream		
6	Egg albumin		

d. Wait 2 minutes and then record the color of the mixture in Table 3-4 in the column "Color Reaction." Also enter your conclusion about the presence of protein in the substance.

D. Testing Unknown Food Substances

Your instructor will provide you with several samples of unidentified food substances. Alternatively, you may be instructed to bring your own food samples so you can attempt to determine their composition.

You may need to grind small quantities of solid substances with a mortar and pestle and then dilute the resulting slurry with water to perform some of the tests. Filter the slurry to obtain a relatively clear liquid for your tests. Use the additional equipment supplied by your instructor to accomplish the preparations.

OPTIONAL MATERIALS

Per student group (4):

- Mortar and pestle
- Funnel and filter paper *or* coffee filter
- Dropping pipettes

PROCEDURE

1. Following the previous procedures, perform tests for carbohydrates (Benedict's and Lugol's tests), lipids (uncoated paper and Sudan IV tests), and proteins (Biuret reagent) on each unknown item.

2. Record your test results (+ or − for each test) in Table 3-5.

3. What kind(s) of biological molecule(s) are found in significant concentrations in each food? Did you find any of the test food substances that appeared to be relatively simple, providing only one of the biological molecules? Did you find any food substances that appeared to be complex, providing multiple kinds of biological molecules? Describe.

_____

_____

_____

_____

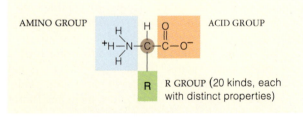

FIGURE 3-4 General structure of an amino acid. The R-group consists of one or more atoms and is unique for each of the 20 amino acids.

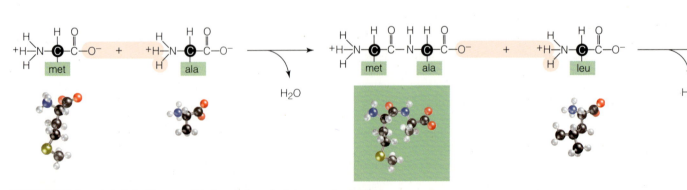

FIGURE 3-5 A protein is built as peptide bonds form between amino acids.

TABLE 3-5 Results of Testing the Compositions of Unknown Food Materials

Unknown	Benedict's Test for Sugars	Lugol's Test for Starch	Brown Paper Test for Lipids	Sudan IV Test for Lipids	Biuret Test for Protein
A					
B					
C					
D					
E					

_____

_____

_____

_____

Note: After completing all experiments, take your dirty glassware to the sink and wash it following the directions given in "Instructions for Washing Laboratory Glassware," page xii. Invert the test tubes in the test tube rack so they drain. Tidy up your work area, making certain all equipment used in this exercise is there for the next class.

3.2 Food Guides and Diet Analysis

Food must be digested before most nutrients are available within the body. In the digestive system, large organic molecules are digested into smaller components (e.g., proteins are broken down into amino acids, polysaccharides are broken down into simple sugars). The resulting smaller molecules are absorbed across the intestinal wall, transported throughout the body, and used to build and power cells. We truly *are* what we eat.

The U.S. Department of Agriculture and health authorities issue dietary guidelines for average people of good health. Several nutrition principles underlie the recommendations, including moderation in portion sizes, balance among food groups to ensure that all key nutrients are included but none are in excess, and emphasis on daily exercise for a multitude of health benefits.

The major food groups are vegetables and fruits; grains; dairy and soy; meat, beans, and alternatives; and vegetable oils. There also may be an allotment of "discretionary calories" for each person, which are calories in excess of basic energy needs. Each type of food provides specific nutrients.

Vegetables and **fruits** are especially important sources of the vitamins and minerals needed for a healthy diet, as well as complex carbohydrates and **phytochemicals,** a catch-all term for a diverse group of molecules whose benefits we are just beginning to understand. Diets that include a colorful variety of vegetables and fruits may help prevent illnesses, including heart disease, stroke, and certain types of cancer.

Grains are high in the polysaccharides (starches and cellulose) that provide ready energy sources plus dietary fiber. Health authorities recommend that you "make half your grains whole," meaning that *at least* half your grain intake should be whole grains, which supply important vitamins and minerals, plus the fiber needed for a healthy digestive system. Whole grains include oatmeal, brown rice, popcorn, and whole-grain or whole wheat flour.

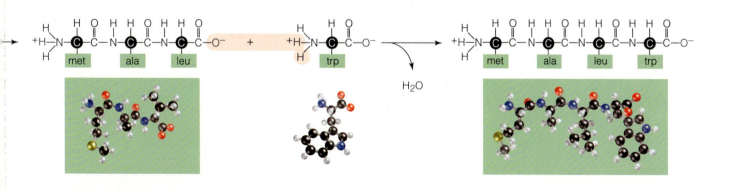

Low-fat **dairy** and **soy** products are important sources of calcium and provide significant amounts of proteins. Protein-rich, vitamin D–fortified dairy and alternative soy foods, along with weight-bearing exercise, build and maintain strong bones.

Smaller quantities are recommended of the **meat, bean, and alternatives** group, which includes poultry, fish, red meats, eggs, beans, and nuts. These protein-rich foods supply essential amino acids, vitamins, and minerals and help maintain the "full" feeling between meals.

Recommended **vegetable oils** include monounsaturated vegetable oils (e.g., olive oil and canola oil) and polyunsaturated oils (e.g., corn oil, soybean oil). These oils are considered healthy in moderation, unlike the solid saturated fats (sour cream, butter, palm oil) and especially the harmful trans fats, which can be identified on food labels as "hydrogenated" or "partially hydrogenated" oils.

A **calorie** is a unit of energy content. Specific calorie guidelines vary depending on your age, gender, and activity level, with more calories of food energy needed to power your cells as your activity level increases. Use Table 3-6 to determine your recommended average daily calorie allowance.

Consult Table 3-7 for guidelines for consumption of each food group for persons of moderate activity level and your age level and gender.

A. Food Diary

You will be on the path toward enjoying a healthy diet that provides balanced nutrients and calories if you follow the guidelines and eat a variety of foods in moderate portions. In this activity, you will analyze *your* diet so you can see in which areas you make good nutrition choices and which areas need improvement.

PROCEDURE (To Be Completed Before Class)

For two typical days during the week preceding this activity, keep a complete diary of *every* food and drink item you consume other than water, unsweetened coffee or tea, or other zero-calorie beverages. In Table 3-8,

---

TABLE 3-6 Approximate Daily Calorie Requirements

	Sedentary (<30 min of Moderate Activity Daily)	Moderate (30–60 min of Moderate Activity Daily)	Active (>60 min of Moderate Activity Daily)
Females			
Ages 16–18 years	1750	2100	2400
Ages 19–30 years	1900	2100	2350
Ages 31–50 years	1800	2000	2250
MALES			
Ages 16–18 years	2400	2800	3200
Ages 19–30 years	2500	2700	3000
Ages 31–50 years	2350	2600	2900

TABLE 3-7 Recommended Daily Food Group Serving Intake for Moderate Activity Levels

	Teens		Adults	
Age in Years	14–18		19–50	
Gender	Females	Males	Females	Males
Vegetables and fruits	7	8	7–8	8–10
Grains	6	7	6–7	8
Dairy and soy	3–4	3–4	2	2
Meat and alternatives	2	3	2	3
Oils and fats	2 to 3 Tablespoons, or 24 grams, of unsaturated or monounsaturated fats for all uses			

TABLE 3-8 Two-Day Food Diary

| Food or Beverage Item | Total Portion Size | Food Group Servings in Each Food Item | | | | | | | Calories |
		Vegetables	Fruits	Grains	Dairy and Soy	Meat, Beans, Alternatives	Fats and Oils (g)	
Example: McDonalds Double Cheeseburger	1	0.2	0	2.4	0.5	1.9	24.4	459
2-day total								
2-day average								

record what you eat and drink and how much (the portion size: ounces, cups, teaspoons, and so on). Enter these data in the first two columns.

It's often difficult to determine portion sizes. To do this accurately, record the information on food product labels regarding serving size. You may also want to use a measuring cup for this activity. For unlabeled foods, use the portion guidelines in Table 3-9.

You should record items immediately after eating, whenever possible. Be as honest, specific, and descriptive as you can. The more specific you can be, the more accurate your diet analysis will be. For example, include brand names, the method of preparation (baked, fried, canned, frozen), and the type of food (whole, 2%, 1%, or skim milk). Record the calorie content of each item whenever possible. Do this for *every* meal and snack.

Eat your usual diet. Don't change your eating habits for this exercise.

B. In-Class or At-Home Food Diary Analysis

MATERIALS

Per student group (4):

- Food diary data
- Diet analysis books or computers with Internet access

PROCEDURE

1. Complete the food diary table (Table 3-8) by using online diet analysis calculators identified by your instructor or reference books. Determine how each food item is allocated as servings of the various food groups: grains, vegetables, fruits, dairy and soy, meat and beans, and vegetable oils. Alternatively, if using online resources such as USDA MyTracker, record summary serving data for an entire day's food intake.

2. Determine and record the calorie content of each item.

3. Average the food group servings and calorie content over the 2 days of data collection and use the averaged data for the rest of this activity.

4. Using Figure 3-6a, create a bar graph to represent the daily food group serving and discretionary calorie *recommendations* for your gender and activity level. Record discretionary calories (those that exceed your recommended calorie intake) in 100-calorie "servings."

5. Now construct a bar graph of your *actual* daily food group serving and discretionary calorie consumption using Figure 3-6b with average 2-day data from your food diary.

6. Answer the following questions *after* you have completed your 2-day food diary and bar graphs.

   (a) What is your typical level of physical activity (sedentary, moderate, active)? _____

   (b) What is your approximate daily calorie requirement? _____

   (c) How does your caloric intake compare with the recommendations for your gender and activity level? If your caloric intake is not in balance, suggest at least two strategies to correct the imbalance.

   _____

   _____

   _____

   _____

   (d) Which, if any, food groups are you eating too much of? What major kinds of nutrients are you thus overconsuming?

   _____

   _____

   (e) Which, if any, food groups are you not eating enough of? What kinds of major nutrients are thus underrepresented in your diet?

   _____

   _____

   _____

   _____

   (f) What proportion of your grain servings included whole grains? _____

   How many different *kinds* of vegetables and fruits did you consume? _____

   (g) Choose one area in which your diet differs from recommendations. Search the Internet for two sites that describe consequences of that dietary choice. List the two sites and briefly summarize their contents. (If your diet fully conforms to recommendations, choose a dietary issue of interest instead and research that topic.)

   http://

   http://

   (h) Given all the above information, describe two *reasonable* changes you could make to improve your diet and your health. (If your diet is currently healthy, describe your two greatest challenges in maintaining that healthy diet plus a successful strategy for overcoming each.)

   _____

   _____

   _____

   _____

TABLE 3-9 Visualizing Portion Sizes

Portion	Approximate Size of Item
1 teaspoon (tsp)	Thumb tip to base of nail
1 tablespoon (Tbsp)	Whole thumb
1 ounce (oz)	Two dice; one golf ball; one slice of processed cheese
3 ounces (oz) (one serving meat)	Palm of hand (thickness and size); one deck of playing cards; checkbook
½ cup	Volume within a cupped hand; one tennis ball; one lightbulb
1 cup	Volume of a woman's fist; one baseball (not the larger softball)
1 oz grains	½ cup of oatmeal or rice; 1 cup of cereal flakes; ½ English muffin; 1 slice of commercial bread loaf

FIGURE 3-6a My *recommended* food group serving bar chart.

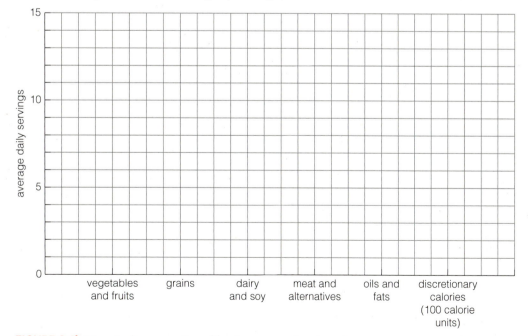

FIGURE 3-6b My *actual* consumption of food group servings bar chart.

_____ **1.** A carbohydrate consists of
(a) amino acid units
(b) one or more sugar units
(c) lipid droplets
(d) glycerol

_____ **2.** A protein is made up of
(a) amino acid units
(b) one or more sugar units
(c) lipid droplets
(d) Biuret solution

_____ **3.** Benedict's solution is commonly used to test for the presence of
(a) proteins
(b) certain carbohydrates
(c) nucleic acids
(d) lipids

_____ **4.** Glycogen is
(a) a polysaccharide
(b) a storage carbohydrate
(c) found in human tissues
(d) all of the above

_____ **5.** To test for presence of starch, one would use
(a) Benedict's solution
(b) uncoated paper
(c) Sudan IV
(d) Lugol's solution

_____ **6.** Rich sources of stored energy that are dissolvable in organic solvents are
(a) carbohydrates
(b) proteins
(c) glucose
(d) lipids

_____ **7.** Rubbing a substance on uncoated paper should reveal if it contains
(a) lipid
(b) carbohydrate
(c) protein
(d) sugar

_____ **8.** Proteins consist of
(a) monosaccharides linked in chains
(b) amino acid units
(c) polysaccharide units
(d) condensed fatty acids

_____ **9.** Biuret reagent will indicate the presence of
(a) peptide bonds
(b) proteins
(c) amino acids units linked together
(d) all of the above

_____ **10.** The largest number of food servings in your daily diet should come from
(a) meats and beans
(b) dairy products
(c) vegetables and fruits
(d) grains

EXERCISE **3** **Chemistry of Life: Biological Molecules and Diet Analysis**

POST-LAB QUESTIONS

A.1 Test for Sugars Using Benedict's Solution

1. Let's suppose you are teaching science in a part of the world without easy access to a doctor and you're worried that you may have developed diabetes. (Diabetics are unable to regulate blood glucose levels, and glucose accumulates in their urine.) What test could you perform to gain an indication of whether you have diabetes?

2. The test tubes in the photograph contain Benedict's solution and two unknown substances that have been heated. What do the results indicate for each substance?

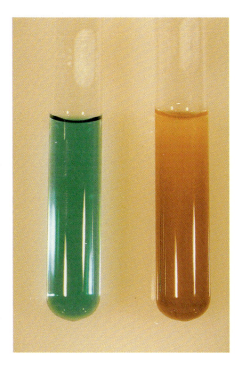

3. How could you verify that a soft drink container contains diet soft drink rather than soft drink sweetened with fructose?

A.2 Test for Starch Using Lugol's Solution

4. Observe the photomicrograph accompanying this question. This thin section of a potato tuber has been stained with Lugol's iodine solution. When you eat French fries, the potato material is broken down in your small intestine into what kind of small subunits?

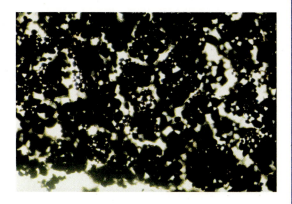

B.2 Sudan IV Test

5. The test tube in the photograph contains water at the bottom and another substance that has been stained with Sudan IV at the top. What is the macromolecular composition of this stained substance?

6. You are given a sample of an unknown food. Describe how you would test it for the presence of lipids.

7. You wish to test the same unknown food for the presence of sugars. Describe how you would do so.

Food for Thought

8. What is the purpose of the distilled water sample in each of the chemical tests in this exercise?

9. Many health food stores carry enzyme preparations that are intended to be ingested orally (by mouth) to supplement existing enzymes in various organs such as the liver, heart, and muscle. Use your knowledge of the digestion process to explain why these preparations are unlikely to be effective as advertised.

10. A young child grows rapidly, with high levels of cell division and high energy requirements. If you were planning the child's diet, which food groups would you emphasize, and why? Which food groups would you deemphasize, and why?

Cell Structure

OBJECTIVES

After completing this exercise, you will be able to

1. define *cell, cell theory, plasma membrane, DNA, cytoplasm, eukaryotic, nucleus, organelle, cytoplasmic streaming, sol, gel, envelope,* and *endomembrane system.*

2. list the structural features shared by all cells.

3. describe the basic structures of the general human cell.

4. identify the cell parts described in this exercise.

5. state the function for each cell part.

6. recognize the structures presented in **boldface** in the procedure sections.

Introduction

Structurally and functionally, all life has one common feature: all living organisms are composed of **cells.** The development of this concept began with Robert Hooke's seventeenth-century observation that slices of cork were made up of small units. He called these units "cells" because their structure reminded him of the small cubicles that monks lived in. Over the next 100 years, the **cell theory** emerged. This theory has three principles: (1) all organisms are composed of one or more cells, (2) the cell is the basic *living* unit of organization, and (3) all cells arise from preexisting cells.

Although cells vary in their organization, size, and function, all share three structural features: (1) all possess a **plasma membrane** defining the boundary of the living material; (2) all contain a region of **DNA** (deoxyribonucleic acid), which stores genetic information; and (3) all contain **cytoplasm,** which is everything inside the plasma membrane that is not part of the DNA region.

Cells of humans, other animals, plants, and many microbes are **eukaryotic** cells, those with a nucleus. (Bacterial cells, by contrast, are *prokaryotic* cells, those without a nucleus.) The nucleus is a double-membrane–bound compartment housing multiple *chromosomes,* each composed of a single molecule of DNA plus associated proteins. Eukaryotic cells also contain many other membrane-bound structures, the **organelles,** which create compartments within the cell in which different cellular functions are carried out.

This exercise will familiarize you with the basics of cell structure and the function of eukaryotic cells, such as those of humans.

4.1 Eukaryotic Cells Observed with a Light Microscope

MATERIALS

Per student:

- Toothpick
- Microscope slides
- Coverslips
- Culture of *Physarum polycephalum*
- Compound microscope

Per student pair:

- Methylene blue in dropping bottle
- Distilled water (dH$_2$O) in dropping bottle

Per student group (table):

- Tissue paper
- Alcohol-containing disposal jar

A. HUMAN CHEEK CELLS

PROCEDURE

1. Using the broad end of a clean toothpick, gently scrape the inside of your cheek. Stir the scrapings into a drop of distilled water on a clean microscope slide and then add a coverslip. Dispose of used toothpicks in the jar containing alcohol.

2. Because the cells are almost transparent, decrease the amount of light entering the objective lens to increase the contrast. (See Exercise 2, page 10.)

Find the cells using the low-power objective of your microscope; then switch to the high-dry objective for detailed study.

3. Find the **nucleus,** a centrally located spherical body within the **cytoplasm** of each cell.

4. Stain your cheek cells with a dilute solution of methylene blue, a dye that stains the nucleus darker than the surrounding cytoplasm. To stain your slide, follow the directions illustrated in Figure 4-1.

 Without removing the coverslip, add a drop of the stain beside one edge of the coverslip. Then draw the stain under the coverslip by touching a piece of tissue paper to the *opposite* side of the coverslip.

5. In Figure 4-2, sketch the cheek cells, labeling the **cytoplasm, nucleus**, and location of the **plasma membrane**. (A light microscope cannot resolve the plasma membrane, but the boundary between the cytoplasm and the external medium indicates its location.) Many of the cells will be folded or wrinkled because of their thin, flexible nature. In your sketch, estimate and

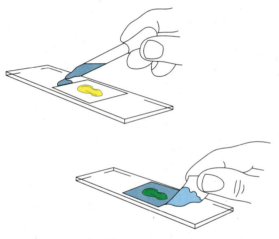

FIGURE 4-1 Method for staining specimens under coverslip on microscope slide.

Labels: cytoplasm, nucleus, plasma membrane

FIGURE 4-2 Drawing of human cheek cells (_____ ×).

record the size of the cells. (The method for estimating the size is found in Exercise 2, page 10.)

B. CELLS OF A SLIME MOLD, *PHYSARUM POLYCEPHALUM*

The slime mold *P. polycephalum* is in the Kingdom Protista, not in the Kingdom Animalia as are humans. *Physarum* is a unicellular organism, so it contains all the metabolic machinery for independent existence. Furthermore, its single cell can grow to relatively gigantic size, allowing easy observation of its cellular organelles, which share many features with human cells.

Physarum absorbs its organic food substances by engulfing particles and digesting them within organelles called *food vacuoles.* Your culture will be growing across the surface of agar gel in a Petri dish. The organism grows branches out across the substrate and then comes together on the surface of a food item. The "food" provided to your culture is bacterial cells naturally occurring on the surfaces of oatmeal flakes.

PROCEDURE

1. Place a plain microscope slide on the stage of your compound microscope. This will serve as a platform on which you can place a culture dish.

2. Obtain a Petri dish culture of *Physarum*, remove the lid, and place it on the platform. Observe initially with the low-power objective and then with the medium-power objective. If you choose to view the culture with the high-dry objective, place a coverslip over part of the organism before rotating the objective into place. (This prevents the agar from getting on the lens.)

 Physarum is *multinucleate,* meaning that more than one nucleus occurs within the cytoplasm. Unfortunately, the nuclei are tiny; you will not be able to distinguish them from other organelles in the cytoplasm.

3. Locate the **plasma membrane**, which is the outer boundary of the cytoplasm. Again, the resolving power of your microscope is not sufficient to allow you to actually view the membrane.

4. Watch the cytoplasm of the organism move. This intracellular motion is known as **cytoplasmic streaming**. Contractile proteins called *microfilaments* are believed to be responsible for cytoplasmic streaming. The cytoplasmic motion carries nutrients, proteins, organelles, and other cytoplasmic components throughout the cell.

5. Note that the outer portion of the cytoplasm appears solid; this is the **gel** state of the cytoplasm. Notice that the granules closer to the interior are in motion within a fluid; this interior portion of the cytoplasm is in the **sol** state. Movement of the organism occurs as the sol-state cytoplasm at the advancing tip pushes against the plasma membrane, causing the region to

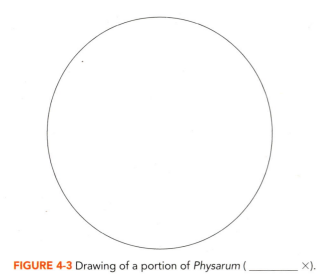

FIGURE 4-3 Drawing of a portion of *Physarum* (_____ ✕).

swell outward. The sol-state cytoplasm flows into the region, converting to the gel state along the margins.

6. In Figure 4-3, sketch the portion of *Physarum* that you have been observing and label it.

4.2 Experiment: Temperature Effects on Cytoplasmic Streaming

Temperature affects many cellular and organismal processes. For example, reptiles and insects are ectotherms (animals that gain heat from the environment), unlike humans, whose body heat comes primarily from cellular metabolism. You may have observed that in nature, these animals are relatively sluggish during cold weather. Is the same true for other organisms, such as the slime mold?

This simple experiment addresses the hypothesis that *cold slows cytoplasmic streaming in P. polycephalum.* Before starting this experiment, you may wish to review the discussion in Exercise 1, "Learning About the Natural World."

MATERIALS

Per experimental group:

• Culture of *P. polycephalum*
• Compound microscope
• Container with ice *or* refrigerator
• Timer *or* watch with second hand
• Celsius thermometer

PROCEDURE

1. Place the *Physarum* culture on the stage of your compound microscope as described in Section I-B.

2. Time the duration of cytoplasmic streaming in one direction and then in the other direction. Do this for five cycles of back-and-forth motion. Calculate the average duration of flow in either direction. Record the temperature and your observations in Table 4-1.

3. Remove your culture from the microscope's stage, replace the cover, and place it and the thermometer in a refrigerator or atop ice for 15 minutes.

4. While you are waiting, in Table 4-1, write a prediction for the effect on the duration of cytoplasmic streaming by reducing the temperature of the culture of *P. polycephalum*.

5. After 15 minutes have elapsed, remove the culture from the cold treatment, record the temperature of the experimental treatment, and repeat the observations in step 2.

6. Record your observations and make a conclusion in Table 4-1, accepting or rejecting the hypothesis.

A logical question to ask at this time is *why* temperature has the effect you observed. If you perform Exercise 6, "Enzymes: The Catalysts of Life," you may be able to make an educated guess (a hypothesis).

4.3 Animal Cells Observed with the Electron Microscope

Studies with electron microscopes have yielded a wealth of information on the structure of eukaryotic cells. Structures too small to be seen with light microscopes have been identified. These include many **organelles**, which are structures in the cytoplasm that have been separated ("compartmentalized") by enclosure in membranes. Examples of organelles are the nucleus, mitochondria, endoplasmic reticulum, and Golgi bodies. Although the cells in each of the six kingdoms have some peculiarities unique to that kingdom, electron microscopy has revealed that all cells are fundamentally similar.

MATERIALS

Per student:

• Textbook

Per classroom:

• Model of animal cell

PROCEDURE

1. Study Figure 4-4, a three-dimensional representation of an animal cell. Use this figure to help you identify the corresponding structures on the model of the animal cell that is on demonstration in your classroom.

2. Figure 4-5 is an electron micrograph of an animal cell (Kingdom Animalia). Study the electron micrograph and with the aid of Figure 4-4 and any electron micrographs in your textbook, label each structure indicated.

TABLE 4-1   Effect of Temperature on Cytoplasmic Streaming

Hypothesis: Cold slows cytoplasmic streaming in *Physarum polycephalum*

Prediction of observations:

Temperature (°C)	Time/Cycle Number	Observations and Duration of Directional Flow (sec)
	1	
	2	
	3	
	4	
	5	
	Average	
	1	
	2	
	3	
	4	
	5	
	Average	
Conclusions about hypothesis		

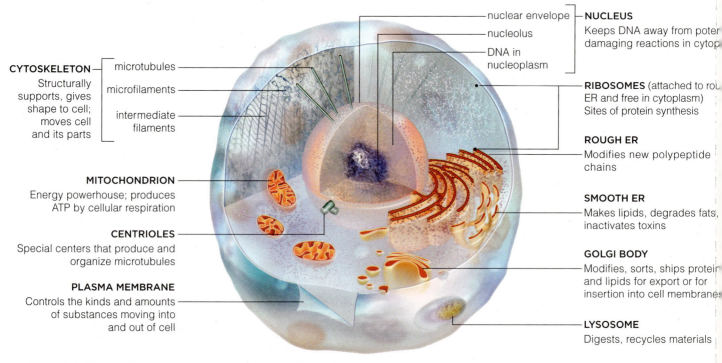

CYTOSKELETON — microtubules
Structurally supports, gives shape to cell; moves cell and its parts — microfilaments — intermediate filaments

MITOCHONDRION
Energy powerhouse; produces ATP by cellular respiration

CENTRIOLES
Special centers that produce and organize microtubules

PLASMA MEMBRANE
Controls the kinds and amounts of substances moving into and out of cell

nuclear envelope — **NUCLEUS**
nucleolus — Keeps DNA away from poten damaging reactions in cytop
DNA in nucleoplasm

RIBOSOMES (attached to rou ER and free in cytoplasm) Sites of protein synthesis

ROUGH ER
Modifies new polypeptide chains

SMOOTH ER
Makes lipids, degrades fats, inactivates toxins

GOLGI BODY
Modifies, sorts, ships protein and lipids for export or for insertion into cell membrane

LYSOSOME
Digests, recycles materials

FIGURE 4-4 Three-dimensional representation of an animal cell.

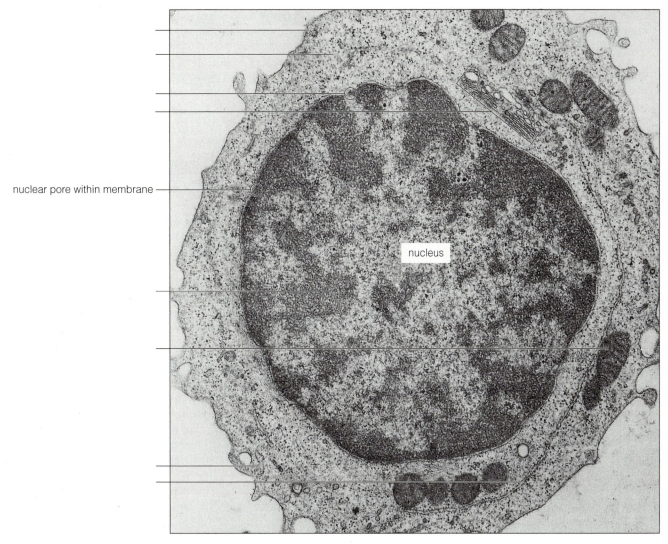

nuclear pore within membrane ————

nucleus

FIGURE 4-5 Electron micrograph of an animal cell (1600X).

Labels: plasma membrane, cytoplasm, nuclear envelope, nuclear pore, chromatin, rough ER, smooth ER, Golgi body, mitochondrion

3. Study closely the membranes surrounding the nucleus and mitochondria. Note that these two types of organelles are each bounded by *two* membranes, which are commonly referred to collectively as their **envelope**.

4. Identify each of the following structures in Figure 4-4 and/or 4-5. Using your textbook as a reference, list the function for each:

(a) Plasma membrane _____

_____

(b) Cytoplasm_____

_____

(c) Nucleus (the plural is *nuclei*) _____

_____

(d) Nuclear envelope _____

_____

(e) Nuclear pores_____

_____

(f) Chromatin_____

_____

(g) Nucleolus (the plural is *nucleoli*) _____

_____

(h) Rough endoplasmic reticulum (RER) _____

_____

(i) Smooth endoplasmic reticulum (SER) _____

_____

(j) Golgi body_____

_____

(k) Mitochondrion (the plural is *mitochondria*) ____

_____

5. Let's examine the parts of the typical animal cell in greater detail. Examine Figure 4-6, a high-magnification electron micrograph of a small portion of a mouse cell. Identify the nucleus in this micrograph. Locate the **nuclear envelope,** which separates the nuclear contents from the cytoplasm. The **nuclear pores** are the ports for information flow between the nucleus and cytoplasm.

6. Within the nucleus, identify the **chromatin**; it appears as aggregated dark granular material. During cell reproduction, the chromatin condenses into

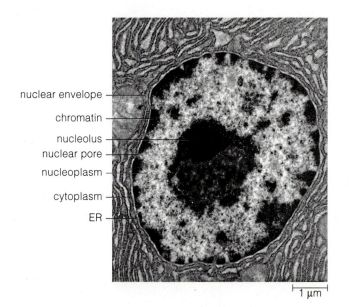

nuclear envelope —
chromatin —
nucleolus —
nuclear pore —
nucleoplasm —
cytoplasm —
ER —

1 μm

FIGURE 4-6 Transmission electron micrograph of a mouse pancreas cell.

chromosomes. Most cells in the human body have 46 chromosomes (23 pairs). One of each pair comes from each parent. Our sex cells (eggs or sperm) have only 23 chromosomes. When the nucleus of a cell has all 46 chromosomes, it is said to be *diploid* (abbreviated *2n*). Sex cells are *haploid* (abbreviated *n*). We will have more to say about this in Exercises 17 and 18.

7. The nucleus is connected to the **endomembrane system,** a series of interacting organelles within the cytoplasm. In Figure 4-6, find the parallel lines with "dots" attached to their surfaces. These are parts of the endoplasmic reticulum, with **ribosomes** on the surface. Endoplasmic reticulum with attached ribosomes is called **rough endoplasmic reticulum** (RER). Polypeptides are synthesized on the ribosomes and then move into the RER for modification. The RER creates the proteins that will be embedded in cell membranes.

Examine Figure 4-7, an electron micrograph (left side) morphing into an artist's rendering of the three-dimensional nature of RER (right side).

8. A second type of endoplasmic reticulum, **smooth endoplasmic reticulum** (SER), is shown in Figure 4-8. Lipids are synthesized in the SER of many cells, and the SER is also the location of many enzymes, especially in the liver. For example, carbohydrates are stored in the liver. Liver SER enzyme activity allows glucose to leave the cell and enter the bloodstream for transport to cells needing the energy supplied by glucose. The SER synthesizes the phospholipids that are the basis of membranes.

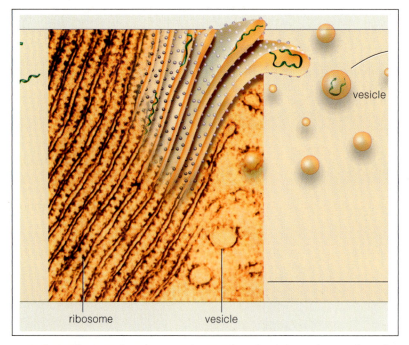

vesicle

ribosome   vesicle

FIGURE 4-7 Transmission electron micrograph and artist's rendering of rough endoplasmic reticulum (RER).

What visible structural feature distinguishes SER from RER? _____

9. Another portion of the endomembrane system is the **Golgi body**, which puts the finishing touches on proteins and lipids and then packages them for shipment to specific locations. Examine Figure 4-9,

an electron micrograph and artist's representation of a Golgi body. Vesicles are budding from the ends of each membrane.

10. Finally, examine Figure 4-10, a high-magnification electron micrograph and three-dimensional sketch of a **mitochondrion**. Observe the envelope (two membranes) bounding the mitochondrion and the numerous folds in the internal membrane, forming numerous **cristae** (the singular is *crista*) in the interior (matrix) of the mitochondrion. Different portions of the process of aerobic cellular respiration take place within the different compartments and membrane areas inside a mitochondrion.

What other cytoplasmic structure is present in this electron micrograph? _____

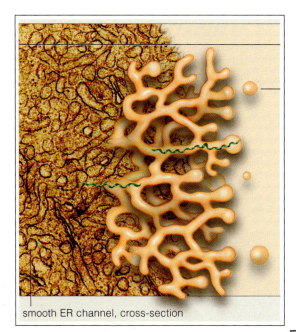

smooth ER channel, cross-section

FIGURE 4-8 Transmission electron micrograph and artist's rendering of smooth endoplasmic reticulum (SER).

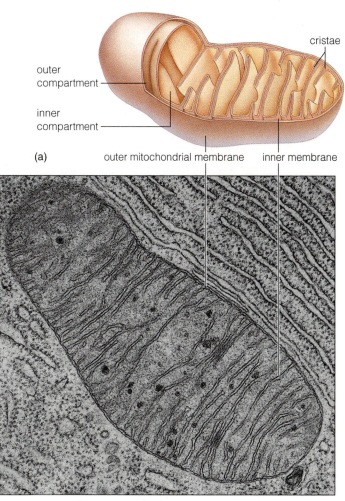

(a)

outer compartment

inner compartment

cristae

outer mitochondrial membrane   inner membrane

(b)

FIGURE 4-10 Mitochondrion (a) Sketch and (b) transmission electron micrograph.

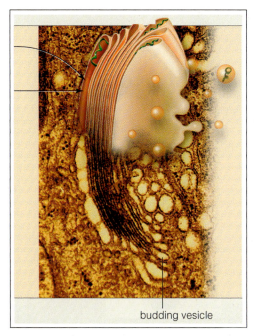

budding vesicle

FIGURE 4-9 Transmission electron micrograph and artist's rendering of Golgi body.

_____ **1.** The person who first used the term *cell* was
(a) Darwin
(b) Leeuwenhoek
(c) Hooke
(d) Watson

_____ **2.** All cells contain
(a) a nucleus, plasma membrane, and cytoplasm
(b) an endomembrane system, nucleus, and cytoplasm
(c) DNA, plasma membrane, and cytoplasm
(d) mitochondria, plasma membrane, and cytoplasm

_____ **3.** The word *eukaryotic* refers specifically to a cell containing
(a) cytoplasm
(b) a nucleus
(c) a plasma membrane
(d) none of the above

_____ **4.** You should expect to be able to clearly see which of the following with a compound light microscope?
(a) cheek cell
(b) cheek cell nucleus
(c) *Physarum* nucleus
(d) all but c

_____ **5.** *Physarum polycephalum* is:
(a) an animal
(b) a multicelled organism
(c) a photosynthetic organism that makes its own food
(d) a protist

_____ **6.** Methylene blue
(a) is used to kill cells that are moving too quickly to observe
(b) detoxifies cells treated with it
(c) provides energy to animal cells
(d) is a biological stain used to increase the contrast of cellular constituents

_____ **7.** Cytoplasmic streaming:
(a) causes mixing of cellular components within a cell
(b) causes *Physarum* sol-state cytoplasm to push against the plasma membrane
(c) happens because of the action of microfilaments
(d) all of the above

_____ **8.** Which of the following is true of the nucleus?
(a) It has no connection to the endomembrane system.
(b) Its membrane has no pores or openings.
(c) Details of its structure can be seen with an electron microscope.
(d) None of the above is true.

_____ **9.** The endomembrane system:
(a) consists of the rough endoplasmic reticulum, smooth endoplasmic reticulum, and Golgi body
(b) synthesizes, modifies, and transports biological molecules
(c) is made up of a series of membranous organelles
(d) all of the above

_____ **10.** An envelope
(a) surrounds the nucleus
(b) surrounds mitochondria
(c) consists of two membranes
(d) does all of the above

EXERCISE 4 Cell Structure

4.1.A Human Cheek Cells

1. Identify the indicated structures.

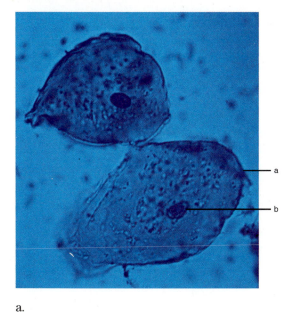

a.

b.

2. List one function for each of the structures you identified in (1).

a.

b.

4.1.B Cells of a Slime Mold, *Physarum polycephalum*

3. Is it possible for a cell to contain more than one nucleus? Explain.

4. One botulinum toxin prevents microfilaments from forming. If a *Physarum* cell were exposed to that chemical, would you expect to see cytoplasmic streaming? Explain.

5. Observe the photo at below. Identify the labeled cell components.

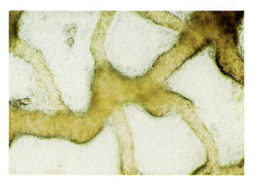

4.2 Experiment: Temperature Effects on Cytoplasmic Streaming

6. Why was it a good idea to time cytoplasmic streaming for more than a single direction cycle?

7. How could this experiment have been improved?

4.3 Animal Cells Observed with the Electron Microscope

8. Observe the electron micrograph to the below. Identify the labeled structures.

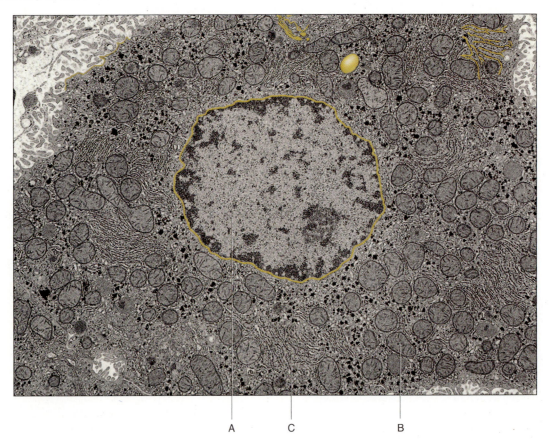

A   C   B

A.

B.

9. What are the numerous "wavy lines" within the cell (labeled C)?

Food for Thought

10. Mitochondrial diseases are common, affecting up to 4000 children per year in the United States. Mild forms cause "exercise intolerance." Search two Internet sites for information on "exercise intolerance," including the role(s) of defective mitochondria and disease symptoms. List the two sites below and briefly summarize their contents.

http://

http://

Cell Membranes and How They Work

OBJECTIVES

After completing this exercise, you will be able to

1. define *solvent, solute, solution, selectively permeable, diffusion, osmosis, concentration gradient, equilibrium, hypertonic, isotonic, hypotonic, plasmolysis, cholera,* and *oral rehydration therapy.*

2. describe the structure of cellular membranes.

3. distinguish between diffusion and osmosis.

4. determine the effects of molecular weight and temperature on diffusion.

5. describe the effects of hypertonic, isotonic, and hypotonic solutions on red blood cells.

6. describe how cholera causes dehydration, as well as how to reverse the effects.

Introduction

Living cells are made up of 75% to 85% water. Virtually all substances entering and leaving cells are dissolved in water, making it the **solvent** most important for life processes. The substances dissolved in water are called **solutes** and include such molecules as salts and sugars. The combination of a solvent and dissolved solute is a **solution.** The cytoplasm of living cells is a solution of many solutes, including ions, sugars, and proteins.

All cells are surrounded by a cell membrane made of a phospholipid bilayer with different kinds of embedded and surface proteins (Figure 5-1). Membranes are boundaries that solutes must cross to reach the cellular site where they are used in the processes of life. These membranes regulate the passage of substances into and out of the cell. They are **selectively permeable,** allowing some substances to move easily while completely or partially excluding others.

The simplest way in which solutes enter a cell is **diffusion,** the movement of solute molecules from a region of high concentration to one of lower concentration. Diffusion occurs spontaneously without the expenditure of cellular energy. Oxygen, carbon dioxide, small nonpolar molecules, and water freely cross biological membranes. Channel and transporter proteins allow the passage of some types of polar, water-soluble molecules and ions across biological membranes. Large molecules, such as proteins and polysaccharides, cannot pass membranes; they must be digested and hydrolyzed into smaller component subunits.

When they are inside the cell, solutes move through the cytoplasm by diffusion, sometimes assisted by cytoplasmic streaming.

Water (the solvent) also moves across the membrane. **Osmosis** is the movement of *water* molecules across selectively permeable membranes. Think of osmosis as a special form of diffusion, with net movement of water molecules from a region of higher *water* concentration (and thus lower solute concentration) to one of lower *water* concentration (and thus higher solute concentration).

The difference in concentration of like molecules between two regions is called a **concentration gradient.** Diffusion and osmosis take place *down* concentration gradients, with net movements of molecules from regions of higher concentration to regions of lower. Over time, the concentration of solvent and solute molecules becomes equally distributed, and the gradient ceases to exist. At this point, the system is said to be at **equilibrium.**

Molecules are always in motion, even at equilibrium. Thus, solvent and solute molecules continue to move because of random collisions. However, at equilibrium, there is no *net change* in their concentrations.

This exercise introduces you to principles of diffusion and osmosis and their application to states of disease and health.

5.1   Rate of Diffusion of Solutes

Solutes move within a cell's cytoplasm largely because of diffusion. Some solutes are able to diffuse across a membrane, either directly through the lipid bilayer *or* through the interior of passive transport protein molecules embedded in the membrane. However, the rate of diffusion (the distance diffused in a given amount of time) is affected by such factors as temperature and the size of the solute molecules. In this experiment, you will discover the effects of these two factors in gelatin (the substance of Jell-O™), a material similar to the cytoplasm and used to simulate it in this experiment.

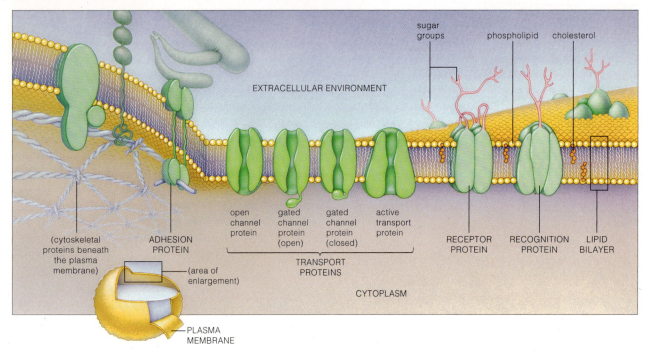

FIGURE 5-1 Artist's rendering of a cutaway view of part of a plasma membrane.

MATERIALS

Per student:

- Metric ruler

Per student group (table):

- One set of three screw-cap test tubes in rack, each half-filled with 5% gelatin, to which the following dyes have been added: potassium dichromate, aniline blue, and Janus green; labeled with each dye and marked "5°C"

- One set of three screw-cap test tubes in rack as above but marked "Room Temperature"

Per lab room:

- 5°C refrigerator

PROCEDURE

Two sets of three screw-cap test tubes have been half-filled with 5% gelatin, and 1 mL of a dye has been added to each test tube. Set 1 is in a 5°C refrigerator; set 2 is at room temperature. Record the time at which your instructor tells you the experiment was started: _____

_____

1. Remove set 1 from the refrigerator and visually compare the distances the dye has diffused in corresponding tubes of each set.

2. Invert and hold each tube vertically in front of a white sheet of paper. Use a metric ruler to measure how far each dye has diffused from the gelatin's surface. Record this distance in Table 5-1.

CAUTION Be certain the cap on each tube is tight!

3. Determine the *rate* of diffusion for each dye by using the following formula:

Rate of diffusion = Distance (cm)/Elapsed time (hours)

Time experiment ended: _____

Time experiment started: _____

Elapsed time: _____ hours

Which of the solutes diffused the slowest (regardless of temperature)? _____

Which diffused the fastest? _____

_____

What effect did temperature have on the rate of diffusion? _____

Write a conclusion about the diffusion of a solute through a gel, relating the rate of diffusion to the molecular weight of the solute and to temperature.

_____

_____

_____

_____

Note: Return set 1 to the refrigerator.

TABLE 5-1   Effect of Temperature on Diffusion Rates of Various Solutes

Solute (dye)	Set 1 (5°C)		Set 2 (Room Temperature)	
	Distance (mm)	Rate	Distance (mm)	Rate
Potassium dichromate (MW = 294)[a]				
Janus green (MW = 511)				
Aniline blue (MW = 738)				

[a]MW = molecular weight, a reflection of the mass of a substance. To determine MW, add the atomic weights of all elements in a compound.

5.2   Experiment: Effect of Solute Size on Diffusion and Selective Permeability of Membranes

One example of a selectively permeable membrane within a living cell is the plasma membrane. In this experiment, you will learn about diffusion using dialysis membrane, a selectively permeable cellulose sheet that permits the passage of water and ions but obstructs passage of larger molecules. If you examined the membrane with a scanning electron microscope, you would see that it has pores; it thus prevents molecules larger than the pores from passing through the membrane.

This activity will demonstrate the effect of solute molecule sizes on their ability to pass through a selectively permeable membrane.

MATERIALS

Per student group (4):

- Bottle of dilute iodine solution (I_2KI)
- Bottle of 10% soluble starch solution
- Two 15-cm lengths of dialysis tubing, soaking in dH_2O
- Four 10-cm pieces of string or waxed dental floss
- Dishpan half-filled with dH_2O
- Two 400-mL graduated beakers
- Ring stand and funnel apparatus (Figure 5-2)
- China marker
- 25-mL graduated cylinder
- Two plastic dropping pipettes
- Spot plate *or* plastic Petri dish half
- Scissors

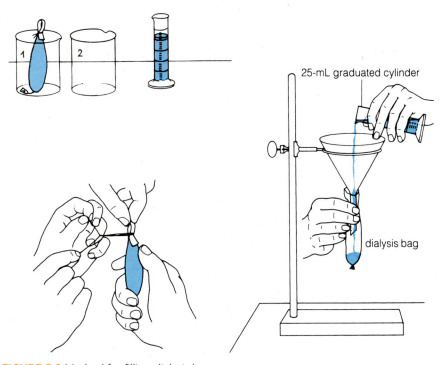

FIGURE 5-2 Method for filling dialysis bags.

PROCEDURE

Work in groups of four.

1. Label one beaker "#1, Starch Bag in Iodine" and the second beaker "#2, Iodine Bag in Starch."

2. Obtain a 15-cm section of dialysis tubing that has been soaked in dH_2O.

3. Fold over one end of the tubing and tie it securely with string or dental floss to form a leakproof bag (Figure 5-2).

4. Slip the open end of the bag over the stem of a funnel.

5. Measure 15 mL of 1% soluble starch and pour it through the funnel to fill the bag.

6. Remove the bag from the funnel; fold and tie the open end of the bag.

7. Rinse the tied bag in a dishpan partially filled with dH_2O; place the bag into Beaker 1.

8. Make and fill a second bag with the iodine solution. Place that dialysis tubing bag into Beaker 2.

9. Pour 200 mL of iodine solution into Beaker 1.

10. Pour 200 mL of starch solution into Beaker 2.

11. Record the time: _____

12. Now and every 15 minutes over the next hour, record your observations of the contents inside each dialysis bag and beaker in Table 5-2.

13. Perform a visual test for the presence of starch by adding 1 drop of iodine solution to several drops of starch solution in the spot plate or Petri dish half. Describe the result of this test. How can you identify the presence of starch?

_____

_____

_____

To which substance(s) was the dialysis tubing permeable?

_____

_____

_____

I2KI has a molecular weight of approximately 420, and that of starch, a large and variable polysaccharide, ranges from 20,000 to 225,000. What physical property of the dialysis tubing might explain its differential permeability to those two substances?

_____

_____

_____

19. Puncture the dialysis bags and discard the contents of the bags and beakers down the sink drain. Wash the glassware by using the technique described on page xii.

20. Discard the dialysis tubing in the wastebasket.

5.3 Osmosis and Red Blood Cells

Recall that the composition of a solution can be described in terms of the proportions of the total made up of solute and solvent. For example, a 5% glucose solution is made of 5% glucose molecules and 95% water molecules.

Fill in the following:

A 10% salt (NaCl) solution is composed of _____ % NaCl and _____ % H_2O.

A 0.9% NaCl solution is composed of _____ % NaCl and _____ % H_2O.

Tonicity describes one solution's solute concentration *compared with that of another solution*. The solution containing the lower concentration of solute molecules than the other is **hypotonic** *relative to the second solution*. Solutions containing equal concentrations of solute are **isotonic** to each other, and one containing a greater concentration of solute relative to another is **hypertonic** to that second solution.

Fill in the following:

A 10% salt solution is _____ relative to a 0.9% salt solution.

A 0.9% salt solution is _____ relative to a 5% salt solution.

Osmosis occurs when different concentrations of water are separated by a selectively permeable membrane.

TABLE 5-2	Observations of Dialysis Bag and Beaker Contents				
	At Start of Experiment	**After 15 Min**	**After 30 Min**	**After 45 Min**	**After 60 Min**
Beaker 1 dialysis bag contents					
Beaker 1 contents					
Beaker 2 dialysis bag contents					
Beaker 2 contents					

Water molecules flow spontaneously (no cellular energy is needed) from the area of higher water concentration to the area of lower water concentration, with a net flow continuing across the membrane until water concentrations are equal on both sides.

Circle the best answers: This means that water exhibits net flow by osmosis from a (hypertonic, hypotonic, *or* isotonic) solution to a (hypertonic, hypotonic, or isotonic) solution.

An animal cell increases in size as water enters the cell. However, because the plasma membrane is relatively fragile, it may rupture when too much water enters the cell. This is because of excessive pressure pushing against the membrane A blood cell that has burst is said to have been **hemolyzed**. Conversely, if water moves out of the cell, it becomes *plasmolyzed* (the cell undergoes the process of **plasmolysis**), shrinking in size. In the case of red blood cells, plasmolysis is given a special term, *crenation;* the blood cell is said to be *crenate.*

In this experiment, you will view red blood cells to see the effects of osmosis in animal cells.

MATERIALS

Per student:

- Compound microscope

Per student group (4):

- Three clean screw-cap test tubes
- Test tube rack
- Metric ruler
- China marker
- Bottle of 0.9% sodium chloride (NaCl)
- Bottle of 10% NaCl
- Bottle of dH_2O
- Three disposable plastic pipets
- Three clean microscope slides
- Three coverslips

Per student group (table):

- Bottle of sheep blood (in an ice bath)

Per lab room:

- Source of dH_2O

PROCEDURE

Work in groups of four for this experiment but do the microscopic observations individually.

1. Observe the scanning electron micrographs in Figure 5-3.

    Figure 5-3a illustrates the normal appearance of red blood cells. They are biconcave disks; that is, they are circular in outline with a depression in the center of both surfaces. Cells in an isotonic solution will appear like these blood cells.

    Figure 5-3b shows cells that have been plasmolyzed.

    Figure 5-3c represents cells that have taken in water but have not yet burst. (Burst red blood cells are said to be *hemolyzed*, and of course they are not visible.) Note the swollen, spherical appearance of the cells.

2. Obtain three clean screw-cap test tubes.

3. Lay test tubes 1 and 2 against a metric ruler and mark lines indicating 5 cm *from the bottom of each tube.*

4. Fill each tube as follows:

    | Tube 1: | 5 cm of 0.9% sodium chloride (NaCl) |
    | | 5 drops of sheep blood |

    | Tube 2: | 5 cm of 10% NaCl |
    | | 5 drops of sheep blood |

5. Lay test tube 3 against a metric ruler and mark lines indicating 0.5 cm and 5 cm *from the bottom of the tube.*

6. Fill tube 3 to the 0.5-cm mark with 0.9% NaCl and to the 5-cm mark with dH_2O. Then add 5 drops of sheep blood. Enter the contents of each tube in the appropriate column of Table 5-3.

7. Replace the caps and mix the contents of each tube by inverting them several times (Figure 5-4a).

8. Hold each tube flat against the printed page of your lab manual (Figure 5-4b). *Only if the blood cells are hemolyzed should you be able to read the print.*

9. In Table 5-3, record your observations in the column "Print Visible?"

10. Number three clean microscope slides.

(a) 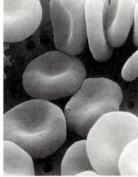 (b) 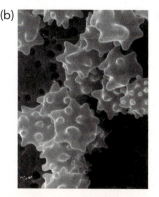 (c)

FIGURE 5-3 Scanning electron micrographs of red blood cells. (*Photos from M. Sheetz, R. Painter, and S. Singer. Reproduced from The Journal of Cell Biology. 1976, 70:193, by copyright permission of the Rockefeller University Press and M. Sheetz.*) **(a)** Red blood cells in an isotonic solution ("normal") **(b)** Red blood cells in a hypertonic solution ("crenate") **(c)** Red blood cells in a hypotonic solution (before hemolysis).

11. With three *separate* disposable pipets, remove a small amount of blood from each of the three tubes. Place 1 drop of blood from tube 1 on slide 1, 1 drop from tube 2 on slide 2, and 1 drop from tube 3 on slide 3.

12. Cover each drop of blood with a coverslip.

13. Observe the three slides with your compound microscope, focusing first with the medium-power objective and finally with the high-dry objective. (Hemolyzed cells are virtually unrecognizable; all that remains are membranous "ghosts," which are difficult to see with the microscope. To see red blood cells placed in a hypotonic solution, you can make a wet mount slide of the preparation immediately after mixing.)

14. In Figure 5-5, sketch the cells from each tube. Label the sketches, indicating whether the cells are normal, plasmolyzed (crenate), or hemolyzed.

15. Record the microscopic appearance in Table 5-3.

16. Record in Table 5-3 the tonicity of the sodium chloride solutions you added to the test tubes, relative to the cytoplasm of the red blood cells.

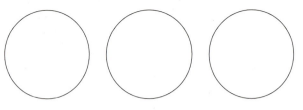

Labels: normal, plasmolyzed (crenate), hemolyzed

FIGURE 5-5 Microscopic appearance of red blood cells in different solute concentrations (____×).

If a hypotonic and a hypertonic solution are separated by a selectively permeable membrane, in which direction will there be net movement of water molecules due to osmosis? _____

If in this same situation, the solute molecules are able to pass through the membrane, in which direction will there be net movement of solute molecules due to diffusion? _____

After completing all experiments, take your dirty glassware to the sink and wash it as directed on page xii. Invert the test tubes in the test tube rack so they drain. Reorganize your work area, making certain all materials used in this exercise are present for the next class.

5.4 When Osmosis Malfunctions: Understanding Cholera

Osmosis and diffusion are homeostatic mechanisms which maintain the water and solute contents of cells and the entire body. If plasma membranes are defective or damaged because of genetic mutations or toxins produced by bacteria or viruses, the consequences can be serious.

Cholera is a disease caused by infection of the small intestine with the bacterium *Vibrio cholerae*, (Figure 5-6) often caused by contaminated drinking water. An infected person experiences severe fluid loss through diarrhea. The dehydration proceeds rapidly and can be fatal if not treated, especially in infants and children.

The cholera bacterium produces a toxin that affects plasma membrane proteins in the cells lining the interior of the small intestine. The toxin attaches to specific cell surface receptors and transfers a portion of the toxin molecule into the cell, which responds by continuously

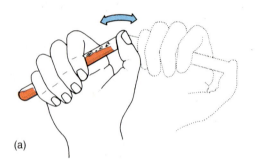

(a)

(b)

FIGURE 5-4 Method for studying effects of different solute concentrations on red blood cells.

TABLE 5-3	Effects of Salt Solutions on Red Blood Cells			
Tube	Contents	Print Visible?	Microscopic Appearance of Cells	Tonicity of External Solution[a]
1				
2				
3				

[a]With respect to that inside the red blood cell at the start of the experiment.

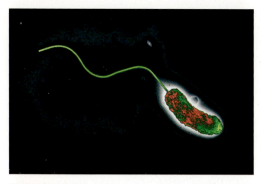

FIGURE 5-6 False color micrograph of *Vibrio cholerae*, the bacterium that causes cholera.

secreting chloride (Cl^-) ions into the intestinal lumen (interior). Water loss into the lumen caused by osmosis follows, producing large volumes of watery diarrhea.

We can visualize this process using a dialysis membrane bag as a stand-in for the small intestine. The contents of the small intestine are normally a watery soup of molecules represented by the liquid filling the dialysis bag.

MATERIALS

Per student group (4):

- Bottle of 80% sucrose
- Two 15-cm lengths of dialysis tubing, soaking in dH_2O
- Four 10-cm pieces of string *or* waxed dental floss
- Two short lengths of plastic tube
- Two large conical tubes standing in rack or larger beakers
- Ring stand and funnel apparatus (Figure 5-2)
- China marker
- 25-mL graduated cylinder
- Source of dH_2O

PROCEDURE

1. Label one tube "#1 Normal" and the second tube "#2, Cholera Affected."

2. Obtain a 15-cm section of dialysis membrane that has been soaked in dH_2O.

3. Fold over one end of the membrane and tie it securely with string or dental floss to form a leakproof bag (Figure 5-2).

4. Slip the open end of the bag over the stem of a funnel.

5. Measure 15 mL of dH_2O and pour through funnel to nearly fill the bag.

6. Remove the bag from the funnel. Press out excess air, insert piece of plastic tubing into the opening, and use dental floss or string to tie the open end of the bag tight around the tubing. The plastic tubing end inside the bag should be immersed in the liquid (Figure 5-7).

7. Place the bag into Tube or Beaker 1.

8. Make and fill a second bag with the sucrose solution. Ensure the end of the tubing is immersed into the sucrose before tying the bag closed. Place the dialysis tubing bag into Tube or Beaker 2.

9. Pour dH_2O into each tube or beaker so that the dialysis bag is covered and the water line is just above the dental floss tie.

   Fill in the following using the terms hypertonic, hypotonic, or isotonic:

   In set #1, the contents of the dialysis bag are _____ relative to the liquid outside the bag.

   In set #2, the contents of the dialysis bag are _____ relative to the liquid outside the bag.

   Describe your expectations and explanation regarding net flow of water in set #1: _____

   _____

   _____

   _____

   Describe your expectations and explanation regarding net flow of water in set #2: _____

   _____

   _____

   _____

10. Observe both tubes every 5 minutes over the next half hour, taking note of apparent changes in bag volume as well as whether or not any liquid is seen overflowing through the tubing. At the end of the time, record your observations and explanations for each set:

FIGURE 5-7 Dialysis bag and tube apparatus for visualizing effects of cholera.

Set 1:

_____

_____

_____

_____

Set 2:

_____

_____

_____

_____

How do your observations relate to the symptoms of cholera?

_____

_____

_____

Although the *Vibrio* bacterium is sensitive to antibiotics, dehydration is too rapid and severe to wait for the time needed for antibiotics to take effect. **Oral rehydration therapy** is a simple, quick, and effective treatment.

A common recipe for oral rehydration solution (ORS) is 1 level tsp of salt (NACL) plus 8 level tsp of sugar dissolved in 1 L of clean drinking water. This inexpensive solution is drunk by the infected person, with the goals of replacing lost fluid and maintaining the internal ion concentrations of body cells.

ORS is isotonic to body fluids. Sodium ions are actively moved (a transport process against a concentration gradient and requiring expenditure of cellular energy) into the intestine cells, replacing ions lost during diarrhea. Maximum water and ion absorption is achieved with the mixture of salt and sugar, though, through a membrane transport mechanism known as "co-transport."

Recall that table sugar, sucrose, is a disaccharide made of fructose and glucose subunits. Sucrose is hydrolyzed into fructose and glucose in the upper small intestine; glucose moves through channel proteins in the plasma membrane of some of the small intestine lining cells. This glucose transport only occurs in the presence of sodium ions, though, with Na^+ ions entering cells from the intestine lumen at the same time as glucose. The ORS composition thus ensures maximal restoration of cytoplasmic tonicity. Sports drinks, such as Gatorade™, are not adequate substitutes for ORS because they contain too high sugar concentration and too low salt.

_____ **1.** If one were to identify the most important compound for sustenance of life, it would probably be
   (a) salt
   (b) $BaCl_2$
   (c) water
   (d) I_2KI

_____ **2.** A solvent is
   (a) the substance in which solutes are dissolved
   (b) a salt or sugar
   (c) one component of a biological membrane
   (d) selectively permeable

_____ **3.** Diffusion
   (a) is a process requiring cellular energy
   (b) is the movement of molecules from a region of higher concentration to one of lower concentration
   (c) occurs only across selectively permeable membranes
   (d) is none of the above

_____ **4.** Cellular membranes
   (a) consist of a phospholipid bilayer containing embedded proteins
   (b) control the movement of substances into and out of cells
   (c) are selectively permeable
   (d) are all of the above

_____ **5.** An example of a solute is
   (a) Janus green B
   (b) water
   (c) sucrose
   (d) both a and c

_____ **6.** Dialysis membrane
   (a) is selectively permeable
   (b) is used in these experiments to simulate cellular membranes
   (c) has pores that allow passage to specific-sized molecules
   (d) all of the above

_____ **7.** Specifically, osmosis
   (a) requires the expenditure of cellular energy
   (b) is diffusion of water from lower to higher concentration
   (c) is diffusion of water across a selectively permeable membrane
   (d) is none of the above

_____ **8.** When the cytoplasm of a red blood cell has lost water to its surroundings, the cell is said to be
   (a) isotonic
   (b) burst
   (c) hemolyzed
   (d) crenate

_____ **9.** If one solution contains 10% NaCl and another contains 30% NaCl, the 30% solution is _____ with respect to the 10% solution
   (a) isotonic
   (b) hypotonic
   (c) hypertonic
   (d) plasmolyzed

_____ **10.** Cholera is a disease whose main symptom is
   (a) increase in water uptake by intestine cells
   (b) rapid dehydration
   (c) uptake of glucose by intestine cells
   (d) none of these

EXERCISE 5 Cell Membranes and How They Work

POST-LAB QUESTIONS

5.1 Rate of Diffusion of Solutes

1. You want to dissolve a solute in water. Without shaking or swirling the solution, what might you do to increase the rate at which the solute would go into solution? Relate your answer to your method's effect on the motion of the molecules.

5.2 Experiment: Effect of Solute Size on Diffusion and Selective Permeability of Membranes

2. If a 10% sugar solution is separated from a 20% sugar solution by a selectively permeable membrane, in which direction will there be a net movement of water? Explain.

3. Based on your observations in this exercise, would you expect dialysis membrane to be permeable to egg albumen, a protein with molecular weight of approximately 36,800? Why or why not?

5.3 Osmosis and Red Blood Cells

4. Explain, using terminology of tonicity and membrane function, why disaster would result if a patient were given intravenous dH_2O rather than 0.9% "physiological saline" solution.

5.4 When Osmosis Malfunctions: Understanding Cholera

5. A human lost at sea without fresh drinking water is effectively lost in an osmotic desert. Why is drinking salt water harmful?

6. How does osmosis differ from diffusion?

7. Explain why a person whose intestinal lining cells continuously "leak" chloride ions into the intestinal lumen experiences dehydration.

Food for Thought

8. What does the word *lysis* mean? (Does the name of the disinfectant product Lysol® make sense?)

9. Oral rehydration therapy is ancient. In 1000 BCE, an Indian physician recommended that his patients with diarrhea drink large amounts of water containing dissolved rock salt and molasses. Describe in your own words how oral rehydration solution acts to prevent death from cholera.

10. Cystic fibrosis is a common genetic disease, affecting about one of every 3200 Caucasian-births in the United States. This disease is recognizable in the ancient Northern European folk saying, "Woe to that child which when kissed on the forehead tastes salty. He is bewitched and soon must die." Search two Internet sites for information on "cystic fibrosis and osmosis" or "cystic fibrosis and membrane transport." List the two sites below and briefly summarize what you learn from them regarding the connections between this disease and malfunctioning plasma membranes.

http://

http://

Enzymes, the Catalysts of Life

OBJECTIVES

After completing this exercise, you will be able to

1. define *enzyme, catalyst, activation energy, substrate, enzyme–substrate complex, product, active site, enzyme specificity,* and *denaturation.*

2. explain how an enzyme operates.

3. identify the substrate and products of the reaction catalyzed by the enzyme catalase.

4. design and carry out experiments to test hypotheses about factors that affect enzyme action.

5. describe how enzyme activity is generally affected by enzyme concentration, pH, and temperature.

Introduction

Life would be impossible without enzymes. **Enzymes** are proteins that function as biological **catalysts.** A catalyst is a substance that lowers the amount of energy necessary for a chemical reaction to occur. You might think of this so-called **activation energy** as a mountain to be climbed. Enzymes decrease the height of the mountain, in effect turning it into a molehill. By lowering the activation energy, an enzyme speeds up the rate at which a reaction occurs.

Enzymes catalyze condensation and hydrolysis reactions in cells, rearrange chemical bonds, and perform many other tasks essential for cellular metabolism and maintenance of the body's homeostasis, or stable internal balance (see Exercise 7).

In an enzyme-catalyzed reaction, the reactant (the substance being acted upon) is called the **substrate.** Substrate molecules combine with enzyme molecules to form a temporary **enzyme–substrate complex.** Products are then formed, and the enzyme molecule is released unchanged. The enzyme molecule is not used up in the process and is capable of catalyzing the same reaction again and again. This can be summarized as follows:

Substrate + Enzyme → Enzyme–substrate complex
→ Products + Enzyme

An enzyme is similar to a key in that it has a specifically shaped "active site" into which only specific substrates fit. After it is bound in the active site, the enzyme–substrate complex changes shape slightly, allowing the chemical reaction to proceed more quickly. The products of the chemical reaction are released, and the enzyme returns to its original shape.

Before we proceed, let's visualize this process (Figure 6-1). Think of an enzyme as the key that unlocks a padlock. Imagine that you have many locks, and all of them use the same key. You would only need to reuse your single key to unlock them all.

Changing the shape of the key just a tiny bit may still allow the key to function in the lock, but you may have to fumble with the key a while to get the lock open. Changing the key shape still further may prevent the key from opening the lock at all.

Most enzymes are specific and catalyze reactions with only one or a very few kinds of substrates and products. This **enzyme specificity** lies in the shape of the protein molecule; anything that changes the shape of an enzyme alters its ability to function. Enzymes function best within a certain range of temperature, pH, or salinity (saltiness), for example. Although thousands of kinds of enzymes are present in living cells, we'll examine the activity of one, catalase.

Catalase is a widespread enzyme found in nearly all aerobic cells (plants, animals, and microbes). It protects cells from the toxic effects of hydrogen peroxide (H_2O_2) generated as a by product of cell metabolism. It does this by catalyzing the decomposition of the substrate hydrogen peroxide into oxygen gas and water products as follows:

$$2H_2O_2 \rightarrow 2H_2O + O_2 \text{ (gas)}$$

Catalase activity can be visualized in Figure 6-2.

In this exercise, you will examine the activity of catalase and design and carry out an experiment to examine the effect of one or more factors on the reaction. You will use liver homogenate (a slurry of liver and water blended together) as a source of catalase. The oxygen produced by the breakdown of hydrogen peroxide will be measured as an indication of catalase activity.

CAUTION Hydrogen peroxide is a reactive material. Wear goggles and avoid eye and skin contact.

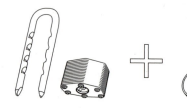

| substrate | enzyme | enzyme–substrate complex | products | enzyme |

FIGURE 6-1 Analogy of enzyme action.

MATERIALS

Per student:

- Safety goggles

Per student group (4):

- Beaker of liver extract containing catalase
- Wash bottle of dH$_2$O
- Bottle or stoppered flask of fresh hydrogen peroxide, H$_2$O$_2$

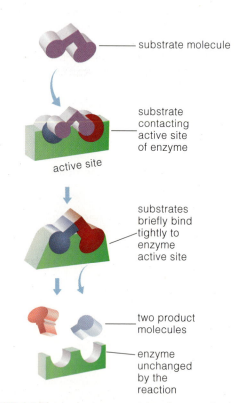

- substrate molecule

active site

- substrate contacting active site of enzyme

- substrates briefly bind tightly to enzyme active site

- two product molecules

- enzyme unchanged by the reaction

FIGURE 6-2 The hydrogen peroxide substrate physically fits into the active site of catalase. The substrate briefly binds to the active site, and an enzyme shape change allows water and oxygen gas to be formed. The products are released and the enzyme resumes its original shape.

- Reaction chamber
- 25-mL graduated cylinder
- Two 10-mL graduated cylinders
- Deep square or rectangular pan
- Filter paper
- Hole punch
- Forceps
- Stopwatch
- Source of distilled water
- Buffers pH 4, 7, and 10 or others as provided by your instructor
- Ice
- Insulated container
- Water bath at 37°C
- Thermometer

Work in groups of four for all sections in this exercise.

6.1 Measuring Catalase Activity

PROCEDURE

1. Punch several filter paper discs with the hole punch. Do not touch the discs with your fingers because skin oils will interfere with the process; handle the discs only with forceps.

2. Fill the pan with water. Lay the 25-mL graduated cylinder on its side in the pan so that it fills with water. Tilt the mouth of the cylinder upward slightly to remove all trapped air bubbles. Keep the mouth of the cylinder underwater while you stand the cylinder upside down in the pan.

3. Pour 10 mL of hydrogen peroxide into the reaction chamber. Take note if the hydrogen peroxide runs down a side wall of the chamber.

4. Hold a filter paper disc with the forceps and hold it in the liver solution for several seconds. Drain the excess solution from the disc by blotting it against the side of the container.

5. Place the disc high on a side wall of the reaction chamber; do not allow the disc to touch the hydrogen peroxide or an area that it flowed over.

6. Insert the stopper in the chamber and close it tightly.

7. Place the reaction chamber on its side in the pan of water, *making sure that the filter paper disc is on the top side.*

8. Tip the graduated cylinder into position so that its mouth lies directly over the tip of the pipette protruding from the reaction chamber. This will allow any oxygen released to rise into the graduated cylinder. Prop or hold the cylinder in position for the duration of data collection (see Figure 6-3.)

9. Rotate the reaction chamber 180 degrees so the paper disc comes in contact with the hydrogen peroxide solution.

10. Allow the reaction to proceed for 10 minutes and then tip the graduated cylinder upside down again with the mouth still under water so that you can measure the accumulated oxygen.

Describe the appearance of the filter paper disc and surrounding solution as the reaction begins and explain this observation in terms of the action of catalase.

_____

_____

_____

_____

How much oxygen was generated during this 10-minute experiment?

_____

11. Discard the reaction chamber contents into the sink or other area designated by your instructor. Rinse the reaction chamber with water and drain it upside down before proceeding with further investigations.

6.2 Effects of Substrate Concentration, pH, and Temperature on Catalase Activity

In the remainder of this exercise, you will work with your team to determine what, if any, changes in catalase activity result from changing various conditions in the reaction chamber. Recall from Exercise 1, The Scientific Method, that you construct a *hypothesis* (tentative, possible general explanation) to answer a *question*. Next, formulate a *prediction* of results from an experimental test of the hypothesis, assuming that the hypothesis is correct; a prediction often takes the form of an "If–then" statement.

Your experiment will allow you to make observations and collect data to test the prediction. You can then compare your predicted results with what you've actually observed and then make a conclusion about whether your hypothesis was supported or falsified.

In an experiment, the factor that you change or manipulate is the *independent variable*. The factor that you measure as a result of the experiment is the *dependent variable*. In every experiment, you will also have

FIGURE 6-3 Equipment to measure catalase activity. Note bubbles in reaction chamber and accumulating oxygen bubble at top of graduated cylinder. *(Photo by Joy B. Perry.)*

a number of *controlled variables*, which are all of the other conditions and factors that are kept the same for all treatment groups. In this way, your experiment will determine if the dependent variable changes based on your manipulations of the independent variable.

A. Experiment: Effect of Enzyme Concentration on Reaction Rate

Is the rate of oxygen production affected by concentration of catalase? In this activity, you will use the method from section 6.1 to design an experiment to test the hypothesis that *reaction rate increases with increasing enzyme concentration*. In this experimental system, placing multiple soaked filter paper discs in the reaction chamber effectively multiplies the enzyme concentration correspondingly.

PREDICTION: If enzyme concentration is (changed how?) _____, then the rate of oxygen production will (change how?) _____
_____ .

EXPERIMENTAL DESIGN:

The independent variable is _____

What levels of the independent variable will you test?
_____

The dependent variable is _____

Controlled variables include _____
_____

Record your data in Table 6-1 below.

CONCLUSION: The observed results (circle one) support *or* falsify the hypothesis.

What sources of error or uncertainty were there in your experiment?

_____

_____

How might another experiment to test this hypothesis be improved?

_____

_____

TABLE 6-1	Effect of Enzyme Concentration on Reaction Rate
PREDICTION:	
Enzyme concentration	Oxygen production at _____ minutes (mL)
1X enzyme concentration	
___X enzyme concentration	
___X enzyme concentration	
CONCLUSION:	

B. Experiment: Effect of pH on Catalase Activity

pH is a measure of the hydrogen ion concentration in a cell on a scale ranging from pH of 1 (extreme acidity) to 7 (neutral) to 14 (extreme base). A change of 1 pH unit reflects a 10-fold change in hydrogen ions and acidity (see Figure 6-4). pH affects the three-dimensional shape of enzymes, thus regulating their function. Most enzymes operate best when the pH of the solution is near neutrality (pH 7). Others, however, have pH optima in the acidic or basic range, corresponding to the environment in which they normally function. When an

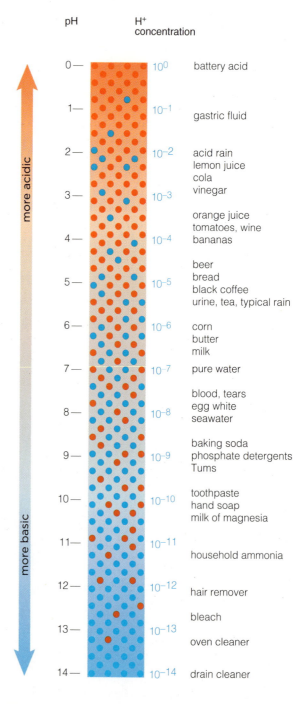

FIGURE 6-4 The pH scale.

TABLE 6-2	Effect of pH on Catalase Activity

PREDICTION:

pH	Oxygen production at ____ minutes (mL)
4	
7	
10	

CONCLUSION:

enzyme's structure is changed sufficiently to destroy its function, the enzyme is said to be **denatured** (has undergone **denaturation**). Most enzymatically controlled reactions have an *optimum* pH and temperature—that is, one pH and temperature at which activity is maximized.

How is catalase activity affected by pH changes? In this activity, you will use the method from section 6.1 and the pH buffers present in your lab to design an experiment to test the hypothesis that *enzyme activity is greatest in a neutral pH range.*

The pH of hydrogen peroxide solution can be changed by adding 8 mL of H_2O_2 to the reaction chamber and then adding 4 mL of the desired buffer solution and swirling to mix well.

PREDICTION: If pH is (changed how?) _____, then the rate of oxygen production will (change how?)

_____.

EXPERIMENTAL DESIGN:

The independent variable is _____

What levels of the independent variable will you test?

_____

The dependent variable is _____

Controlled variables include _____

_____

Record your data in Table 6-2 above.

CONCLUSION: The observed results (circle one) support *or* falsify the hypothesis.

What sources of error or uncertainty were there in your experiment? _____

_____

How might another experiment to test this hypothesis be improved?

_____

_____

C. Experiment: Effect of Temperature on Enzyme Activity

The rate at which chemical reactions take place is largely determined by the temperature of the environment. *Generally, for every 10°C increase in temperature, the reaction rate doubles.* Within a rather narrow range, this is

true for enzymatic reactions as well. However, because enzymes are proteins, excessive temperature alters their structure, interferes with activity, and may denature them if too hot.

In this activity, you will use the methods in section 6.1 to design an experiment to determine the *effect of different temperatures on the activity of catalase.* This may help you determine the best (optimum) temperature for the reaction. Ice is available to create low temperature conditions surrounding the reaction chamber, and a heated waterbath will increase the water temperature to approximately body temperature, 37°C.

PREDICTION: If temperature is (changed how?)

_____, then the rate of oxygen production will (change how?) _____

_____.

EXPERIMENTAL DESIGN:

The independent variable is _____

What levels of the independent variable will you test?

_____

The dependent variable is _____

Controlled variables include _____

Record your data in Table 6-3 below.

CONCLUSION: The observed results (circle one) support *or* falsify the hypothesis.

What sources of error or uncertainty were there in your experiment? _____

_____

_____

How might another experiment to test this hypothesis be improved? _____

_____

_____

 NOTE After completing all experiments, take your dirty glassware to the sink and wash it following the directions on page xii. Invert the glassware so it drains. Tidy up your work area, making certain all materials used in this exercise are there for the next class.

TABLE 6-3	Effect of Temperature on Enzyme Activity

PREDICTION:

Temperature (°C)	Oxygen production at ____ minutes (mL)

CONCLUSION:

_____ **1.** Enzymes are
  (a) biological catalysts
  (b) agents that speed up cellular reactions
  (c) proteins
  (d) all of the above

_____ **2.** Enzymes function by
  (a) being consumed (used up) in the reaction
  (b) lowering the activation energy of a reaction
  (c) combining with otherwise toxic substances in the cell
  (d) adding heat to the cell to speed up the reaction

_____ **3.** The substance that an enzyme combines with is
  (a) another enzyme
  (b) a product
  (c) an active site
  (d) the substrate

_____ **4.** *Enzyme specificity* refers to the
  (a) specific name of each kind of enzyme
  (b) fact that enzymes catalyze one particular substrate or a small number of structurally similar substrates
  (c) effect of temperature on enzyme activity
  (d) effect of pH on enzyme activity

_____ **5.** For every 10°C increase in temperature, the rate of most chemical reactions will
  (a) double
  (b) triple
  (c) increase by 100 times
  (d) stop

_____ **6.** When an enzyme becomes denatured, it
  (a) increases in effectiveness
  (b) loses its requirement for a substrate
  (c) forms an enzyme–substrate complex
  (d) loses its ability to function

_____ **7.** An enzyme may lose its ability to function because of
  (a) excessively high temperatures
  (b) a change in its three-dimensional structure
  (c) a large change in the pH of the environment
  (d) all of the above

_____ **8.** pH is a measure of
  (a) an enzyme's effectiveness
  (b) enzyme concentration
  (c) hydrogen ion concentration
  (d) none of the above

_____ **9.** Catalase
  (a) is an enzyme found in most cells, including liver
  (b) catalyzes the production of hydrogen peroxide
  (c) has as its substrate water and oxygen gas
  (d) is a substance that encourages the growth of microorganisms

_____ **10.** The oxygen production measured in the experiments of this exercise
  (a) is a consequence of the breakdown of hydrogen peroxide
  (b) is an index of enzyme activity
  (c) may depend on pH, temperature, or enzyme concentration
  (d) is all of the above

EXERCISE **6** Enzymes: Catalysts of Life

POST-LAB QUESTIONS

INTRODUCTION

1. Explain the difference between *substrate* and *active site*.

6.2.A. Experiment: Effect of Enzyme Concentration on Reaction Rate

2. Given your observations in this activity, predict the effect on the catalase reaction rate if the hydrogen peroxide solution were diluted by half. Explain.

6.2.B. Experiment: Effect of pH on Catalase Activity

3. Explain what happens to the catalase molecule and activity when the pH is on either side of the optimum level.

4. a. Whereas the pH of saliva ranges from 6.5 and 11.5, that in the stomach is about pH 2. What would you expect the optimum pH to be for the enzymes secreted into your stomach that digest proteins into amino acids?

b. What would you expect the optimum pH to be for the salivary enzyme amylase, which digests starches into sugars?

6.2.C. Experiment: Effect of Temperature On Enzyme Activity

5. Eggs can contain bacteria such as *Salmonella*. Considering what you've learned in this exercise, explain why cooking eggs makes them safe to eat.

6. Why do you think high fevers alter cellular functions?

7. Some surgical procedures involve lowering a patient's body temperature during periods when blood flow must be restricted. What effect might this have on enzyme-controlled cellular metabolism?

Food for Thought

8. Is it necessary for a cell to produce one enzyme molecule for every substrate molecule that needs to be catalyzed? Why or why not?

9. What happened to the composition of the solution inside the reaction chamber as a catalase reaction proceeded? Why was it a good practice to start with fresh hydrogen peroxide for each experimental trial?

10. Hydrogen peroxide solution is often used to sterilize cuts and scrapes. Why does hydrogen peroxide bubble when it is poured on an open wound but not when it is poured on intact skin?

EXERCISE 7

Homeostasis

OBJECTIVES

After completing this exercise, you will be able to

1. define *homeostasis, intracellular fluid, extracellular fluid, plasma, interstitial fluid, homeostatic mechanisms, pH scale, neutral solution, alkaline solution, acidic solution, buffer, systolic blood pressure, diastolic blood pressure,* and *sounds of Korotkoff.*

2. explain the logarithmic nature of the pH scale.

3. describe the role of sensory receptors, integrating centers, and effectors in homeostatic mechanisms using the regulation of body temperature and body fluid volume as examples.

4. use a sphygmomanometer and stethoscope to estimate blood pressure.

5. compare negative and positive feedback and feedforward homeostatic mechanisms.

Introduction

The world outside ourselves (our external environment) is constantly changing. For example, consider the seasonal and daily range of air temperature. However, the body must maintain an internal environment that is very constant, not only in its temperature but in all of its other functional and structural characteristics (variables) as well. Maintenance of a stable internal environment is called **homeostasis.**

The internal environment is mostly fluid. An alien in a science fiction story once described us as "dirty bags of water." Body fluid plays a central role in homeostasis. Its major compartments are summarized in Table 7-1. Body fluid is divided into two major compartments, (1) fluid Inside of cells, or **intracellular fluid,** and (2) body fluid external to cells, or the **extracellular fluid.** The extracellular fluid is further subdivided into the fluid portion of blood or **plasma** and into the body fluid outside of blood vessels or **interstitial fluid.**

7.1 Homeostatic Mechanisms

Homeostatic mechanisms function throughout the body to keep the chemistry of the body fluids constant and to maintain the various structures of the body.

MATERIALS

Per student:

- Text

PROCEDURE

A. Vital Functions of Organ Systems

Each organ system has functions vital to homeostasis. Use your textbook to complete Table 7-2.

B. Regulation of Body Temperature

All homeostatic mechanisms involve feedback loops (Figure 7-1). A typical feedback loop consists of three steps. First, a **sensory receptor** receives a stimulus—a rise or fall in body temperature, for example. In the case

of temperature regulation, temperature-sensitive receptors are located in the skin and in the brain. Second, the sensor relays the information that a change has occurred to an **integrating center.** This control center for temperature regulation is also located in the brain. An integrating center compares the intensity of a stimulus with a preset value. Third, the integrating center activates an **effector,** which produces a response that reverses the direction of the original change back toward the preset value. Effectors are usually muscles or glands. Normally, over time, variables such as body temperature are maintained by homeostatic mechanisms within a range of values. The midpoint of the range is the preset point (Figure 7-2).

1. From your own experience, identify one effector that is activated when body temperature gets too hot— and its response to cool the body down. Also identify the organ system to which the effector belongs.

    (a) Effector _____

    (b) Response _____

    (c) Organ system _____

TABLE 7-1 Fluid Compartments

Body Fluid	Definition	Function	Average Volume in the Human Body (L)
Intracellular	All of the fluid in cells	Contains dissolved substances	29
Interstitial	All of the fluid between the cells and the blood vessels	Diffusion of dissolved substances between cells and blood	12
Plasma	All of the fluid portion of the blood	Transport of dissolved substances throughout body	3

TABLE 7-2 Organ Systems of Mammals

Systems	Vital Functions
Integumentary	
Nervous	
Endocrine	
Skeletal	
Muscular	
Circulatory	
Lymphatic	
Respiratory	
Digestive	
Urinary	
Reproductive	

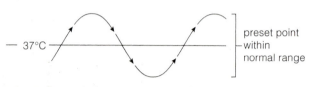

FIGURE 7-2 Body temperature trace (line with arrowheads) over several seconds showing that it stays within its normal range due to the action of homeostatic mechanisms.

2. From your own experience, describe one effector that is activated when temperature gets too cold—and its response to warm up the body. Also identify the organ system to which the effector belongs.

(a) Effector _____

(b) Response _____

(c) Organ system _____

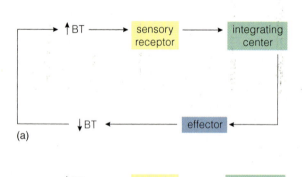
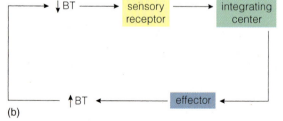
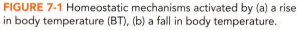

FIGURE 7-1 Homeostatic mechanisms activated by (a) a rise in body temperature (BT), (b) a fall in body temperature.

7.2 Regulation of pH

The **pH scale** (Figure 7-3) measures the concentration of chemically active hydrogen ions (H^+) in solutions. It is logarithmic in nature, and each full unit is 10 times more or less than the next full unit. A **neutral solution** has a pH of 7 and depending on the temperature has or almost has an equal number of H^+ and OH^- (hydroxyl) ions. Pure water and intracellular fluid are examples of neutral solutions. Extracellular fluid is a little more alkaline at a pH of 7.3 to 7.5. An **alkaline solution** (basic solution) has a pH higher than 7, which means a lower concentration of H^+. An **acidic solution** has a pH lower than 7 and a higher concentration of H^+.

The regulation of the pH of our bodies involves homeostatic mechanisms at two levels—molecular and organ system (respiratory and excretory). Molecules of some of the compounds dissolved in our body fluids minimize pH changes because of the addition of acids (H+ donors) or bases (H+ acceptors). A compound that does this is called a **buffer.**

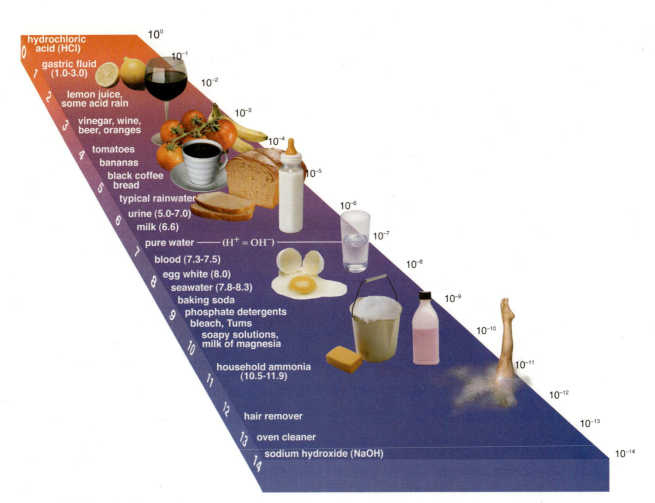

FIGURE 7-3 The pH scale.

MATERIALS

Per student group (4):

- Plastic tray
- Five 100-mL beakers labeled, respectively, W, M, OJ, D, and V
- 1-mL pipet and safety bulb or filling device
- Glass stirring rod
- Wide-range pH test paper and dispenser with scale
- Pencil

Per lab room:

- 500-mL container of nonphosphate detergent solution (1/4 teaspoon/500 mL dH_2O)
- Cartons of milk
- Orange juice
- 500-mL bottle of white vinegar

PROCEDURE

1. Pour 20 mL of water, milk, orange juice, detergent solution, and vinegar separately into their own 100-mL beakers, labeled W, M, OJ, D, and V, respectively.

2. Tear off 11 short pieces (about 1 cm in length) of pH indicator paper and label them with a pencil W, W+V, W+2V, W+3V, M, M+V, M+2V, M+3V, OJ, D, and V. Place them on the plastic tray.

3. Dip a clean glass stirring rod into the water and transfer a drop of water to the pH paper labeled W.

4. Wait for the paper to absorb the drop. Find the closest match between the color of the paper and the colored scale of the pH paper dispenser and read the corresponding pH value. Record this value in the second row (Water) of the second column (pH) of Table 7-3.

5. Wash and dry the glass stirring rod.

6. Repeat steps 4 and 5 for the remaining fluids using the appropriately labeled piece of pH paper (M, OJ, D, or V) and record the pH values you read in the third to sixth rows of the second column of Table 7-3.

7. Pipet 1 mL of vinegar into the beaker containing water and stir it with the glass rod to make the W+V solution.

8. As before, measure the pH and record it in the second row of the third column of Table 7-3.

TABLE 7-3 pH Values of Common Fluids

Fluid	pH	pH After Addition of:		
		1 mL of Vinegar	2 mL of Vinegar	3 mL of Vinegar
Water				
Milk				
Orange juice				
Detergent				
Vinegar				

9. Pipet another 1 mL of vinegar into the W+V solution and stir with the glass rod to make the W+2V solution.

10. Measure the pH and record it in the second row of the fourth column of Table 7-3.

11. Pipet one last 1 mL of vinegar into the W+2V solution and stir with the glass rod to make the W+3V solution.

12. Measure the pH and record it in the second row of the fifth column of Table 7-3.

13. Wash and dry the glass stirring rod.

14. Repeat steps 7 to 12 for the beaker containing milk (M+V, M+2V, and M+3V) except record the data in the appropriate columns of the third row. The milk will curdle as it sours.

15. Plot the four data points for W, W+V, W+2V, and W+3V and then plot them for milk (M, M+V, M+2V, M+3V) in Figure 7-4.

16. Which fluid, water or milk, is more likely to contain buffers?

17. Explain any differences that you observe between the slopes of the two lines in Figure 7-4.

_____

_____

_____

_____

_____

_____

7.3 Simulation of Homeostatic Mechanisms Regulating Body Fluid Volume

In this section, you will be part of model homeostatic feedback loops that regulate the volume of fluid in the body. Specifically, you will be the sensor, integrating center, and controller of the effectors. Your instructor has prepared a vertically stacked double-buret system, as illustrated in Figure 7-5, for each group of students. The lower buret has been half-filled with alkaline buffer solution, containing the pH indicator phenol red. Imagine that this solution is blood.

MATERIALS

Per student group (4):

- Vertically stacked double-buret system
- 1-L Erlenmeyer flask of water
- Empty 1-L Erlenmeyer flask

Per lab room:

- 1-L container of red alkaline buffer solution (pH 10) containing phenol red
- 1-L container of colorless solution acid buffer (pH 6)

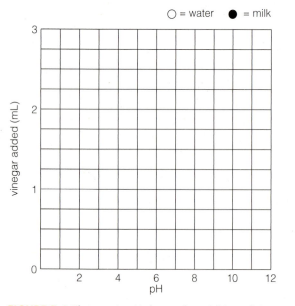
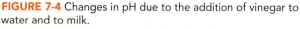

FIGURE 7-4 Changes in pH due to the addition of vinegar to water and to milk.

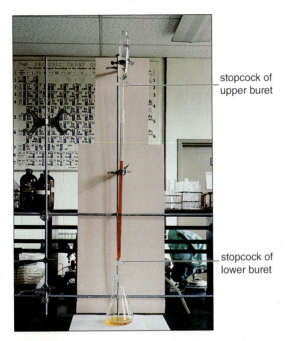

stopcock of upper buret

stopcock of lower buret

FIGURE 7-5 Double buret apparatus used to simulate homeostasis. *(Photo by D. Morton.)*

PROCEDURE

1. Fill about three-quarters of the reservoir of the upper buret with water. Imagine that this is drinking water, which, after you drink it, adds to the body fluid volume.

2. Pour 100 mL of acid buffer solution into a 1-L Erlenmeyer flask and place it under the lower buret. Imagine its contents are urine.

3. In the body, your kidneys function to filter the blood to make urine. This is a continuous process that results in the obligatory loss of fluid from the body. To simulate this fluid loss, turn the stopcock of the lower buret so that it slowly drips into the flask at about 24 drops per minute.

4. Adjust the stopcock of the upper buret to keep the fluid (blood) level in the lower buret constant. When the dripping of both burets is stable, read and record on the line below the number on the side of the lower buret that corresponds to the fluid level.

_____

This number represents the normal value for body fluid volume with water intake equal to that lost in urine.

5. On a hot day, water is lost from the body during temperature regulation and thus is not available for urine formation. Your kidneys produce a smaller amount of more concentrated urine. To simulate this situation, reduce the drop rate from the upper buret to 12 drops per minute. Readjust the drop rate of the lower buret to hold body fluid volume constant, albeit at a lower level.

6. Figuratively, take a drink. That is, open the stopcock on the upper buret to return the fluid level in the lower buret to the number that corresponds to the normal body fluid volume. When this level is reached, reduce the drip rate to maintain this level.

7. When you drink excess fluid, the kidneys increase urine production to maintain the correct amount of fluid in the body. To simulate this situation, increase the drip rate from the upper buret. Adjust urine production to return to and maintain normal body fluid volume.

8. When alcoholic beverages are consumed, regulation of kidney function is disturbed. Urine production is higher than it should be until the alcohol is metabolized. Simulate this by slightly increasing the drip rate of the lower buret for 30 seconds. Read and record on the line below the new number for body fluid volume.

_____

The morning after excessive alcohol consumption, a person feels dehydrated and thirsty.

9. In carrying out this activity, you used different body parts to simulate the homeostatic feedback loop that maintains water balance. In Table 7-4, match the body part used with the correct listed part of the **homeostatic feedback loop. Use each answer only once.**

10. When red fluid drops from the lower buret into the flask, it turns yellow, simulating the color of urine. On the lines below, write a hypothesis as to the mechanism that accomplishes this color change.

_____

_____

_____

11. Dispose of solutions as directed by your instructor.

7.4 Blood Pressure

The force that causes blood to flow through the heart and vessels of the circulatory system is blood pressure. By means of fast-circulating blood, the effects of homeostatic

TABLE 7-4	Parts of Homeostatic Feedback Loops
Body Part	**Homeostatic Feedback Loop**
_____ Hands	a. Integration center
_____ Eyes	b. Effector
_____ Brain	c. Sensor

mechanisms are spread first to the interstitial fluid and then to the intracellular fluid throughout the body. Various effectors regulate the blood pressure so that the right amount of blood is delivered to the body organs. For example, blood pressure in the arteries (relatively large vessels delivering blood to capillary beds in organs) is partly controlled by the number of times the heart contracts in a minute (heart rate).

The most common test done when you visit your physician is the measurement of blood pressure in the artery of the upper arm usually with the body at rest in the sitting position. This blood pressure is reported as **systolic blood pressure** (blood pressure when the heart contracts) over **diastolic blood pressure** (blood pressure when the heart relaxes). The diastolic blood pressure is caused by the elastic rebound of the walls of the largest arteries.

The most common unit of blood pressure is the number of millimeters in a column of mercury (mm Hg) needed to produce a force equal to a particular blood pressure. Normal blood pressure at rest for an 18-year-old young woman is 116/72 mm Hg. Normal blood pressure is slightly higher in males and increases with age for both genders. Many homeostatic mechanisms function to maintain normal blood pressure. Values that are significantly different from the average blood pressure for a person's age, height, and weight or from his or her normal blood pressure may indicate a medical problem that needs immediate attention.

The instrument used to measure blood pressure is the **sphygmomanometer** (Figure 7-6). Blood pressure measurements by a sphygmomanometer are approximations of true blood pressure. Tissue composition of the arm, the size of the sphygmomanometer cuff, and

operator experience are three factors that can cause inaccurate measurements.

The specific steps taken to measure blood pressure depend on the type of sphygmomanometer used. The following procedure lists the steps for the measurement of blood pressure. Make sure that you read and understand the complete instructions before you begin.

MATERIALS

Per student pair:

- Aneroid sphygmomanometer
- Stethoscope

PROCEDURE

CAUTION   Do not leave a fully inflated cuff on a subject's arm for more than a few seconds.

1. Have your lab partner sit with one arm resting on the lab bench.

2. Wrap the cuff around that arm and position it 2 to 3 cm above the elbow.

3. Feel for the pulse of the brachial artery in the hollow between the muscles of the upper arm, just above the crease of the elbow, and place the membrane end of a stethoscope over this location (Figure 7-7).

4. The sounds you hear as the pressure rises above the diastolic level are called the Korotkoff sounds and are caused by the turbulence created as blood squeezes through the artery. With the valve closed, pump up the cuff pressure to 20 mm Hg above the point where these sounds cease.

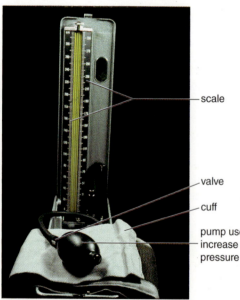

scale

valve

cuff

pump used to increase cuff pressure

FIGURE 7-6 Sphygmomanometer with a mercury column to illustrate the basis of blood pressure units (i.e., mm Hg) . You will be using a much safer aneroid instrument but the principle of how it works is the same. *(Photo by D. Morton.)*

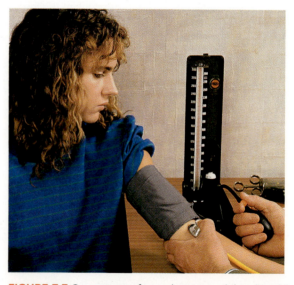

FIGURE 7-7 Correct use of a stethoscope while taking blood pressure. *(Photo by Sheila Terry/SPL/Photo Researchers.)*

5. While watching the scale, open the valve to allow the blood pressure to decrease at a rate of about 2 mm Hg per second.

6. Listen carefully for the muffled return of the Korotkoff sounds. The reading at the instant these sounds are first heard is the systolic blood pressure, and the reading at the instant they disappear is the diastolic blood pressure.

7. Measure and record your lab partner's blood pressure: _____ mm Hg

8. Have your lab partner measure and record your blood pressure: _____ mm Hg.

7.5 Negative Feedback, Positive Feedback, and Feedforward

All of the examples given so far are negative feedback homeostatic mechanisms in that the response reverses the direction of the change. That is, if a change in the value of the variable occurs in either a positive or negative direction away from the preset value, then the value of the variable is turned around back toward the preset point. Buffers help to maintain the correct pH by releasing H^+ when the pH increases and consumes H^+ when the pH decreases. A decrease in body fluid volume is compensated for by producing more concentrated urine and taking a drink. A decrease in blood pressure in an artery delivering blood to an organ or an increase in an organ's activity that requires greater blood flow is compensated partly by an increase in blood pressure elsewhere in the circulatory system.

In a few situations, positive feedback contributes to homeostasis. Positive feedback occurs when the response intensifies the change from a variable's preset point. One example is the increasing contractions of the uterus just before childbirth. In all cases of positive feedback, they are smaller parts of larger negative feedback homeostatic situations. In the case of childbirth, the mother's body is returned to the "not pregnant" condition.

Some other homeostatic mechanisms are feedforward rather than feedback, meaning the response occurs before the change. An example is the small increase in stress hormones that occurs before taking a test. (i.e., a little anxiety is good for your performance, but, of course, too much is detrimental).

Fill in Table 7-5.

TABLE 7-5	Review of Types of Homeostatic Mechanisms	
Homeostatic Mechanism	**Homeostatic Feedback Loop**	**Explanation**
_____ Negative feedback	a. Blood clotting (liquid to solid state)	
_____ Positive feedback	b. Visualization of performance just before an athletic event	
_____ Feedforward	c. Removal of calcium from bones because of low levels in the blood	

_____ 1. The portion of the body fluid contained in cells is
  (a) interstitial fluid
  (b) plasma
  (c) intracellular fluid
  (d) extracellular fluid

_____ 2. All the body fluid located outside of cells is
  (a) interstitial fluid
  (b) plasma
  (c) intracellular fluid
  (d) extracellular fluid

_____ 3. Extracellular fluid includes
  (a) interstitial fluid
  (b) plasma
  (c) intracellular fluid
  (d) both a and b

_____ 4. In the case of temperature regulation, an example of an effector is
  (a) a receptor in the brain
  (b) a center in the brain
  (c) a shivering skeletal muscle
  (d) none of the above

_____ 5. In the case of temperature regulation, integration is accomplished by
  (a) receptors in the brain
  (b) centers in the brain
  (c) muscles and glands
  (d) none of the above

_____ 6. A neutral solution has a pH of
  (a) 3
  (b) 5
  (c) 7
  (d) 9

_____ 7. The instrument used to measure blood pressure is called
  (a) a sphygmomanometer
  (b) a hemocytometer
  (c) a hematocrit centrifuge
  (d) none of the above

_____ 8. When measuring blood pressure, the Korotkoff sounds are heard
  (a) below the diastolic pressure
  (b) above the systolic pressure
  (c) between the diastolic and systolic pressures
  (d) at none of the above times.

_____ 9. Normal blood pressure for an 8-year-old boy is 100/67 mm Hg. The systolic blood pressure is
  (a) 33 mm Hg
  (b) 100 mm Hg
  (c) 67 mm Hg
  (d) 167 mm Hg

_____ 10. In positive feedback homeostatic mechanisms
  (a) the response reverses the change
  (b) the response intensifies the change
  (c) the response occurs before the change
  (d) none of the above occurs

EXERCISE 7 Homeostasis

POST-LAB QUESTIONS

Introduction

1. Define homeostasis.

2. List and describe the various body fluid components.

7.1 Homeostatic Mechanisms

3. Which organ systems are involved with the acquisition of oxygen and its delivery to the cells of the body?

7.2 Regulation of pH

4. Explain the pH scale. How many hydrogen ions are present in a solution at pH 5 compared with those in a solution at pH 6?

5. Describe how neutral, alkaline, and acidic solutions differ from each other.

7.3 Simulation of Homeostatic Mechanisms Regulating Body Fluid Volume

6. List the components of a homeostatic feedback loop. Use the regulation of fluid volume to illustrate your answer.

7.4 Blood Pressure

7. A normal blood pressure for a man in his late forties is 130/82 mm Hg. Explain the two numbers and what they mean.

7.5 Negative Feedback, Positive Feedback, and Feedforward

8. Describe how negative feedback, positive feedback, and feedforward are similar and dissimilar when it comes to homeostatic mechanisms.

Food for Thought

9. The lowest body pH is found in the fluid of the working stomach. How does a pH of about 3 in this particular location benefit the body as a whole?

10. What is hypertension? List some possible causes of this condition.

How Mammals Are Constructed

OBJECTIVES

After completing this exercise, you will be able to

1. define *tissue, organ, system, organism, histology, basement membrane, goblet cells, cilia, brush border of microvilli, keratinization, keratin, collagen fibers, elastic fibers, fibroblast, fat cells, lumen, chondrocytes, lacunae, Haversian systems, osteocytes, actin filaments, myosin filaments, intercalated disk, neuron, cell body, axon, dendrites, neuroglia, body cavities—thoracic, abdominal, pleural, pericardial, pelvic, cranial,* and *spinal.*

2. discuss the high degree of organization present in animal structure and explain its significance.

3. recognize the four basic tissues and their common mammalian subtypes.

4. list the functions of the four basic tissues and their common mammalian subtypes.

5. explain how the four basic tissues combine to make organs.

6. list each system of a mammal and its vital functions.

7. describe the basic plan of the mammalian body.

8. explain the layout of the body's major cavities.

9. locate the major organs in a mammal's body.

Introduction

Each animal cell, including those of mammals, is specialized to emphasize certain activities, although most continue to carry on basic functions such as cellular respiration. Groups of similarly specialized cells, along with any extracellular material, associate together to form **tissues.** Different subtypes from each of the four basic tissue types are combined much like quadruple-decker sandwiches to make the body's **organs.** Groups of related organs are strung together functionally, and usually structurally, to form **systems.** Taken together and arranged correctly, systems form the entire body. The levels of organization of an animal with systems can be summarized as follows:

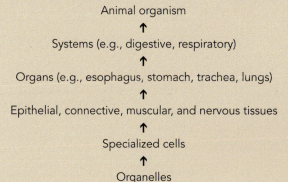

Animal organism
↑
Systems (e.g., digestive, respiratory)
↑
Organs (e.g., esophagus, stomach, trachea, lungs)
↑
Epithelial, connective, muscular, and nervous tissues
↑
Specialized cells
↑
Organelles

8.1 Tissues and Organs

The study of normal tissues is called **histology.** The four basic tissues types are **epithelial tissue, connective tissue, muscle tissue,** and **nervous tissue.** Each type has subtypes. In your study of these subtypes, you will examine a variety of tissues and organs, all permanently mounted on glass microscope slides. Most slides have sections of tissues and organs, usually 6 to 10 μm thick. Other slides have whole pieces of organs that are either transparent or are gently pulled apart (teased) and spread on the surface of the slide until they are thin enough to see

through. The tissues of some organs are simply smeared onto the surface of slides. Differences in tissue structure go with the particular functions they perform.

MATERIALS

Per student:

- Compound microscope, lens paper, a bottle of lens-cleaning solution (optional), a lint-free cloth (optional), a dropper bottle of immersion oil (optional)

- Prepared slide of a whole mount of mesentery (simple squamous)

- Prepared slide of a section of the cortex of the mammalian kidney
- Prepared slide of a section of trachea
- Prepared slide of a small intestine (preferably a cross-section of the ileum)
- Prepared slide of a section of esophagus
- Prepared slide of a section of mammalian skin
- Prepared slide of sections of contracted and distended urinary bladders
- Prepared slide of a teased spread of loose (areolar) connective tissue
- Prepared slide of a tendon, longitudinal section
- Prepared slide of a compact bone ground cross-section
- Prepared slide of a section of white adipose tissue
- Prepared slide of sections of the three muscle types
- Prepared slide of a smear of the spinal cord of an ox (neurons)
- Colored pencils
- Computer with Internet access (optional)

Per lab room:

- Demonstration of intercalated disks in cardiac muscle tissue
- 50-mL beaker three-fourths full of water
- 50-mL beaker three-fourths full of immersion oil
- Two small glass rods

PROCEDURE

A. Epithelial Tissues

Epithelial tissues are widespread throughout the body, covering both the body's outer surface (epidermis of skin) and inner surfaces (e.g., ventral body cavities. inner surface of the small intestine). Their main functions are *protection* and *transport*—secretion, absorption, and excretion. Specialized functions include sensory reception and the maintenance of the body's gametes (eggs and sperm).

Epithelial cells carry on rapid cell division in adults, and various stages of mitosis are often seen in this basic tissue type. As a consequence, epithelia have the highest rates of cell turnover among tissues. For example, the lining of the small intestine replaces itself every 3 to 4 days. Epithelial tissues do not have blood vessels and are attached to the underlying connective tissue by an extracellular **basement membrane** that is difficult to see if it is not specially stained. Use the following questions to identify most subtypes of epithelial tissue:

QUESTION 1. What is the shape of the outermost cells?

There are three possible answers: *squamous* (scalelike or flat), *cuboidal*, or *columnar*. These choices describe the shape of the cells when they are viewed in a section

perpendicular to the surface. In surface view, all three types are multi-sided (polygonal.)

QUESTION 2. How many layers of cells are there?

Epithelial tissues are *simple* if there is one layer of cells; they are *stratified* if there are two or more. Figure 8-1 applies these definitions to squamous epithelia. *Pseudostratified* epithelial subtypes appear stratified but really are not because their cells all rest on the basement membrane (Figure 8-2). As only the tallest columnar-shaped cells reach the free surface, cell nuclei lie at different levels, giving the false appearance of several layers of cells.

In addition, the epithelium that lines the inside of the urinary bladder (and parts of other nearby urinary system organs) changes its subtype as it fills or empties with urine. This epithelium is called *transitional* to avoid confusion.

There are six common subtypes of epithelia: *simple squamous, simple cuboidal, simple columnar, pseudostratified columnar, stratified squamous,* and *transitional.*

1. *Whole mount of mesentery* (Figure 8-3). The mesentery is a fold of the abdominal wall and holds the intestines in place. This slide is a surface view of the **simple squamous epithelium** (Figure 8-4), which

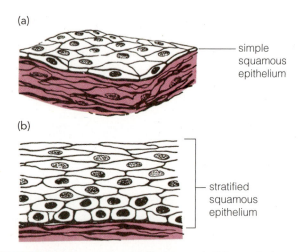

(a)

simple squamous epithelium

(b)

stratified squamous epithelium

FIGURE 8-1 Squamous epithelia (a) simple (b) stratified.

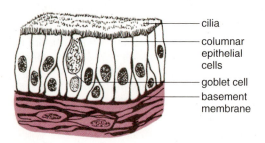

cilia

columnar epithelial cells

goblet cell

basement membrane

FIGURE 8-2 Pseudostratified columnar epithelium. It is columnar because of the shape of the cells that reach the free surface.

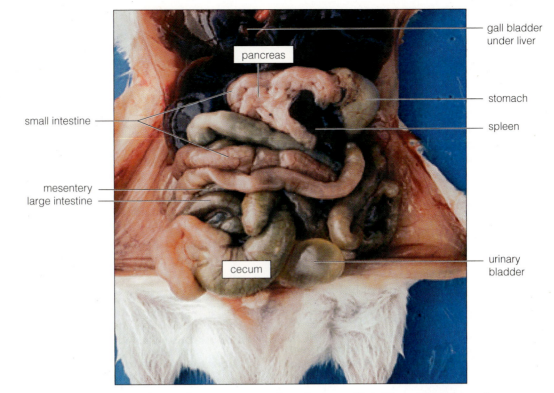

FIGURE 8-3 Ventral view of lower digestive system of a male mouse (2✕). *(Photo by D. Morton.)*

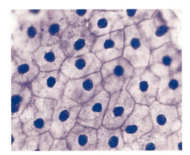

FIGURE 8-4 Surface view of simple squamous epithelium of the mesentery (300✕). *(© RAY Simmons/Photo Researchers.)*

lines the inside of the ventral body cavities and the organs within.

What is the shape of the surface cells?

_____

Draw what you see in Figure 8-5 and fill in the total magnification of the compound microscope you are using. Repeat this procedure for each subsequent drawing. Alternatively, if computers and Internet access are available, construct an atlas of links to images of sections, and so on, of the various tissues and organs under study.

2. *Kidney* (Figure 8-6). With the help of Figure 8-7a, find a *renal corpuscle*. A renal corpuscle is composed of a tuft of capillaries surrounded by a hollow capsule formed from the cup-shaped end of one of the kidney's functional units, the nephron. Identify a side view of the simple squamous epithelium that comprises the outer wall of the capsule (Figure 8-7b). Around the renal corpuscle, you can see a number of transverse and oblique sections of the tubular portion of the nephron. Their walls are composed of **simple cuboidal epithelium.** The main function of the nephrons and the entire kidney is the production

FIGURE 8-5 Drawing of the surface of simple squamous epithelium (_____ ✕).

of urine. Draw simple squamous epithelium in Figure 8-8. In Figure 8-9, draw simple cuboidal epithelium.

3. *Trachea* (Figure 8-10). The trachea is a tubular organ that conveys air to and from the lungs. Note that its inner surface is lined by **pseudostratified columnar epithelium** (Figure 8-11). Locate the unicellular glands, which are called **goblet cells** because of their shape. They secrete mucus that is difficult to see unless it is specifically stained. Using high power, do you see the numerous hair-like structures that project from the surface of the columnar epithelial cells? These are

cilia, which in life move together, sweeping mucus, trapped bacteria, and debris up the trachea to the throat. When you clear your throat, you collect this mucus and swallow it (a form of recycling). Draw pseudostratified columnar epithelium in Figure 8-12.

4. *Small intestine* (Figure 8-3). The small intestine is a tubular organ that connects the stomach to the large intestine. Its inner surface is lined with **simple columnar epithelium** (Figure 8-13). Do not miss the *goblet cells* present in the epithelium. Locate the **brush border of microvilli** situated along the free surface of the epithelium. Do not confuse this relatively thin structure with

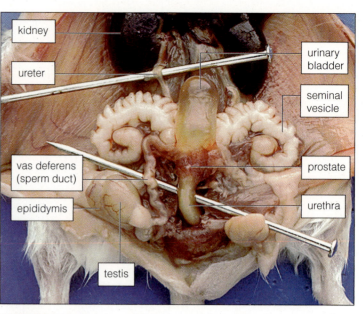

FIGURE 8-6 Ventral view of the urinary and reproductive systems of a male mouse after removal of the digestive system (3✕). *(Photo by D. Morton.)*

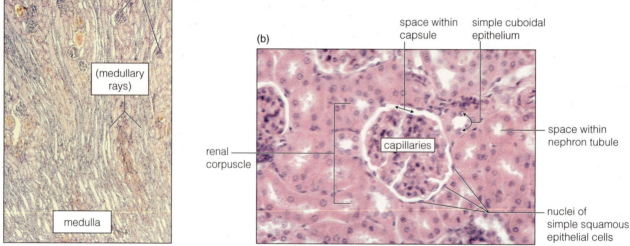

(a)

(b)

FIGURE 8-7 Sections of the cortex of the kidney, (a) Low power (25✕) and (b) medium power (183✕) views. *(Photos by D. Morton.)*

FIGURE 8-8 Drawing of a side view of simple squamous epithelium (_____ ✕).

FIGURE 8-9 Drawing of a side view of simple cuboidal epithelium (_____ ✕).

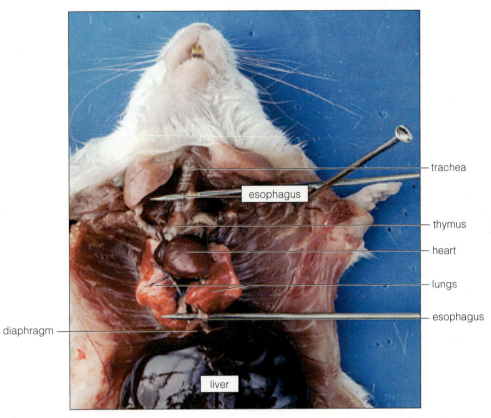

FIGURE 8-10 Ventral view of the organs in the thoracic cavity of a dissected mouse (2×). *(Photo by D. Morton.)*

(a)

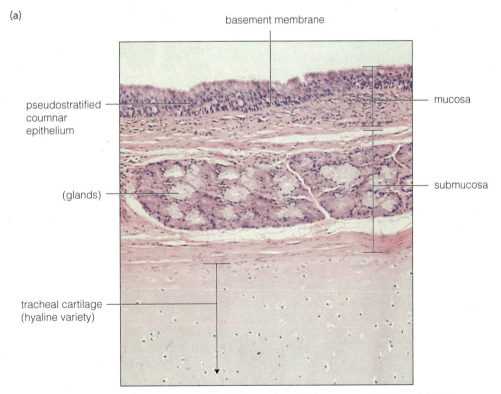

FIGURE 8-11 Sections of the inner surface of the trachea, (a) low power (100×) and (b) oil immersion (1000×) views. *(Photos by D. Morton.)*

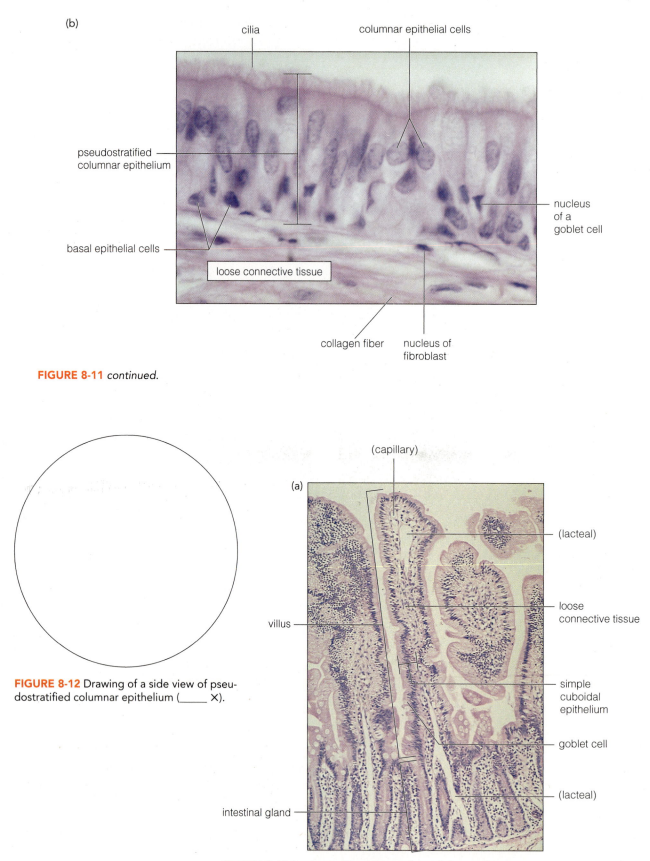

(b)

cilia

columnar epithelial cells

pseudostratified
columnar epithelium

basal epithelial cells

loose connective tissue

nucleus
of a
goblet cell

collagen fiber

nucleus of
fibroblast

FIGURE 8-11 *continued.*

FIGURE 8-12 Drawing of a side view of pseudostratified columnar epithelium (_____ ✕).

(a)

(capillary)

(lacteal)

loose
connective tissue

villus

simple
cuboidal
epithelium

goblet cell

(lacteal)

intestinal gland

FIGURE 8-13 Sections of the inner surface of the small intestine, (a) medium-power
(250✕) and (b) high-power (400✕) views. *(Photos by D. Morton.)*

the cilia from the previous slide. Microvilli are primarily responsible for the large surface area of the small intestine. Individual microvilli can best be seen at the electron-microscopic level (Figure 8-14). In Figure 8-15, draw simple columnar epithelium.

5. *Esophagus* (Figure 8-10). The tubular esophagus connects the throat to the stomach. Examine the epithelium lining the inner surface of the esophagus (Figure 8-16).

Is the shape of the outermost cells squamous, cuboidal, or columnar?

_____

Are there one or many layers of cells in this tissue?

_____

Name this subtype of epithelial tissue.

_____

In Figure 8-17, draw simple columnar epithelium.

6. *Skin.* The skin is divided into three layers (Figure 8-18). The *epidermis* is composed of stratified squamous epithelium. The *dermis* and *hypodermis* (or subcutaneous layer) are connective tissue and will be studied later.

The epidermis is the most extreme example of a protective epithelium. The process of keratinization, whereby the cells transform themselves into bags of the protein keratin, causes strata in the epidermis. This substance gives skin its tough, flexible, and water-resistant surface.

At one of the free edges of your section of mammalian skin, locate the **keratinized stratified squamous epithelium.** It will be stained bluer than the predominately pink connective tissue layers. Hair follicles and multicellular sweat glands may be present in the connective tissue layers. These structures grow into the connective tissue layers from the epidermis during the development of the skin. Draw keratinized stratified squamous epithelium in Figure 8-19.

(b)

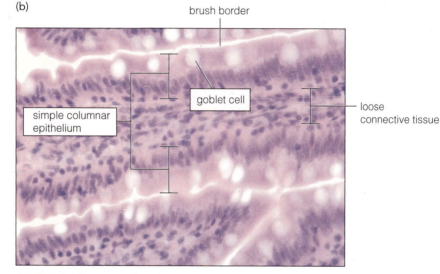

brush border

goblet cell

simple columnar epithelium

loose connective tissue

FIGURE 8-13 *continued.*

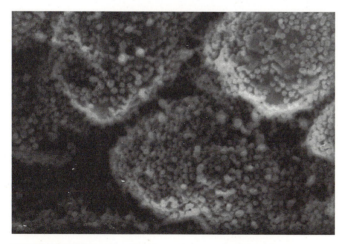

FIGURE 8-14 Scanning electron micrograph of a surface view of microvilli of simple columnar epithelial cells of the fundic cecum (stomach) of a vampire bat (10,000✕). *(Photo by D. Morton.)*

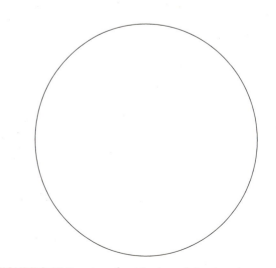

FIGURE 8-15 Drawing of a side view of simple columnar epithelium (_____ ✕).

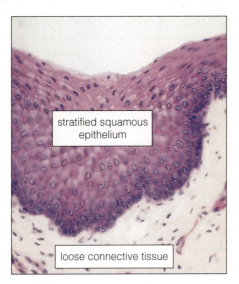

FIGURE 8-16 Section of the epithelial lining of the inner surface of the esophagus (300✕). *(Photo by D. Morton.)*

FIGURE 8-17 Drawing of a side view of stratified squamous epithelium (_____ ✕).

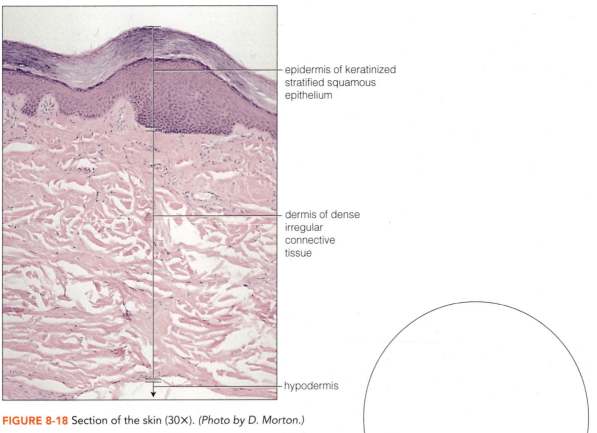

FIGURE 8-18 Section of the skin (30✕). *(Photo by D. Morton.)*

epidermis of keratinized stratified squamous epithelium

dermis of dense irregular connective tissue

hypodermis

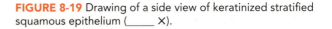

FIGURE 8-19 Drawing of a side view of keratinized stratified squamous epithelium (_____ ✕).

7. *Urinary bladder* (Figure 8-6). Figure 8-20 shows two sections, one from a contracted bladder and the other from a distended bladder. Locate the **transitional epithelium** at the surface of each section. The transitional epithelium from the contracted bladder appears like stratified cuboidal epithelium. In the distended bladder, the transitional epithelium is thinner and looks like stratified squamous epithelium. Draw a high-power view of these two extremes in Figure 8-21.

B. Connective Tissue

Connective tissues occur in all parts of the body. They contain a large amount of material external to the cells, called the *extracellular matrix*. This matrix consists of *fibers* embedded in *ground substance*. Two basic kinds of fibers exist: collagen and elastic. Whereas collagen fibers are tough, flexible, and inelastic, elastic fibers stretch when pulled, returning to their original length when the pull is removed. Except for cartilage, connective tissues contain blood vessels. Similar to epithelia, connective tissue cells are capable of cell division in adults but at a reduced rate. Table 8-1 lists most connective tissue subtypes along with their matrix characteristics.

The hard and special groups are sometimes combined as a specialized category of connective tissue subtypes. Blood is studied in Exercise 12. Find and examine subtypes in the following slides.

1. *Teased spread of loose connective tissue.* **Loose connective tissue** forms much of the packing material of the body and fills in the spaces between other tissues. Identify **collagen fibers** and **elastic fibers** (Figure 8-22). Collagen fibers are stained light pink and are variable in diameter, but they are generally wider than the elastic fibers. Elastic fibers are darkly stained, thin, and branched. Areolar connective tissue contains a number of different cell types. To see a **fibroblast**—the cell that produces the matrix of loose and other soft connective tissues—look for an elongated, oval-shaped nucleus associated with a fiber. Many kinds of cells that play important roles in the body's immune system live in this tissue. The spaces between fibers and cells contain ground substance in the living tissue. The ground substance of all soft connective tissues consists of a viscous soup of carbohydrate/protein molecules and is usually extracted from tissue sections

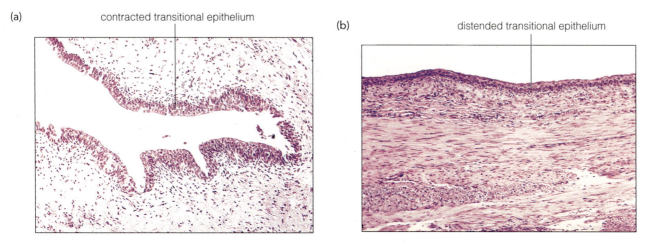

(a) contracted transitional epithelium

(b) distended transitional epithelium

FIGURE 8-20 Sections of (a) contracted and (b) distended urinary bladders (89✕). *(Photos by D. Morton.)*

TABLE 8-1	Connective Tissue Groups and Common Subtypes	
Group	**Subtype**	**Matrix Contains**
Soft	Loose (also called areolar)	Few collagen and elastic fibers arranged apparently randomly
	Dense irregular	Many collagen or elastic fibers arranged apparently randomly
	Dense regular	Many collagen or elastic fibers arranged in a parallel fashion
Hard	Cartilage	Many collagen or elastic fibers and polymerized (or jellylike) ground substance
	Bone	Many collagen fibers and mineralized ground substance
Special	Adipose	Delicate collagen fibers
	Blood	Plasma

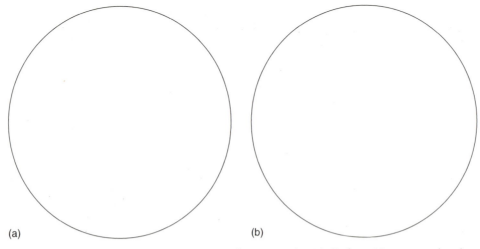

(a)　　　　　　　　　　　　　　　　　　(b)

FIGURE 8-21 Drawings of high power views of transitional epithelia from (a) contracted and (b) distended urinary bladders. (_____ ✕).

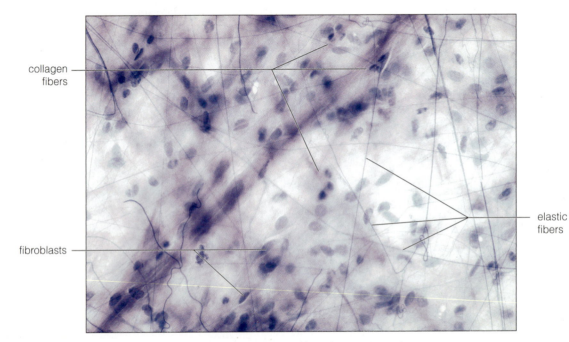

collagen fibers

elastic fibers

fibroblasts

FIGURE 8-22 Spread of loose connective tissue (450✕). *(Photo by D. Morton.)*

during processing. Draw loose connective tissue in Figure 8-23.

2. *Skin.* Reexamine the section of skin. Look at the second layer of the skin, the dermis, which is primarily composed of **dense irregular collagenous connective tissue** (Figure 8-18). The dermis cushions the body from everyday stresses and strains. Note that the looser irregular connective tissue of the hypodermis has a lower concentration of collagen fibers and islands of fat cells. It functions as a shock absorber, an insulating layer, and a site for storing water and energy. In mammals that move their skin independently of the rest of the body (e.g., cats), skeletal muscle tissue is also found in the hypodermis. In Figure 8-24, draw dense irregular collagenous connective tissue.

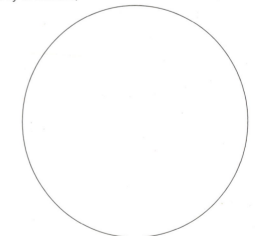

FIGURE 8-23 Drawing of a spread of loose connective tissue (_____ ✕).

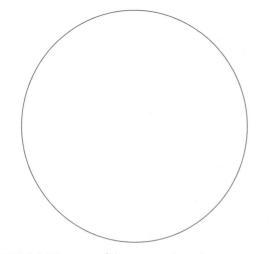

FIGURE 8-24 Drawing of dense irregular collagenous connective tissue (_____ ×).

FIGURE 8-26 Drawing of dense regular collagenous connective tissue (_____ ×).

3. *Tendon*. A tendon connects a skeletal muscle organ to a bone organ. Tendons are composed largely of **dense regular collagenous connective tissue.** Examine a longitudinal section of a tendon (Figure 8-25).

Describe the density and arrangement of the fibers.

_____

_____

_____

This design makes tendons very strong, much like a rope composed of braided strings, which in turn are made of even smaller fibers. How are the fibroblasts oriented relative to the arrangement of the fibers?

_____

_____

Draw dense regular collagenous connective tissue in Figure 8-26.

4. *Trachea* (Figures 8-10 and 8-11a). Reexamine the section of the trachea. Find a portion of one of the supporting cartilage, which prevents the wall from collapsing and closing the **lumen** (the space within any hollow organ). These cartilaginous structures are composed of **hyaline cartilage** tissue (Figure 8-27). Look for the **chondrocytes** (cartilage cells) that are located in small holes in the matrix called **lacunae** (singular, *lacuna*).

Although invisible, many collagen fibers are embedded in the polymerized ground substance. They cannot be seen because the indices of refraction of these two matrix components are similar. Your instructor has set up a demonstration of this phenomenon. Observe the two labeled beakers, one filled with immersion oil and the other with water. Look at the glass rod placed in each of them. The index of refraction is about 1.58 for glass, about 1.33 for water, and about 1.52 for immersion oil. In which fluid is it easier to see the glass rod?

_____

The difference between the index of refraction of the glass rod and immersion oil is less than that between the glass rod and water. This is exactly why

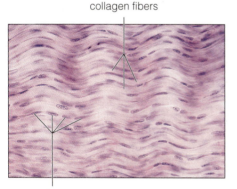

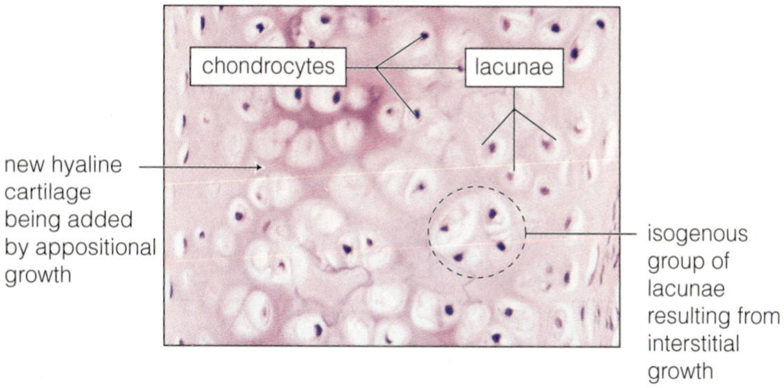

collagen fibers

nuclei of fibroblasts
aligned in a row
between collagen fibers

FIGURE 8-25 Dense collagenous connective tissue from a longitudinal section of a tendon (250×). *(Photo by D. Morton.)*

chondrocytes | lacunae

new hyaline cartilage being added by appositional growth

isogenous group of lacunae resulting from interstitial growth

FIGURE 8-27 Section of hyaline cartilage in the wall of the trachea (130×). *(Photo by D. Morton.)*

FIGURE 8-28 Drawing of hyaline cartilage (_____ ✕).

collagen fibers are hard to see in the ground substance of hyaline cartilage.

In locations where cartilage has to be more durable (intervertebral disks, for example), the collagen content is higher and the fibers are visible. In other sites (such as outer ear flaps), large numbers of elastic fibers are present. Elastic cartilage is deformed by a small force and returns to its original shape when the force is removed. Draw hyaline cartilage in Figure 8-28.

5. *Bone.* Living bones are amazingly strong. Bone organs are mostly composed of hard yet flexible **bone** tissue. Its hardness is attributable to minerals (predominantly calcium–phosphate salts called *hydroxyapatites*) deposited in the matrix. Its flexibility comes from having the highest collagen content of all connective tissues.

Examine the cross-section of compact bone tissue (Figure 8-29). Grinding a piece of the shaft of a long bone with coarse and then finer stones until a thin wafer remains has produced this preparation. Although only the mineral part of the matrix is present, the basic architecture has been preserved. Compact bone tissue is primarily composed of longitudinally arranged **Haversian systems** (also called *osteons*). Locate a Haversian system and identify its *central canal, lamellae* (singular, *lamella*), *lacunae, and canaliculi*—little canals that you see connecting lacunae with each other and with the central canal.

In living bone, blood vessels and nerves are present in central canals, concentric layers of matrix form the lamellae, and the intervening rings of *lacunae* contain bone cells called **osteocytes.** In young living bone, the canaliculi contain the cytoplasmic processes of osteocytes. Thus, the osteocytes can easily exchange nutrients, wastes, and other molecules with the blood. By comparison, substances in

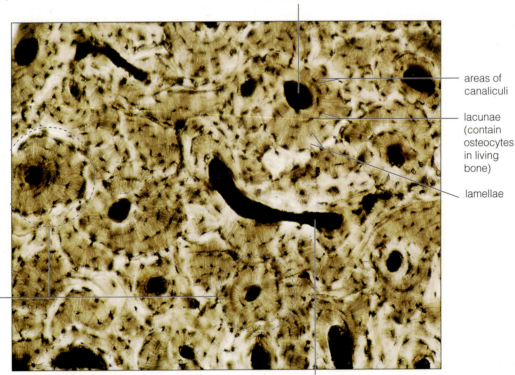

central canal

areas of canaliculi

lacunae (contain osteocytes in living bone)

lamellae

Haversian systems

canal connecting two central canals

FIGURE 8-29 Ground cross-section of a long bone showing compact bone tissue (100✕). *(Photo by D. Morton.)*

FIGURE 8-30 Drawing of compact bone tissue (_____ ✕).

cartilage have to diffuse across the matrix between chondrocytes and blood vessels in the surrounding soft connective tissue. Which tissue, bone or cartilage, will heal quicker? Explain why.

_____

_____

In Figure 8-30, draw a cross-section of compact bone tissue.

6. *White adipose tissue.* As you see in Figure 8-31, **white adipose tissue** consists mainly of **fat cells.** However, careful examination of the section using high power shows them to be surrounded by delicate collagen fibers, fibroblasts, and capillaries. Because the fat has been lost during the slide preparation, the fat cells look empty. The primary function of this tissue is energy storage. Draw white adipose tissue in Figure 8-32.

C. Muscle Tissue

Muscle tissue is contractile. Its cells (fibers) can shorten and produce changes in the position of body parts or they try to shorten and produce changes in tension. Contraction results from interactions between two types of protein filaments: **actin** and **myosin.** Similar to epithelial tissue, muscle tissue is primarily cellular; but unlike both epithelial and connective tissues, its cells (fibers) do not normally divide in adults. Therefore, the repair of damaged and dead fibers is limited. The three subtypes of muscle tissue are *skeletal*, *cardiac*, and *smooth*.

Skeletal muscle tissue is under *voluntary control*. This means that your conscious mind can order skeletal muscles to contract, although most of their contractions are actually involuntary. Cardiac muscle and smooth muscle tissues are always under *involuntary control*. Involuntary control means that contractions are directed at the unconscious level.

Examine your slide with sections of all three subtypes (Figure 8-33). Identify their characteristics (Table 8-2). In skeletal and cardiac muscle fibers, actin and myosin filaments overlap to produce the alternating pattern of light and dark bands (cross- or transverse striations) seen in these tissues. Only cardiac muscle cells are branched. Where the branch of one fiber joins another, the cells are stuck together and in direct communication through a complex of cell-to-cell junctions. The complex is called an **intercalated disk.** Find an intercalated disk in your section, but if you have trouble seeing it, look at the demonstration set up by your instructor. Draw the three muscle tissue subtypes in Figure 8-34.

D. Nervous Tissue

Nervous tissue is found in the brain, spinal cord, nerves, and all of the body's organs. Its function is the point-to-point transmission of information. Messages are carried

capillaries   fat cells

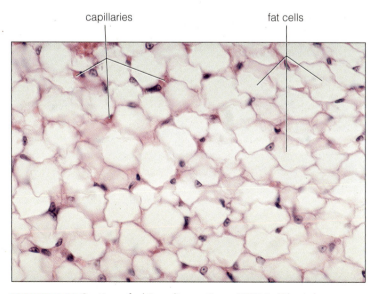

FIGURE 8-31 Section of white adipose tissue (450✕). *(Photo by D. Morton.)*

FIGURE 8-32 Drawing of white adipose tissue (_____ ✕).

(a)

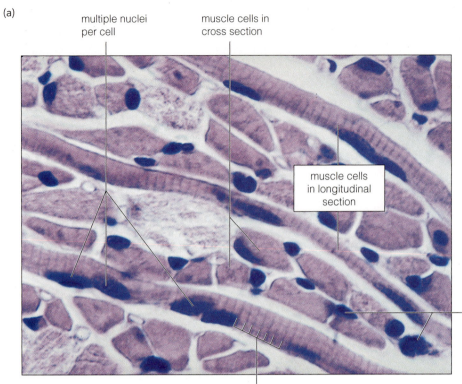

multiple nuclei
per cell

muscle cells in
cross section

muscle cells
in longitudinal
section

fibroblasts in
connective tissue
between fibers

striations

(b)

intercalated disks

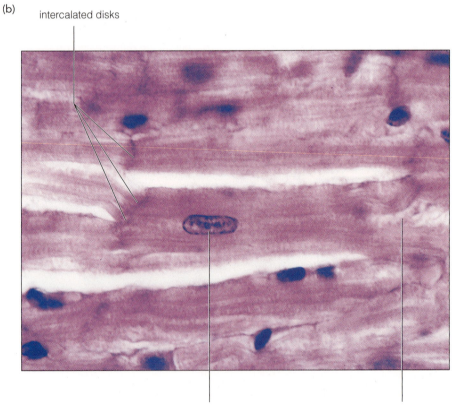

nucleus of cardiac
muscle cell

branch

FIGURE 8-33 Sections of muscle tissue: (a) skeletal (400×), (b) cardiac (1000×), (c) smooth (400×). *(Photos by D. Morton.)*

(c)

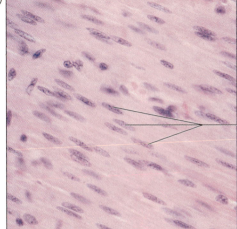

nuclei of smooth muscle cells

FIGURE 8-33 *continued.*

by impulses that travel along the functional unit of the nervous system, the **neuron** (or nerve cell). Similar to muscle cells, neurons do not divide in adults, so natural replacement is impossible.

The largest cells in Figure 8-35 are *motor neurons,* which connect the spinal cord to muscle fibers or glands. Find a similar cell in your smear of an ox spinal cord. The motor neuron has a **cell body** and a number of slender cytoplasmic extensions called *neuron processes,* including one long **axon** and several shorter **dendrites.** The dendrites and cell body are stimulated within the spinal cord, and the axon conducts impulses out of it. The cells with smaller nuclei are accessory cells called **neuroglia.** Accessory cells help neurons function and make up about half the mass of nervous tissue. In Figure 8-36, draw a smear of nervous tissue.

8.2 Systems

There are 11 systems in mammals, and each one contains a number of organs. In this portion of the exercise, you will identify the vital functions of these systems and many of their constituent organs. Many of these organs are located in or near the major **body cavities.** The *ventral* (located toward the belly surface) and *dorsal* (located toward the back surface) *body cavities* of humans, which are typical of mammals, are shown in Figure 8-37. A muscular *diaphragm* separates the **thoracic cavity**

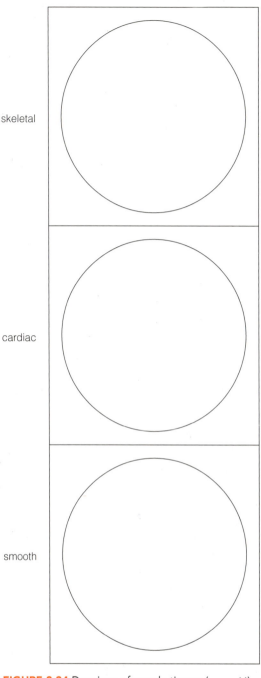

skeletal

cardiac

smooth

FIGURE 8-34 Drawings of muscle tissues (_____ ×).

TABLE 8-2	Characteristics of Muscle Tissue Subtypes			
Subtypes	**Organ Location**	**Position and Number of Nuclei in Fibers**	**Cross-Striated Fibers**	**Special Fiber Features**
Skeletal	Skeletal muscles	Peripheral and many	Yes	Long, cylindrical shape
Cardiac	Heart	Central and usually one	Yes	Branches and intercalated disks
Smooth	Walls of internal organs	Central and one	No	Shape tapers at both ends

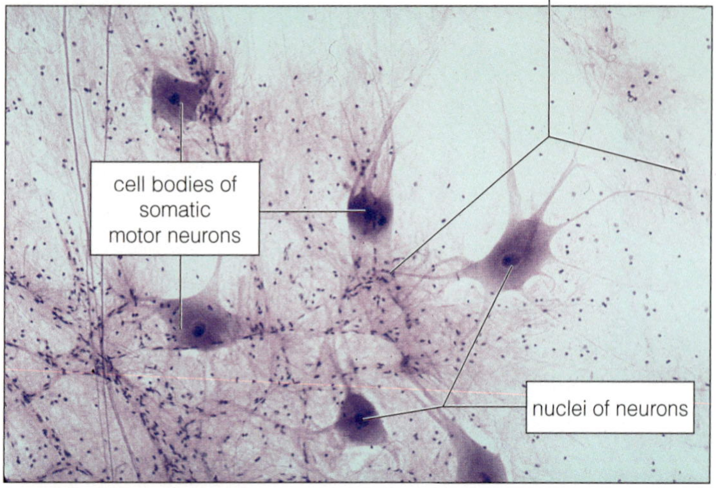

smaller nuclei of numerous
or neuroglial accessory cells

cell bodies of
somatic
motor neurons

nuclei of neurons

FIGURE 8-35 Smear of spinal cord, showing nervous tissue (100×). *(Photo by D. Morton.)*

FIGURE 8-36 Drawing of nervous tissue (_____ ×).

and **abdominal** cavities. The thoracic cavity is further subdivided into two lateral (away from the midline) **pleural cavities** and a medial (at or near the midline) **pericardial cavity.** The abdominal cavity is continuous with a lower **pelvic cavity,** and together they are alternatively referred to as the *abdominopelvic cavity.* The continuous **cranial** and **spinal cavities** make up the dorsal body cavity. Table 8-3 summarizes the layout of the body cavities.

MATERIALS

Per student:

- Textbook
- Colored pencils—green, yellow, black, red, brown, pink, and blue

Per lab room:

- Labeled demonstration dissection of a mouse (optional)
- Demonstration dissection of a sheep brain

PROCEDURE

1. *Vital functions of organ systems.* Using your text, list the vital functions of each system in Table 8-4.

2. *Major organs and systems of the ventral body cavities.* Look at the photos of dissected mice (Figures 8-3, 8-6, and 8-10), which show the organs of the ventral body cavities. In addition, your instructor may have

TABLE 8-3	Layout of Mammalian Body Cavities		
Cavity			**Contains**
Ventral	Thoracic	Pleural (right and left)	Lungs
		Pericardial	Heart
	Abdominopelvic	Abdominal	Stomach, small intestine, liver, and so on
		Pelvic	Urinary bladder, and so on
Dorsal	Cranial		Brain
	Spinal		Spinal cord

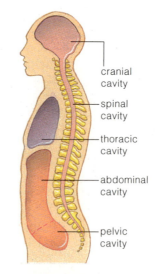

cranial cavity

spinal cavity

thoracic cavity

abdominal cavity

pelvic cavity

FIGURE 8-37 Major body cavities of humans. *(After Starr and Taggart 2001.)*

TABLE 8-4 The Organ Systems of Mammals

Systems	Vital Functions
Integumentary	
Nervous	
Endocrine	
Skeletal	
Muscular	
Circulatory	
Lymphatic	
Respiratory	
Digestive	
Urinary	
Reproductive	

prepared a labeled demonstration dissection of a mouse. Use your text to identify the systems the labeled organs belong to, underlining each name with a colored pencil: green for the respiratory system, yellow for the digestive system, black for the lymphatic system, red for the circulatory system, brown for the urinary system, pink for the female reproductive system, and blue for the male reproductive system. Use Figure 8-38 to see the urinary and reproduction systems of a dissected female mouse.

3. *Major organs and systems of the dorsal body cavities.* The organs of the nervous system are located in the dorsal body cavities—the brain in the cranial cavity and the spinal cord in the spinal cavity. Examine the demonstration dissection of a sheep brain and identify the structures labeled in Figure 8-39.

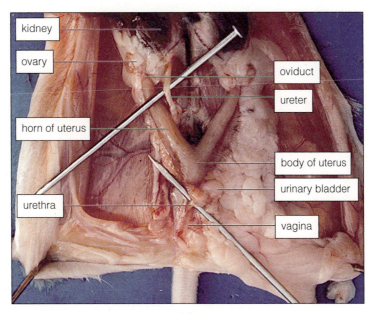

FIGURE 8-38 Ventral view of the urinary and reproductive systems of a female mouse (3✕). *(Photo by D. Morton.)*

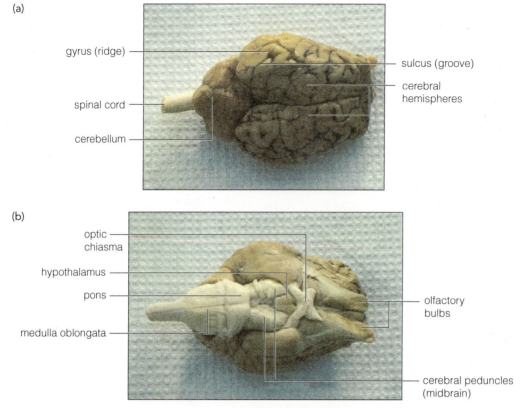

(a)

gyrus (ridge)

sulcus (groove)

cerebral
hemispheres

spinal cord

cerebellum

(b)

optic
chiasma

hypothalamus

pons

medulla oblongata

olfactory
bulbs

cerebral peduncles
(midbrain)

FIGURE 8-39 Sheep brain: dorsal (a) and ventral (b) views. *(Photos by D. Morton.)*

_____ **1.** Histology is the study of
(a) cells
(b) organelles
(c) tissues
(d) organisms

_____ **2.** A collection of similarly specialized cells and any extracellular material they secrete and maintain describes
(a) an organ
(b) a system
(c) a tissue
(d) organelles

_____ **3.** Organs strung together functionally and usually structurally form
(a) organs
(b) systems
(c) tissues
(d) organelles

_____ **4.** _____ are constructed of all four basic tissue types.
(a) organs
(b) systems
(c) cells
(d) organelles

_____ **5.** An epithelial tissue formed by more than one layer of cells and with columnlike cells at the surface is called
(a) simple squamous
(b) stratified squamous
(c) simple columnar
(d) stratified columnar

_____ **6.** The middle layer of the skin
(a) is called the dermis
(b) is primarily connective tissue
(c) contains collagen fibers
(d) is all of the above

_____ **7.** To which subtype of muscle tissue does a fiber with cross-striations and many peripherally located nuclei belong?
(a) skeletal
(b) cardiac
(c) smooth
(d) none of the above

_____ **8.** In which tissue would you look for cells that function in point-to-point communication?
(a) connective
(b) epithelial
(c) muscle
(d) nervous

_____ **9.** The ventral body cavities include
(a) the thoracic cavity
(b) the cranial cavity
(c) the abdominopelvic cavity
(d) both a and c

_____ **10.** The cranial cavity contains
(a) the lungs and heart
(b) the spinal cord
(c) the brain
(d) both b and c

EXERCISE **8** **How Mammals are Constructed**

POST-LAB QUESTIONS

Introduction

1. In the correct order from smallest to largest, list the levels of organization present in most animals.

8.1 Tissues and Organs

2. Describe the main structural characteristics of the four basic tissues.

    a.   Epithelial tissue

    b.   Connective tissue

    c.   Muscle tissue

    d.   Nervous tissue

3. Describe the main functions of the four basic tissues.

    a.   Epithelial tissue

    b.   Connective tissue

    c.   Muscle tissue

    d.   Nervous tissue

4. Identify the following tissues.

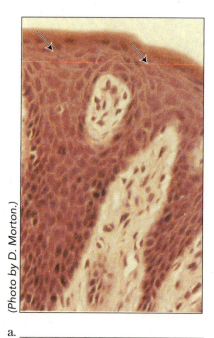

(Photo by D. Morton.)

a. _____

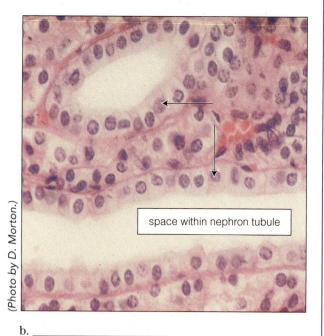

space within nephron tubule

(Photo by D. Morton.)

b. _____

8.3 Systems

5. Choose any organ. Describe how its construction allows for its specific function.

6. Briefly describe the ventral body cavities and the major organs they contain.

7. Briefly describe the dorsal body cavities and the major organs they contain.

8. Identify the following structures.

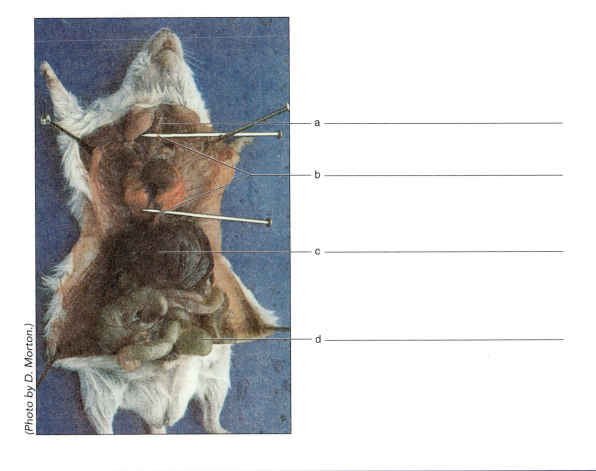

(Photo by D. Morton.)

a ——————————————

b ——————————————

c ——————————————

d ——————————————

Food for Thought

9. Bone is a subtype of connective tissue, and bones are organs. How are the two related yet different from each other? Can you think of another tissue/organ pair that potentially creates a similar confusing situation?

10. Search for the phrases "tissue engineering" and "organ engineering" on the Internet. List two websites and briefly summarize their content.

http://

http://

Support and Movement: Human Skeletal and Muscular Systems

OBJECTIVES

After completing this exercise, you will be able to

1. define *bones, ligaments, joints, skeletal muscles, tendons, muscle tone, posture, sutures, synovial joints, diaphysis, epiphyses, compact bone, spongy bone, marrow cavity,* and *lever.*

2. identify the major bones of the human skeleton.

3. describe the structure of a typical bone.

4. define *origin, insertion,* and *action* as these terms apply to skeletal muscles and their tendons.

5. distinguish between isometric and isotonic contractions of skeletal muscles.

6. give everyday and anatomical examples of the three classes of levers.

7. present a simple biomechanical analysis of walking.

Introduction

The skeletal system and muscular system are often considered together to stress their close structural and functional ties. They determine the basic shape of the body, support other systems, and provide the means by which people move in the external environment.

Bones are the main organs of the skeletal system. They are composed mostly of bone tissue, although all four basic tissue types are present. The places in the body where two or more bones are connected are called **joints.** The joints you are most familiar with are the *movable synovial joints,* including the shoulder, elbow, wrist, hip, knee, and ankle. However, there are other kinds of joints that are less movable. Around many synovial joints, bones are held together by straplike structures called **ligaments.** Ligaments are primarily dense connective tissue that is more or less elastic. Elastic ligaments around mobile joints stretch to allow movement.

Skeletal muscles are the main organs of the muscular system and are largely composed of skeletal muscle tissue. Skeletal muscles are connected to bones by dense fibrous connective tissue structures called **tendons.** Tendons are inelastic, so all of the force of skeletal muscle contraction is transferred to the bones of the skeleton.

When a skeletal muscle contracts, movement may or may not occur. If the skeletal muscle is allowed to shorten, the bone moves, and in doing so, it moves some body part. On the other hand, if the skeletal muscle does not shorten, the tension in that muscle and in its tendons increases. All skeletal muscles exhibit tension or **muscle tone** except when a person is asleep. This tension maintains **posture**—the ability to hold the body erect and to keep the position of its parts, all against the pull of gravity.

The organs of the skeletal and muscular systems have other functions as well. Bones protect internal organs (e.g., the skull protects the brain, the eyes, and the middle and inner parts of the ears). Bones also store minerals and produce blood cells in the bone marrow. When body temperature drops below a certain level, the skeletal muscles produce heat by shivering.

9.1 Adult Human Skeleton

An articulated human skeleton is prepared by joining the degreased and bleached bones of an individual so that many of the bones can be moved as they were in life. Often, plastic casts of the original bones are used.

Although bone tissue predominates, fresh bones are composed of all four basic tissue types. However, when they are prepared for study, their organic portion is lost; these bones consist only of the mineral portion of bone tissue. Original details remain, but the bones are brittle.

Therefore, you must handle bones gently. Use a pipe cleaner to point out details and never use a pencil or pen because it is very difficult to remove marks.

MATERIALS

Per student:

- Pipe cleaner

- Compound microscope, lens paper, a bottle of lens-cleaning solution, a lint-free cloth

- Prepared slides of:
  - A synovial joint
  - A ground cross-section of compact bone (optional)
  - A cross-section of a skeletal muscle

Per student group:

- Articulated adult human skeleton (natural bone or plastic)
- Skull (natural bone or plastic)
- Femur (natural bone or plastic)
- Femur that has been sawed into two halves lengthwise

Per lab room:

- Labeled chart and illustrations of the adult human skeleton

PROCEDURE

A. Identification of Some Bones

There are 206 separate bones in the adult human skeleton. Using the labeled chart and illustrations of the human skeleton, identify the following bones on the articulated human skeleton and label them in Figure 9-1.

1. Axial skeleton
   a. **Skull** (28 separate bones, including the middle ear bones) (Figure 9-2)
   b. **Vertebrae** (singular, *vertebra;* 26 separate bones, including the **sacrum,** which is composed of five fused vertebrae, and the **coccyx,** which is usually composed of four fused vertebrae) (Figure 9-3)
   c. **Ribs** (12 pairs of ribs for a total of 24 separate bones) (Figure 9-4)
   d. **Sternum** (three fused bones) (Figure 9-4)
   e. **Hyoid** (only bone that does not form a joint with another bone) (Figure 9-4)

2. Appendicular skeleton (these are all paired bones found on the right and left sides of the body)

Pectoral girdle (shoulder) (Figure 9-5)
   a. **Scapula**
   b. **Clavicle**

Arm (Figures 9-5 and 9-6)
   a. **Humerus**
   b. **Radius**
   c. **Ulna**
   d. **Carpals** (8)
   e. **Metacarpals** (5)
   f. **Phalanges** (14)

Pelvic girdle (hip) (Figure 9-7)
   a. **Coxal bone** (three fused bones—**pubis, ischium, ilium**)

Leg (Figure 9-8)
   a. **Femur**
   b. **Tibia**

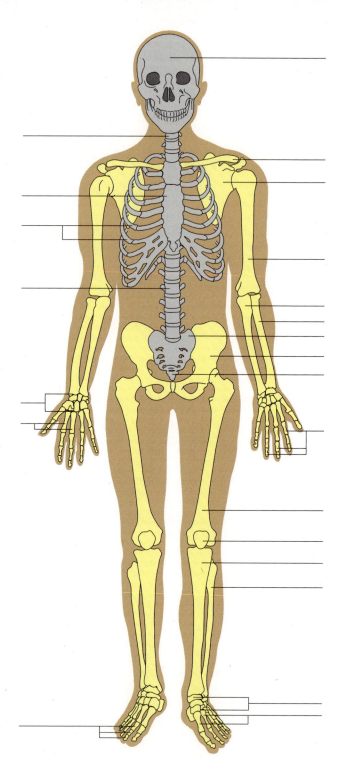

FIGURE 9-1 Label this front view of adult human skeleton (axial portion shaded gray and appendicular portion colored yellow).

   c. **Fibula**
   d. **Patella**
   e. **Tarsals** (7)
   f. **Metatarsals** (5)
   g. **Phalanges** (14)

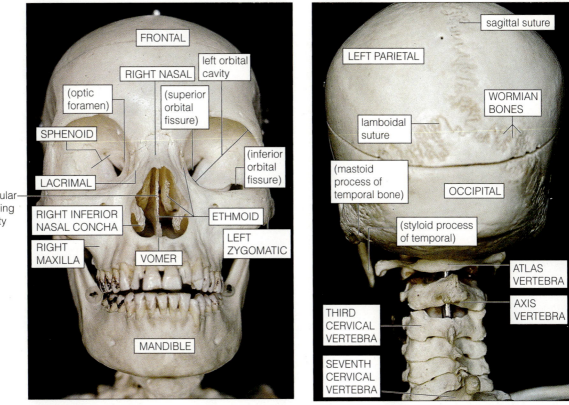

(a)

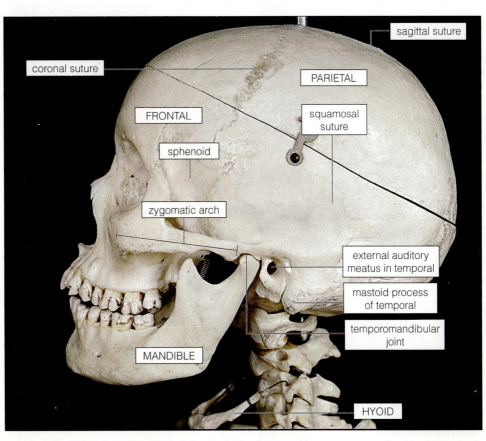

(b)

(c)

FIGURE 9-2 (a) Front, (b) back, and (c) side views of the skull. The names of bones are capitalized in this and subsequent figures to clearly separate them from other features. *(Photos by D. Morton.)*

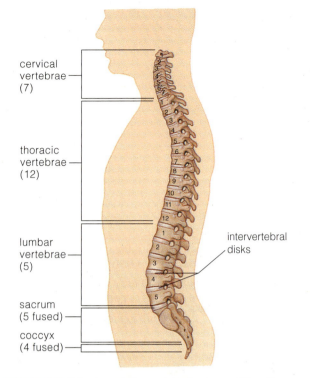

cervical
vertebrae
(7)

thoracic
vertebrae
(12)

lumbar
vertebrae
(5)

intervertebral
disks

sacrum
(5 fused)

coccyx
(4 fused)

FIGURE 9-3 Side view of the vertebral column. *(After Starr and McMillan, 2010.)*

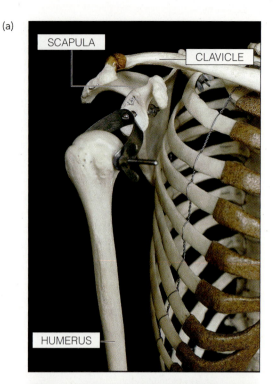

(a)

SCAPULA

CLAVICLE

HUMERUS

(b)

CLAVICLE SCAPULA

HUMERUS

FIGURE 9-5 (a) Anterior and (b) posterior views of pectoral girdle and shoulder. *(Photos by D. Morton.)*

B. Joints

The degree of movements allowed at different joints ranges from none to freely movable (Figure 9-9). Examples of immovable joints are the **sutures** that connect the bones of the roof of the skull of young adults. Joints that allow the freest movement are **synovial joints,** such as the ones listed in Table 9-1.

1. Examine a prepared section of a synovial joint with your compound microscope. Identify the structures

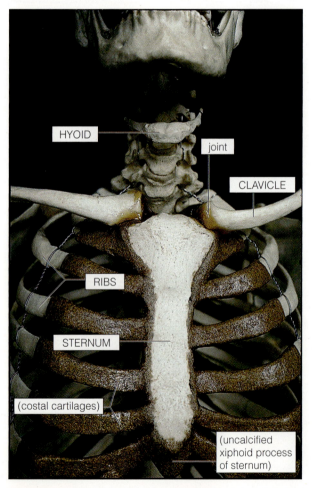

HYOID

joint

CLAVICLE

RIBS

STERNUM

(costal cartilages)

(uncalcified xiphoid process of sternum)

FIGURE 9-4 Anterior view of skeleton of upper trunk. *(Photo by D. Morton.)*

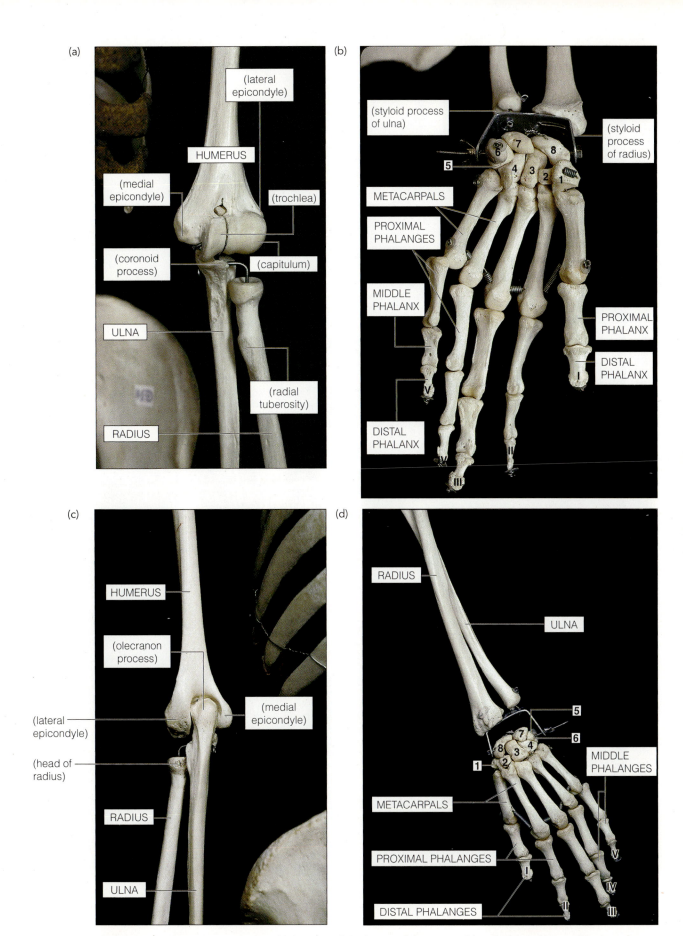

FIGURE 9-6 Anterior (a and b) and posterior (c and d) views of the bones of the arm and hand. The eight carpal bones are numbered and the five digits in (b) and (d) are labeled with roman numerals. *(Photos by D. Morton.)*

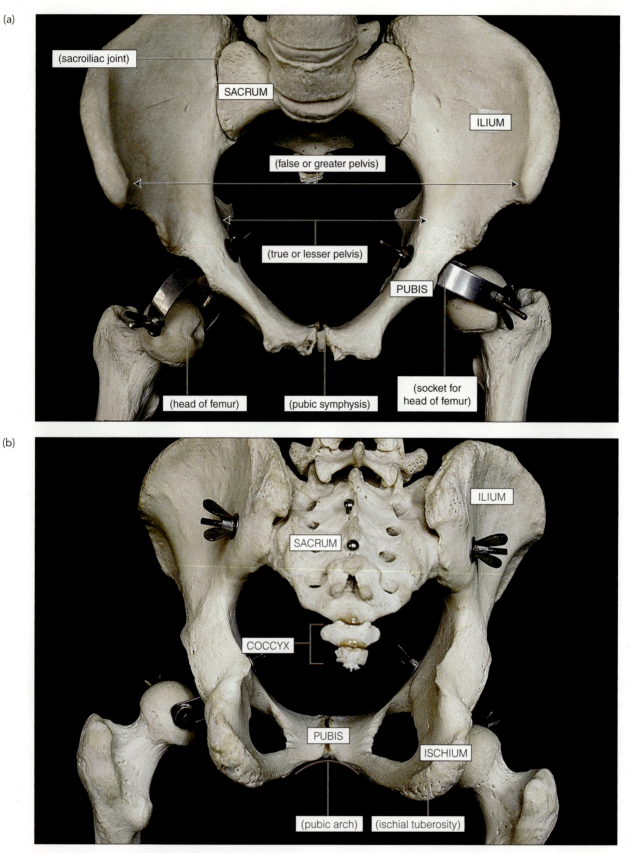

FIGURE 9-7 (a) Anterior and (b) posterior views of pelvic girdle and hip. *(Photos by D. Morton.)*

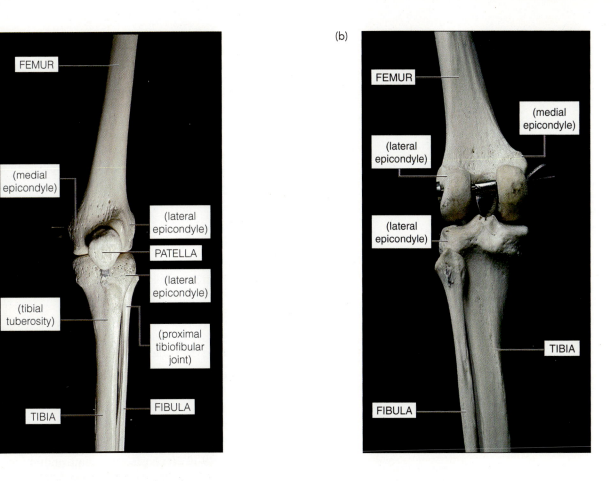

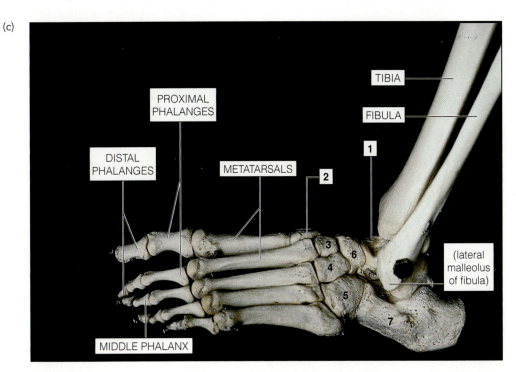

FIGURE 9-8 (a) Anterior and (b) posterior views of knee and (c) lateral view of foot. The seven tarsal bones are numbered. *(Photos by D. Morton.)*

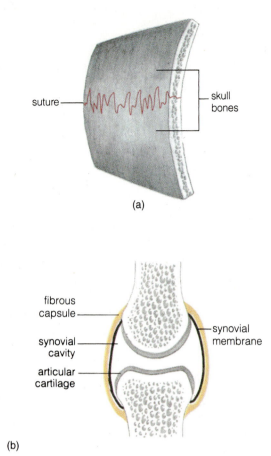

(a)

(b)

FIGURE 9-9 Diagrams of (a) suture and (b) synovial joint.

labeled in Figure 9-9. The *fibrous capsule* of synovial joints is lined by a *synovial membrane*, which secretes lubricating *synovial fluid*. The fibrous capsule and ligaments function to stabilize synovial joints. Ligaments can be located outside and inside the capsule, and they may be thickenings of its wall.

2. Identify the synovial joints listed in Figure 9-1. List the adjacent bones that form them in Table 9-1.

C. Surface Features

There are many places on your body surface where bones can be felt. However, it is often difficult to tell specifically which bone you are feeling. Some are easy.

1. Feel the bone supporting your lower jaw, the *mandible*. This is the only bone of the skull that forms a synovial joint with another skull bone.

2. Let us try a harder example, the piece of bone that projects from the point of the elbow joint. Touch it and alternately extend and flex the forearm, increasing and decreasing the angle between the forearm and upper arm, respectively.

   Which part of the arm does the projection move with, the forearm or upper arm?_____

   _____

   While still touching this projection, alternately turn the hand palm down and up. Does the projection move? _____ (yes or no)

   Which bone belongs to this projection? _____

   _____

   In general, to identify a portion of a bone near a joint, move the body parts adjacent to the joint while touching the bone.

3. Identify the bones that have the surface features listed in Table 9-2.

D. Structure of a Bone

1. Look at a femur, the longest bone of the skeleton (Figure 9-10). It consists of a shaft, or **diaphysis,** with two knobby ends, or **epiphyses** (singular, *epiphysis*). One end has a narrow neck and a round head. Which fused bone does the femur join? _____

   To which bone of the skeleton does the other end join? _____

   Note the other surface features on the femur, such as projections of various sizes and lines. These surface features are attachment sites for tendons and ligaments.

   Are there small tunnels opening onto the surface of the femur? _____ (yes or no)

   These tunnels are best seen in natural specimens. In life, these nutrient canals serve as routes for blood vessels and nerves.

2. Examine a femur that has been sawed in half lengthwise. There are two kinds of adult bone tissue: **compact**

TABLE 9-1	Bones That Form the Major Synovial Joints	
Joint		**Adjacent Bones**
Wrist		
Elbow		
Shoulder		
Hip		
Knee		
Ankle		

TABLE 9-2 Surface Features and Bones

Surface Feature	Bone
Knuckles	_____
Bump next to the wrist and on the same side of the upper appendage as the little finger	_____
Smaller bump next to the wrist and on the same side of the upper appendage as the thumb	_____
Bump next to and outside the ankle	_____
Bump next to and inside the ankle	_____

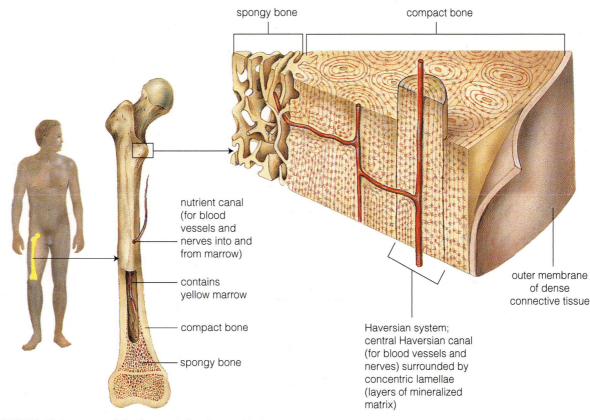

spongy bone   compact bone

nutrient canal
(for blood
vessels and
nerves into and
from marrow)

contains
yellow marrow

compact bone

spongy bone

Haversian system;
central Haversian canal
(for blood vessels and
nerves) surrounded by
concentric lamellae
(layers of mineralized
matrix)

outer membrane
of dense
connective tissue

FIGURE 9-10 Structure of the femur. *(After Starr, 1991.)*

bone and **spongy bone.** Compact bone is solid and dense and is found on the surface of the femur. Spongy bone is latticelike and is found on the inside of the femur, especially in the epiphyses and surrounding the **marrow cavity.**

Which kind of bone tissue looks denser? _____

Comparing pieces of equal size, which kind of bone tissue looks lighter? _____

E. Structure of a Skeletal Muscle

Similar to bones, skeletal muscles are composed of all four basic tissue types. Skeletal muscles are mostly skeletal muscle tissue with the individual skeletal muscle fibers arranged parallel to the axis along which the muscle

shortens when contracting. A substantial amount of connective tissue surrounds the fibers and connects them to the tendons.

Use your compound microscope to examine a section of a skeletal muscle. Look for fibers, fiber bundles (the more or less loose connective tissue located between fibers, the loose connective tissue between bundles of fibers), and the fibrous connective tissue that surrounds the entire organ (Figure 9-11).

9.2   Leverage and Movement

Much of the skeletal system is a system of levers, in which each bone is a lever and the joints are fulcrums. During a typical movement, one end of a skeletal muscle, the **origin,**

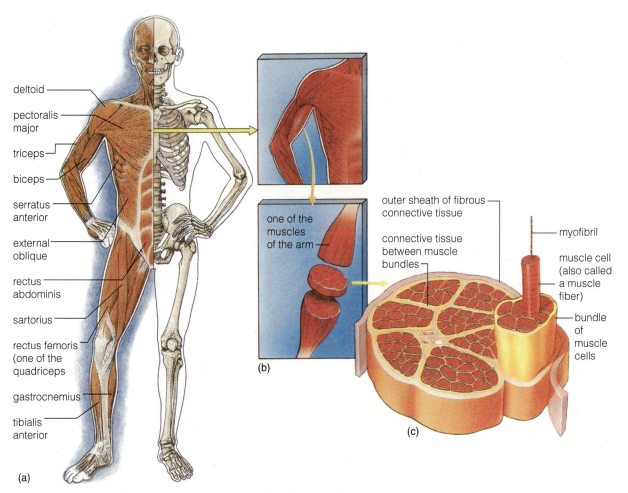

deltoid

pectoralis major

triceps

biceps

serratus anterior

external oblique

rectus abdominis

sartorius

rectus femoris (one of the quadriceps

gastrocnemius

tibialis anterior

(a)

one of the muscles of the arm

(b)

outer sheath of fibrous connective tissue

connective tissue between muscle bundles

myofibril

muscle cell (also called a muscle fiber)

bundle of muscle cells

(c)

FIGURE 9-11 (a) Some of the major skeletal muscles of the human. (b) A closer look at the structure of the skeletal muscle organ. (c) Cross section of a skeletal muscle organ. *(After Starr, 1991.)*

remains stationary. The other end, the **insertion,** moves along with the bone and surrounding body part. The movement produced by the contraction is the **action** of the skeletal muscle. Most insertions are close to their joints, and the advantage gained by this is that the muscle has to shorten a small distance to produce a large movement of the corresponding body part.

MATERIALS

Per student pair:

- Pair of scissors
- Toggle switch mounted on a board (alternatively, you can use any light switches present in the room)
- Pair of forceps
- Pencil
- Textbook

PROCEDURE

A. Classes of Levers

Levers are simple machines. When a pulling force or effort is applied to a lever, it moves about its fulcrum, overcoming a resistance or moving a load.

1. There are three classes of levers (Figure 9-12):
   a. Class I. The fulcrum is located between the effort and the load.
   b. Class II. The load is located between the fulcrum and the effort.
   c. Class III. The effort is located between the fulcrum and the load. Class III levers are the most common in the skeletal system.

2. Test your understanding of the three classes of levers by examining the objects listed in the following matching question and then matching them with the appropriate class of lever.

Lever	Object
Class I. _____	a. Scissors
Class II. _____	b. Toggle switch or light switch
Class III. _____	c. Forceps

3. To remember the relative position of the fulcrum, load, and effort for each class of lever, use this mnemonic

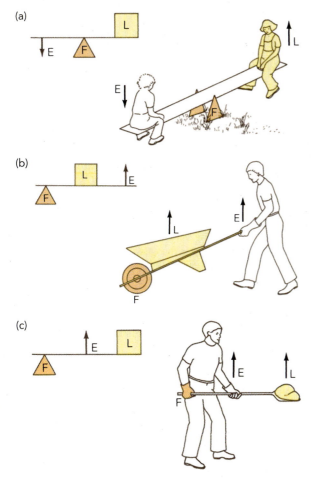

(memory device): "1, 2, 3; F, L, E." For example, 2 has the same relative position as L in the mnemonic. For the wheelbarrow in Figure 9-12b, having the L in the center tells you that it is a class II lever. In a class II lever, the load is at one of the ends, and the fulcrum is at the other. Try it for the other two classes of levers.

B. Analysis of Simple Movements

Let us analyze three simple movements: flexion of the forearm at the elbow, extension of the forearm at the elbow, and plantar flexion of the foot (Figure 9-13).

1. *Flexion of the forearm.* While sitting, turn your hand so your palm is up and place it under the lab bench. Try to flex the forearm (decrease the angle between the forearm and upper arm). Because the skeletal muscle that is attempting to flex the forearm cannot shorten, the tension in it will increase. A contraction of a skeletal muscle in which tension increases but no movement results is called an **isometric contraction.** While repeating this movement, feel with your other hand the front surface of the upper arm. The large, tense muscle is the biceps brachii. Its origin is the scapula, and its insertion is the radius.

Which joint is the fulcrum? _____

Now place a pencil in the palm of your hand and try to flex your forearm at the elbow. A contraction of a skeletal muscle that results in movement is called an **isotonic contraction.** There is no increase in tension during the movement. Feel the tension in the biceps brachii as you make this movement. Repeat this procedure but replace the pencil with a textbook. Both the pencil and the book add to the load being lifted, the forearm.

In which case—lifting the pencil or the textbook—was the tension in the biceps brachii the greatest?

_____

When you lift any object, the tension in the muscle must slightly exceed the weight of that object before

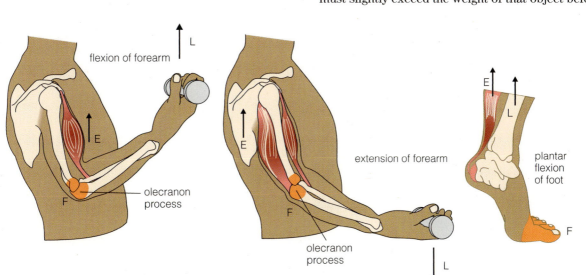

FIGURE 9-13 Some simple actions of skeletal muscles.

movement can occur. Therefore, normal movements have an isometric phase followed by an isotonic phase.

Where is the pulling force applied? (insertion, origin, or both the insertion and origin)

_____

Even simple movements require the coordination of a group of muscles. For example, the origin does not move because other skeletal muscles hold the scapula stationary.

What class of lever (I, II, or III) is illustrated by the preceding example? _____

2. *Extension of the forearm.* Place your hand, still palm up, on the top of the lab bench and try to extend your forearm (increase the angle between the forearm and the upper arm). Feel for a tense muscle on the back surface of your upper arm. This is the triceps brachii. The origin of the triceps brachii is the scapula and the upper humerus; its insertion is the *olecranon process* of the ulna (Figure 9-13). The fulcrum is the same as the previous example except that it has shifted position relative to the effort and the load.

What class of lever is illustrated by this movement?

_____

Extension of the forearm is the opposite movement to flexion of the forearm. Hold the textbook, palm still up, halfway between full flexion and full extension. Feel the tension in the biceps brachii and triceps brachii. Repeat this procedure without the book.

Is the tension in the biceps brachii greater with or without the book? _____

Is the tension in the triceps brachii greater with or without the book? _____

The state of contraction of a group of skeletal muscles has to be coordinated to accomplish a particular movement or posture. Both the tendons of the biceps brachii and the triceps brachii pull on their insertions on the bones of the forearm to keep the forearm stationary. Other muscles keep the shoulder stationary.

3. *Plantar flexion of foot.* You need to stand up to do this movement. A lab partner should stand behind you and watch that you do not fall during this procedure. With one hand on the lab bench to steady your balance, stand on the tips of your toes (Figure 9-12). With your other hand, feel one of the very large, tense muscles on the back of each calf. This is the gastrocnemius. The origin of the gastrocnemius is the femur, and its insertion is a tarsal—the calcaneus or heel bone. The fulcrum is the metatarsal-phalangeal joints, and the weight is the weight of the body transmitted through the tibia.

What class of lever does this movement exemplify?

_____

9.3 Walking

Walking is a complex activity that requires many movements and the coordinated contractions of several groups of skeletal muscles. For each leg, walking involves two phases, which together make up the *step cycle.* The *stance phase* occurs when the leg bears weight, and the *swing phase* occurs when the leg is in the air.

MATERIALS

Per lab room:

- Safe place to walk

PROCEDURE

1. *Follow your instructor's directions as to where to walk safely.* Walk a few normal steps, concentrating on one leg.

What part of the foot (toe or heel) strikes the ground first? _____

What part of your foot leaves the ground last?

_____

Does it leave passively, or does it push off?

_____

2. Now put your hands on your hips and concentrate on what your pelvic girdle is doing while you walk. First take short strides and then long ones.

Does the pelvic girdle rotate more during short or long strides? _____

Rotation of the pelvic girdle can be demonstrated in a different way. Find a lab partner of about equal height. Walk right next to each other but out of step, that is, with opposite feet leading. First take short steps and then long ones. What happens?

_____

_____

This sideways movement is called *lateral displacement.* Incidentally, females in general have to rotate their pelvic girdles a little more than males for a given length of stride. This is because of differences in the proportions of the female and male pelvic girdles.

3. *Vertical displacement* also occurs during walking. From the side, observe two individuals of equal height walking out of step and next to each other.

Do their heads remain at the same level, or do they bob up and down? _____

_____ **1.** Ligaments connect
  (a) bones to bones
  (b) skeletal muscles to bones
  (c) tendons to bones
  (d) skeletal muscles to tendons

_____ **2.** Tendons connect
  (a) bones to bones
  (b) skeletal muscles to bones
  (c) ligaments to bones
  (d) skeletal muscles to tendons

_____ **3.** Which bone is part of the axial skeleton?
  (a) clavicle
  (b) radius
  (c) coxal bone
  (d) sternum

_____ **4.** The two kinds of bone tissue are
  (a) compact and loose
  (b) compact and spongy
  (c) dense and spongy
  (d) loose and dense

_____ **5.** There are _____ classes of levers.
  (a) two
  (b) three
  (c) four
  (d) more than four

_____ **6.** The class of lever in which the effort is located between the fulcrum and the load is called
  (a) class I
  (b) class II
  (c) class III
  (d) class IV

_____ **7.** The end of the skeletal muscle that remains stationary during a movement is
  (a) the action
  (b) the origin
  (c) the insertion
  (d) none of the above

_____ **8.** In an isotonic contraction of a skeletal muscle,
  (a) the tension in the muscle increases
  (b) movement occurs
  (c) no movement occurs
  (d) both a and c occur

_____ **9.** In an isometric contraction of a skeletal muscle,
  (a) the tension in the muscle increases
  (b) movement occurs
  (c) no movement occurs
  (d) both a and c occur

_____**10.** The step cycle of walking consists of
  (a) a stance phase
  (b) a swing phase
  (c) both a and b
  (d) none of the above

EXERCISE **9**

Support and Movement: Human Skeletal and Muscular Systems

POST-LAB QUESTIONS

9.1 Adult Human Skeleton

1. Match the following bones to their location in the body.

Bone	Location
_____ Radius	a. Pectoral girdle
_____ Coxal bone	b. Leg
_____ Ribs	c. Axial skeleton
_____ Scapula	d. Arm
_____ Fibula	e. Pelvic girdle

2. Label this photo of a femur that has been sawed in half lengthwise with compact bone and spongy bone.

marrow cavity

(Photo by D. Morton.)

3. Identify the bones indicated in this photo (top of next page).
4. Label the fibrous capsule and synovial membrane of this joint (middle of next page).

9.2 Leverage and Movement

5. Define the following terms:

   a. The *insertion* of a skeletal muscle

   b. The *origin* of a skeletal muscle

   c. The *action* of a skeletal muscle

6. Explain the difference between isometric and isotonic contractions. How are both important to normal body movements?

7. Draw and label the structures of a typical skeletal muscle organ.

9.3 Walking

8. In your own words, describe one step in the walking cycle.

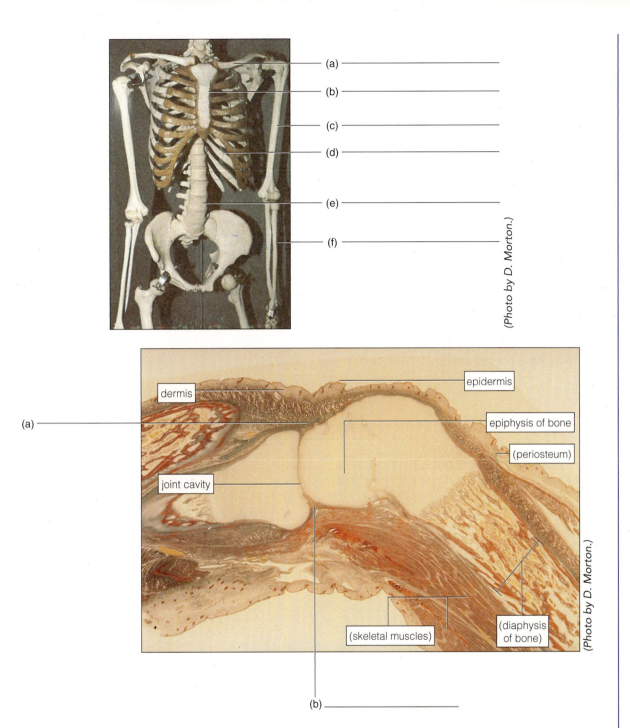

(a) _____

(b) _____

(c) _____

(d) _____

(e) _____

(f) _____

(Photo by D. Morton.)

dermis

epidermis

(a) _____

epiphysis of bone

(periosteum)

joint cavity

(skeletal muscles)

(diaphysis of bone)

(Photo by D. Morton.)

(b) _____

Food for Thought

9. The skeletal muscle that flexes the forearm after pronation (palm down position as in a pull-up) is the brachialis. Its origin is the humerus, and the insertion is the upper front of the ulna. Identify the class of lever involved and explain why you made this choice.

10. Search the Internet for websites that describe diseases of bones (e.g., osteoporosis) and of skeletal muscles (e.g., muscular dystrophy). List two websites and briefly summarize their contents.

http://

http://

External Anatomy and Organs of the Digestive and Respiratory Systems

OBJECTIVES

After completing this exercise, you will be able to

1. define *digitigrade locomotion, biped, plantigrade locomotion, thoracic cavity, abdominopelvic cavity, exocrine gland, endocrine gland,* and *digestive tract.*

2. locate and describe the external features of a fetal pig and human.

3. determine the sex of a fetal pig.

4. describe the function of the umbilical cord.

5. locate the organs of the digestive and respiratory systems in a fetal pig and human.

6. describe and give the functions of the organs of the digestive and respiratory systems in a fetal pig and human.

7. explain the importance of the digestive and respiratory systems to a living mammal.

8. trace the pathway of ingested food through the digestive tract.

9. trace the pathway of oxygen and carbon dioxide into and out of the lungs.

Introduction

In this and the following exercise, you will study the organs of some of the systems of mammals. As you do this, keep in mind the functional interrelationship between systems (Figure 10-1). You will dissect a fetal pig, observe a previously dissected fetal pig, or examine these organs in a human torso model. Many aspects of the structural and functional organization of a fetal pig are identical to those of humans. Thus, a study of the fetal pig is in a very real sense a study of us.

Fetal pigs are purchased from a biological supply house, which obtained them from a plant where pregnant sows are slaughtered for food. At the slaughterhouse, fetuses are quickly removed from the sow and then cooled and embalmed with preservative, which is injected through one of the umbilical arteries. After this, the arterial and venous blood vessels are injected under pressure with latex or a rubberlike compound. Red latex is injected into the arteries, and blue latex is injected into the veins.

While dissecting the fetal pig, keep several points in mind. First, be aware that *to dissect* does not mean "to cut up" but rather "to expose to view." To understand the dissection directions, you need to be familiar with the terms used in virtually all anatomical work (Appendix 2). Spend a few minutes relating each of these terms to the fetal pig body and to your body as well. Figure 10-2 will aid you in this activity. If you are dissecting a fetal pig, follow the instructions below. You must wear a pair of protective lab gloves and safety goggles. If you are going to observe pigs prepared by your instructor, do not forget to wear lab gloves and safety goggles. If you are working with human torso models, identify the bold-faced structures, regions, and so on in the text below.

MATERIALS

Per student pair:

- One preserved fetal pig injected with red and blue latex
- Plastic bag to store fetal pig
- Dissection pan
- Dissecting kit
- Dissecting pins
- Bone shears
- Four large rubber bands or two pieces of string, each 60 cm long
- Piece of string 20 cm long

Per student group (4–6):

- Human torso model (optional)
- Fetal pigs prepared by instructor (optional)

Per lab room:

- Liquid waste disposal bottle
- Boxes of different sizes of lab gloves
- Box of safety goggles

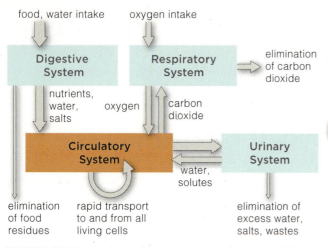

FIGURE 10-1 Inter-relationships between the functions of four of the systems of the body. *(After Starr and McMillan, 2010.)*

10.1 External Anatomy

The external surface of an organism has the greatest amount of contact with the environment. Thus, the greatest differences between humans and pigs may be their external features rather than the internal ones.

A. Preparing the Fetal Pig for Examination

1. The fluid used to preserve fetal pigs may be irritating to the hands and eyes. Therefore, wear lab gloves during dissection and put on protective goggles.

CAUTION Preserved specimens are kept in preservative solutions. Use lab gloves whenever you handle a specimen. Wash any part of your body exposed to this solution with lots of water. If preservative solution is splashed into your eyes, wash them with the safety eyewash for 15 minutes. Even if you wear contact lenses, you should wear safety goggles during dissection and when observing or studying a dissection. Eyeglasses should suffice in a situation when the goggles do not fit over them.

---

2. Obtain a fetal pig and place it on a dissection pan lined with paper towels. If the plastic bag supplied with your pig contains any excess preservative, pour the liquid into the waste bottle provided by your instructor and save the bag. As you will be using the same fetal pig for several days, you should place it in the plastic bag at the end of each day's exercise so it does not dry out.

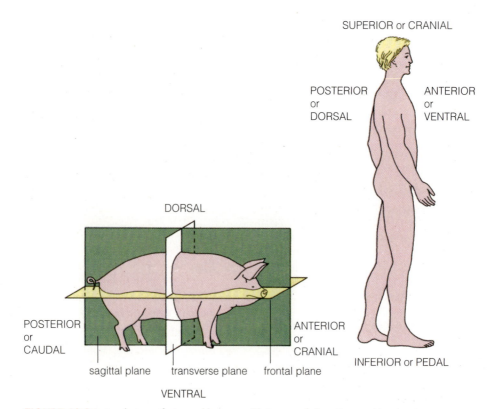
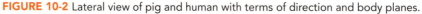

FIGURE 10-2 Lateral view of pig and human with terms of direction and body planes.

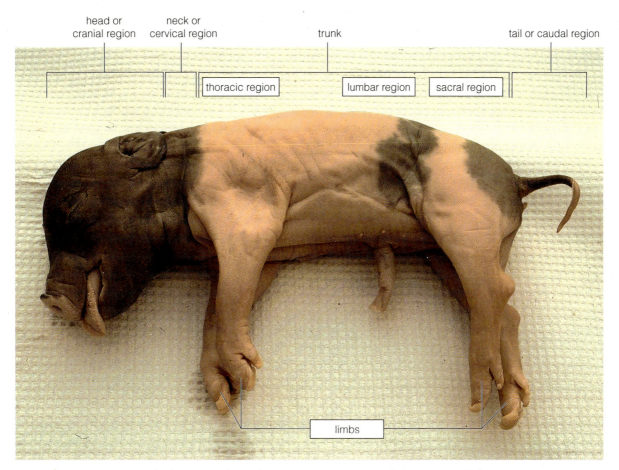

head or
cranial region

neck or
cervical region

trunk

tail or caudal region

thoracic region

lumbar region

sacral region

limbs

FIGURE 10-3 Lateral view of a fetal pig with the four major body regions indicated. *(Photo by D. Morton.)*

3. For easy identification, tie a nametag to a hind leg, the bag, or both, *according to the instructor's instructions.* Use a pencil to fill in the tag.

B. Body Regions and Their Features

1. Identify the four regions of the fetal pig's body: the large, compact **head;** the **neck;** the **trunk** with four **limbs** (or appendages); and the *tail* (Figure 10-3).

2. Examine the head in more detail (Figure 10-4) and identify the **eyes** with **upper** and **lower eyelids,** the **external ears,** the **mouth,** and the characteristic **nose** or *snout.* Note the position of the **nostrils,** or *external nares,* on the snout. Feel the texture of the snout. It is composed of bone, cartilage, and other tough connective tissue and as such allows the pig to root and push soil and debris in its search for food.

3. Open the pig's mouth and note the **tongue** with its covering of **papillae,** which contain *taste buds.* Papillae are especially concentrated and seen easily along the posterior edges and tip of the tongue. Also notice if any *baby teeth* are present. Similar to humans, pigs are omnivores; that is, they eat both animal and plant matter.

4. Note that the trunk is divided further into the **thoracic region, lumbar region,** and **sacral region** (Figure 10-3). These regions, along with the cervical region in the neck, also describe the corresponding regions of the vertebral column or spine.

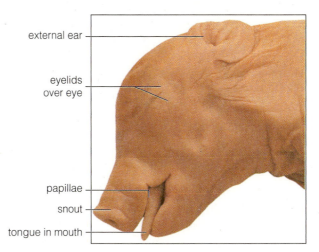

external ear

eyelids
over eye

papillae

snout

tongue in mouth

FIGURE 10-4 External features of the fetal pig head. *(Photo by D. Morton.)*

5. Place the pig on its back (dorsal surface) and examine its *abdomen* (belly). Identify on the ventral surface the **umbilical cord** seen near the posterior end of the abdomen (see Figure 10-4).

Is an umbilical cord present in the adult pig and human? _____ (yes or no)

During its development, the fetus was connected to the placenta on the uterine wall of its mother's reproductive system via the umbilical cord. The cord contains two *arteries* (red), a large *vein* (blue), and a fourth vessel, usually collapsed, the *allantoic duct.* The blood in the umbilical vein carries nutrients and oxygen from the mother to the fetus, and blood in the umbilical arteries carries waste materials and carbon dioxide from the fetus to the mother.

6. Note on the ventral surface of the pig the pairs of **nipples** or *teats* (Figure 10-5). Both male and female pigs may have from five to eight pairs of these structures situated in two parallel rows on the **thoracic region** (chest) and **abdominal region** of the body. Finally, locate the **anus,** the posterior opening of the digestive tract. The anus is situated immediately under (ventral to) the tail.

7. Locate and identify the following: the **wrist, lower forelimb, elbow, upper forelimb, shoulder, ankle, shank (lower leg), knee, thigh,** and **hip** (Figure 10-6).

8. Examine the toes. The first *toe*, or *digit*, which corresponds to your big toe or thumb, is absent in both forelimbs and hind limbs of the pig. Furthermore, the second and fifth digits are reduced in size, and the middle two digits, the third and fourth, are flattened or *hoofed.*

Pigs and other hoofed animals walk with the weight of their body borne on the tips of the digits. This type of walking is referred to as **digitigrade locomotion.** By contrast, humans are **bipeds**—walk on two feet—and use the entire foot for walking and have **plantigrade locomotion.** Compare the structure of your hands and feet with the foot of a pig.

C. Determining the Sex of Your Fetal Pig

1. Find the *urogenital opening* immediately ventral to the anus in females (Figure 10-5a) and just behind the umbilical cord in males (Figure 10-5b). A small fleshy

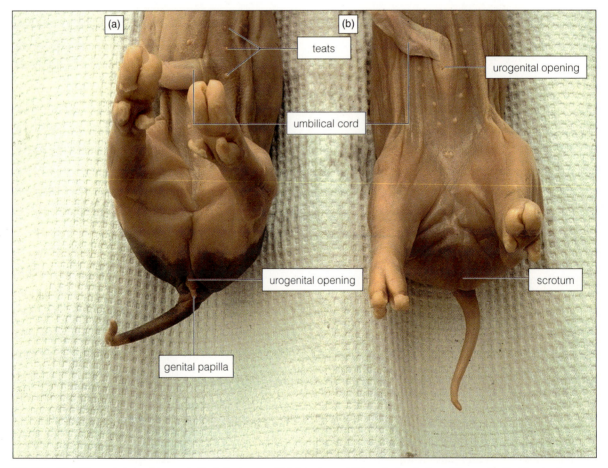

FIGURE 10-5 Ventral view of lower half of the body of (a) female and (b) male fetal pigs. *(Photos by D. Morton.)*

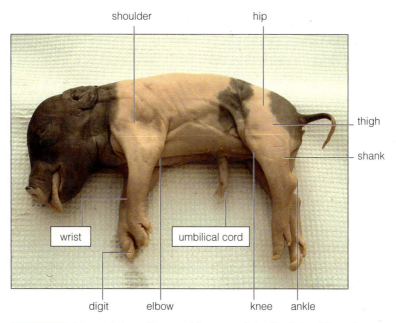

shoulder   hip

thigh

shank

wrist

umbilical cord

digit   elbow   knee   ankle

FIGURE 10-6 Lateral view of external features of the fetal pig. *(Photo by D. Morton.)*

genital papilla projects from the urogenital opening of female fetal pigs.

All fetal pigs have a common urogenital opening shared by the urinary and the reproductive systems. This situation persists in adult male pigs and humans. In adult female pigs and humans, however, there are separate openings to the urinary and reproductive systems.

2. In your or another group's male fetal pig, find a swelling on the posterior portion of the abdomen between the upper ends of the hind limbs. The swelling is the **scrotum,** which contains the sperm-producing *testes*, a pair of small, oval structures that are part of the male reproductive system. The testes are generally easy to locate in older fetuses because during late development, they descend into the scrotum. Identify the *penis*, a large, tubular structure immediately under the skin just posterior to the urogenital opening.

10.2   Opening the Ventral Body Cavities

For the following dissection, use Figure 10-7 as a guide for making the various incisions. In the following directions, numbers in the figure correspond to the numbers in parentheses in the text.

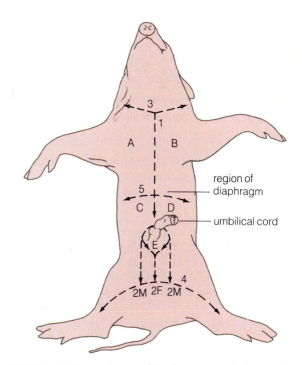

region of diaphragm

umbilical cord

FIGURE 10-7 Ventral view of the fetal pig, with the positions of incisions indicated. Note that the second incision is different for male (M) and female (F) pigs.

PROCEDURE

1. Place the pig, ventral side up, in the dissection pan and hold it down using rubber bands or string as you did in the previous exercise. Begin an incision at the small tuft of hair on the upper portion of the neck (1), and continue it posteriorly to approximately 1.5 cm anterior to the umbilical cord. Cut through the muscle layer but not too deeply, or you may damage the internal organs.

2. *If your pig is a male, move on to step 3.* If your pig is a *female*, make a second incision (2F) completely around the umbilical cord and continue it posteriorly for about 3 cm, stopping at a point between the hindlimbs.

3. If your fetal pig is a *male*, make the second incision (2M) as a half circle anterior to the umbilical cord and then proceed with two parallel incisions posteriorly to a region between the hind limbs. The two incisions are necessary to avoid cutting the *penis*, which lies under the skin just posterior to the umbilical cord. The incisions made in the region of the *scrotum* should be made carefully so as not to damage the testes, which lie just under the skin.

4. Carefully deepen incisions 1 and 2 until the body cavity is exposed. To make lateral flaps of the muscle tissue, which can be folded out of the way, make a third (3) and fourth (4) incision, as illustrated in Figure 10-7. Carefully open the body cavity. If it is filled with fluid, pour the fluid into the waste container provided (not into the sink!) and carefully rinse out the cavity with a little water.

5. Use your fingers to locate the lower margin of the *rib cage.* Just below it, make a fifth (5) incision laterally in both directions from the first incision (1). In this region is the **diaphragm,** a skeletal muscular sheet connected to the body wall and separating the two major ventral body cavities: an anterior **thoracic cavity** and a posterior **abdominopelvic cavity.** Use your scalpel to free the diaphragm from the body wall (do not remove it, however).

6. Carefully peel back flaps A, B, C, and D (see Figure 10-7) and pin them beneath your pig. It may be necessary to cut through the ventral part of the rib cage with a pair of bone shears to separate body wall flaps A and B. Do so carefully, so as not to damage the heart and lungs, which are located in the thoracic cavity.

7. To free the umbilical cord and the flesh immediately surrounding it (flap E), first locate the umbilical vein, which is a dark, tubular structure extending from the umbilical cord forward (anteriorly) to the liver. Then tie small pieces of string around the vein in two places (approximately 1.5 cm apart) and with your scissors cut through the vein between the pieces of string. Now pull back the umbilical cord to a position between the hind legs and pin the flesh surrounding it to the body. The pieces of string around the umbilical vein will aid in identifying this structure during the dissection of the circulatory system.

8. When the body cavities are fully exposed, carefully remove any excess red or blue latex that may be present. (This occurs when some veins and arteries burst when injected with latex.) Remove large pieces with forceps. Remove your pig from the dissection pan and rinse out any smaller pieces of latex in a sink equipped to screen out debris.

10.3 Digestive System

The digestive system of a vertebrate consists of the **digestive tract** (mouth, oral cavity, pharynx, esophagus, stomach, small intestine, large intestine or colon, cecum, rectum, and anus) and associated structures and glands (salivary glands, gallbladder, liver, and pancreas) (Figure 10-8). In addition to these digestive system organs, you will locate and identify the thymus and thyroid, two endocrine system glands. The digestive system digests the complex molecules in food, absorbs the useful end products of digestion along with water and most everything else that the body adds to digesting food, and processes indigestible remnants for defecation.

PROCEDURE

A. Mouth

1. Place your pig in the dissection pan. Observe the area between the **lips** and **gums;** this is called the **vestibule.** The larger area behind the gums is referred to as the **oral cavity.**

2. Carefully cut through the corners of the mouth and back toward the ears with bone shears until the lower jaw can be dropped and the oral cavity exposed (Figure 10-9).

3. If teeth are present, carefully extract a **tooth** and examine it. A tooth consists of the *crown,* the *neck* (surrounded by the gum), and the *root* (embedded in the jawbone). If your specimen does not have exposed teeth, cut into the gums and look for developing teeth.

4. Feel the roof of the oral cavity and determine the position of the **hard palate** and **soft palate.** What is the difference between the two regions?

_____

_____

_____

_____

_____

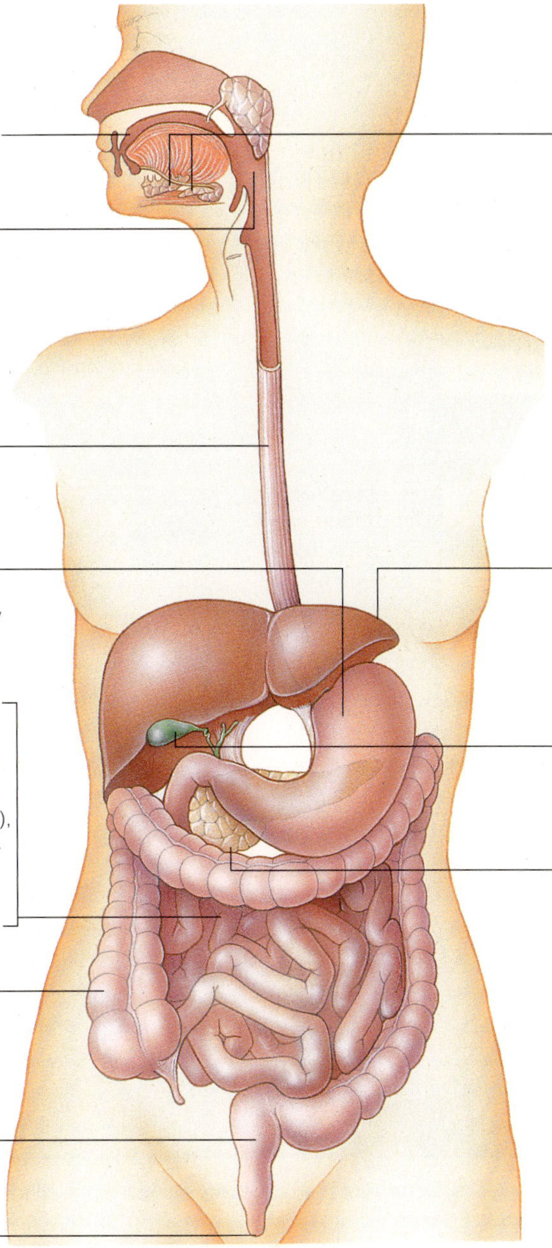

Major Components:

Mouth (Oral Cavity)
Entrance to system; food is moistened and chewed; polysaccharide digestion starts.

Pharynx
Entrance to tubular part of system (and to respiratory system); moves food forward by contracting sequentially.

Esophagus
Muscular, saliva-moistened tube that moves food from pharynx to stomach.

Stomach
Muscular sac; stretches to store food taken in faster than can be processed; gastric fluid mixes with food and kills many pathogens; protein digestion starts. Secretes ghrelin, an appetite stimulator.

Small Intestine
First part (duodenum, C-shaped, about 10 inches long) receives secretions from liver, gallbladder, and pancreas.

In second part (jejunum, about 3 feet long), most nutrients are digested and absorbed.

Third part (ileum, 6–7 feet long) absorbs some nutrients; delivers unabsorbed material to large intestine.

Large Intestine (colon)
Concentrates and stores undigested matter by absorbing mineral ions, water; about 5 feet long: divided into ascending, transverse, and descending portions.

Rectum
Distension stimulates expulsion of feces.

Anus
End of system; terminal opening through which feces are expelled.

Accessory Organs:

Salivary Glands
Glands (three main pairs, many minor ones) that secrete saliva, a fluid with polysaccharide-digesting enzymes, buffers, and mucus (which moistens food and lubricates it).

Liver
Secretes bile (for emulsifying fat); roles in carbohydrate, fat, and protein metabolism.

Gallbladder
Stores and concentrates bile that the liver secretes.

Pancreas
Secretes enzymes that break down all major food molecules; secretes buffers against HCl from the stomach. Secretes insulin, a hormonal control of glucose metabolism.

FIGURE 10-8 Human digestive system. *(After Starr and McMillan, 2010.)*

5. Posterior to the soft palate is the **pharynx** (Figure 10-9). In humans, the portion of the pharynx just posterior to the oral cavity is also referred to as the *throat*. Note that unlike humans, pigs do not have a fingerlike piece of tissue, the **uvula,** projecting from the posterior region of the soft palate. Confirm its presence in humans by looking into the throat of your lab partner.

6. Carefully close the pig's jaws. In the neck locate the **trachea** (Figure 10-11), a tube that is kept open throughout its length by a series of cartilaginous supports. Although the trachea is actually a part of the respiratory system, its identification aids in finding the **esophagus** (Figure 10-11), which lies behind it on its

dorsal surface. Carefully slit the esophagus and insert a blunt probe into it and run it back toward the mouth. Open the mouth and note where the probe emerges. This is the opening of the esophagus (Figure 10-9).

If you would run your probe posteriorly through the esophagus, where would it emerge?

_____

7. Continue your study of the pharynx by locating the opening to the *larynx*, the **glottis.** It can be identified by the presence of a small white cartilaginous flap, the **epiglottis,** on its ventral surface (Figure 10-9). The epiglottis covers the glottis when a mammal swallows.

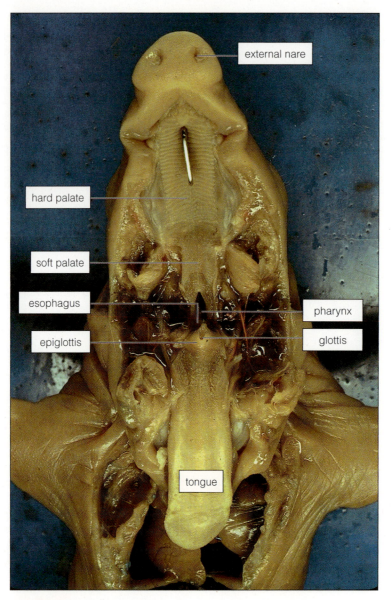

FIGURE 10-9 Structures of the oral cavity and pharynx. *(Photo by D. Morton.)*

B. Salivary Glands

1. Place your pig on its right side and, proceeding from the base of the ear, carefully cut through the skin to the corner of the eye, then ventrally toward the chin, and finally continue the incision posteriorly toward the forelimb. Carefully remove the skin.

2. Tease away the muscle tissue below the ear to reveal a large, relatively dark, triangular **parotid gland.** This salivary gland extends from the edge of the ear posteriorly to halfway down the neck (Figure 10-10). Note that the parotid appears to be composed of many small nodules compared with the fibrous large masseter muscle lying underneath and anterior to it.

3. Cut through the middle of the parotid gland to expose the somewhat lobed **mandibular gland** (Figure 10-10). Do not confuse this second salivary

gland with the small, oval lymph nodes present in the head and neck region.

4. The third salivary gland is the **sublingual gland.** It is located under the tongue, and there is not enough time to dissect it today. The fluids secreted by the mandibular and sublingual glands are more viscous than that secreted by the parotid. Collectively, the secretions by the three salivary glands maintain the oral cavity in a moist condition, ease the mixing and swallowing of food, and sometimes contain enzymes that begin the breakdown of starch to sugars.

C. Thymus and Thyroid Glands

1. Work from the ventral side of your pig with the legs secured by string or rubber bands and the body wall flaps pinned to the sides of the body.

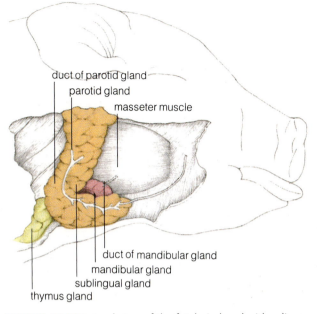

duct of parotid gland
parotid gland
masseter muscle

duct of mandibular gland
mandibular gland
sublingual gland
thymus gland

FIGURE 10-10 Lateral view of the fetal pig head with salivary and thymus glands. *(After Gilbert, 1966.)*

2. Identify the **thymus gland,** a whitish structure that is divided into two lobes. It extends from the neck, where it covers the trachea and larynx to the upper thoracic cavity, where it partially covers the anterior portion of the heart (Figures 10-10 and 10-11). The thymus plays important roles in the development and maintenance of the body's immune system.

3. Immediately beneath the thymus in the neck region, find the **thyroid gland,** a small, reddish, oval mass with a relatively solid consistency (Figure 10-11). Thyroid hormones function in the regulation of metabolism, growth, and development.

D. Liver, Gallbladder, and Pancreas

1. Identify the brownish colored **liver** (Figure 10-11), which is largest organ of the abdominopelvic cavity. Count and carefully determine the extent of all of its four lobes. The liver has many functions, including secreting *bile.*

The liver also plays a very important role in maintaining the stable composition of the blood. The nutrients from digested meals are absorbed into the blood of the small intestine. This blood, which contains high concentrations of sugars such as glucose and amino acids, is transported from the small intestine to the liver (via the hepatic portal vein), and there the excess glucose is converted to glycogen for storage. If the liver has stored a full capacity of glycogen, it converts the glucose into fat, which is stored in other parts of the body. The liver also removes excess amino acids from the blood by converting them to carbohydrates and fats. During this process, an amino group ($-NH_2$) is removed from the amino acid and converted into ammonia (NH_3). Ammonia is a very toxic substance,

and the liver combines it with carbon dioxide to form urea. The urea, which is less toxic than ammonia, is then eliminated from the body in urine.

2. Lift the right central lobe of the liver to expose the **gallbladder.** This saclike organ stores the bile secreted by the liver. When food enters the small intestine, the gallbladder contracts to deliver bile to the small intestine. Bile contains bile salts that function to stabilize tiny fat droplets in digesting food.

3. Carefully move the small intestine and locate the **pancreas,** an elongated globular mass lying between the *stomach* and small intestine.

The pancreas is both an **exocrine gland** (whose secretions are released into a duct) and an **endocrine gland** (whose hormones are released into the blood). The exocrine portion delivers enzymes for digestion along with other substances. The endocrine portion of the pancreas secretes insulin and other hormones involved with controlling the levels of glucose in the blood of mammals.

E. Stomach, Small Intestine, Large Intestine or Colon, Rectum, and Anus

1. Push the lobes of the liver to one side to fully expose the bean-shaped **stomach** (Figure 10-11). Earlier in this dissection, you made a small slit in the esophagus. Return to this incision and insert a blunt probe through the slit only this time posteriorly until you feel its tip in the stomach. Note that the esophagus passes through the diaphragm before joining the stomach. Two smooth muscular rings, the *cardiac sphincter* and the *pyloric sphincter,* control the movement of food through the stomach. Feel for these sphincters by gently squeezing the stomach between your index finger and thumb at the stomach's entrance and exit.

2. Cut the stomach lengthwise with your scissors. Describe any contents of the stomach.

_____

_____

The contents of the fetal pig's digestive tract are called *meconium* and are composed of a variety of substances, including amniotic fluid swallowed by the fetus, epithelial cells sloughed off from the digestive tract, and hair.

3. Clean out the stomach and note the folds, or *rugae,* on its internal surface. What role might the rugae play in digestion?

_____

_____

4. The **small intestine** (Figure 10-11) is divided into three regions: the *duodenum,* the *jejunum,* and the *ileum.* The first portion, the duodenum, leaves the pyloric end of the stomach and runs along the edge of

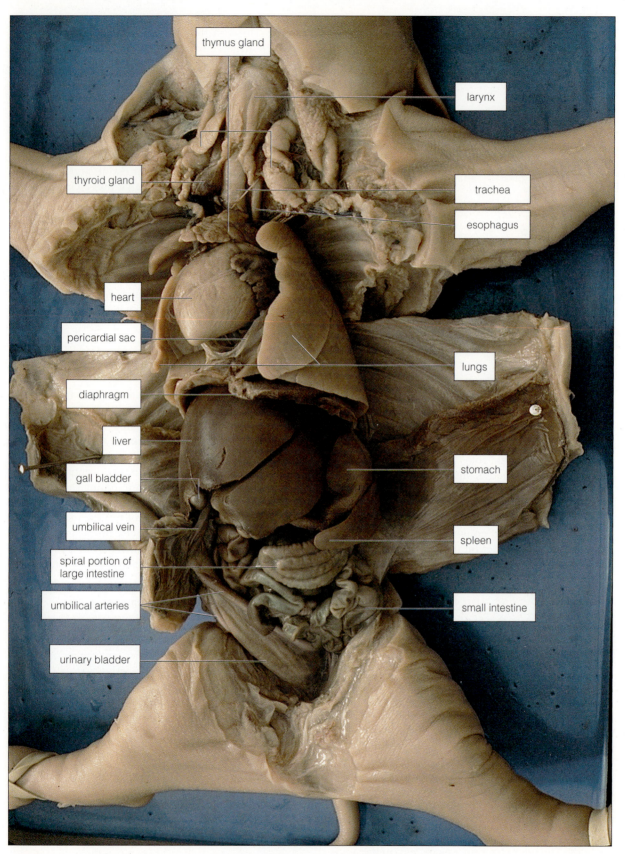

FIGURE 10-11 Ventral view of the internal anatomy of the fetal pig. *(Photo by D. Morton.)*

the pancreas. The junctions of the duodenum and ileum with the jejunum cannot be easily distinguished.

5. A thin membrane, the **mesentery**, holds the coils of the small intestine together. Cut the mesentery from the dorsal body wall and from between the coils of the small intestine. Uncoil the small intestine.

    Measure the length with a meter stick and record it: _____ cm

    A rule of thumb is that the small intestine in both pigs and humans is about five times the length of the individual.

6. Using your scissors, cut a 0.5-cm section of the intestine, slit it lengthwise, and place it in a clear, shallow dish filled with water. Then examine it using a dissection microscope.

    How does the inner surface appear?

    _____

    _____

    _____

    Locate the *villi*. Most of the nutrients provided by the digestive process are absorbed by these small projections from the wall of the small intestine.

7. Locate the juncture of the **large intestine** (Figure 10-11), or **colon**, and the ileum. This may be more difficult in a pig than in a human because in a pig, there is not such a noticeable difference in the size of the small and large intestines. However, this juncture is marked by the presence of a blind pouch, the *cecum*, which is relatively large in pigs. In humans, the cecum is very short and bears a small, fingerlike projection known as the **appendix**.

8. As with other junctures in the digestive tract, the region where the ileum joins the large intestine is the site of a sphincter of smooth muscle, sometimes called the *ileocecal valve*. Feel for it by rolling the junction between your index finger and thumb.

9. The coiled large intestine stretches from the cecum to the straight **rectum**, which opens to the outside at the **anus**. The anus is the site of the final muscle in the digestive tract, the *anal sphincter*. Locate the rectum, anus, and anal sphincter but do not dissect these structures at this time.

10. Review the digestive system by tracing the pathway of an ingested indigestible fiber in and out of the body.

10.4 Respiratory System

The respiratory system of a mammal consists of various organs and structures (Figure 10-12) associated with the exchange of gasses between the internal and external environment. Air rich in oxygen is inhaled into the air sacs of the lungs. Oxygen diffuses from this air into the capillaries of the lungs; carbon dioxide moves in the other direction. Carbon dioxide–rich air is then exhaled from the body.

PROCEDURE

A. Nose

1. Remove your pig from the dissection pan. In the pig and other mammals, molecules of air enter the body through the *nostrils* and pass through a pair of **nasal cavities** dorsal to the hard palate and into the nasal portion of the pharynx or *nasopharynx* (Figure 10-9). Examine the nostrils and hard and soft palates and then carefully cut the soft palate longitudinally to examine the nasopharynx of your specimen.

2. From the nasopharynx, air passes through the *glottis* into the **larynx** (Figures 10-11 and 10-13a and b). In humans, the front of the larynx is often referred to as the *Adam's apple* or *voice box*.

B. Trachea, Bronchial Tubes, and Lungs

1. Place the pig, ventral side up, in the dissection pan and hold it down using rubber bands or string as you did in the first section of this exercise. Pin back the body flaps to the sides of the body.

2. Slit the larynx longitudinally to expose the *vocal cords* (Figure 10-13c).

3. Locate the trachea (Figure 10-13). The trachea extends from the larynx and divides into two major branches, the **bronchi** (singular, *bronchus*), to the lungs. Note again the series of cartilaginous structures that prevent the trachea from collapsing. These apparent rings of cartilage are actually incomplete on their dorsal side.

4. Note that the thoracic cavity is divided into two **pleural cavities,** which contain the lungs, and a **pericardial sac** that is located between them. Inside this sac are the *pericardial cavity* and the *heart*. Carefully examine the lungs and note the thin, transparent **pleural membrane** that lines the inner surface of the thoracic cavity and the outer surface of the lungs. The right lung consists of four lobes and the left of two or three. Are the lungs of the fetal pig filled with air? _____ (yes or no)

5. Carefully push the *heart* to one side and gently tease away some of the lung tissue to expose the bronchi. Notice that the bronchi divide into smaller and smaller branches.

    These branches continue to divide and branch into smaller and smaller tubes, eventually ending as microscopic air sacs called **alveoli** (singular, *alveolus*). The thin walls of the alveoli are well supplied with blood capillaries, and it is here that the exchange of carbon dioxide and oxygen occurs after birth.

    Where does this exchange occur in a fetus?

    _____

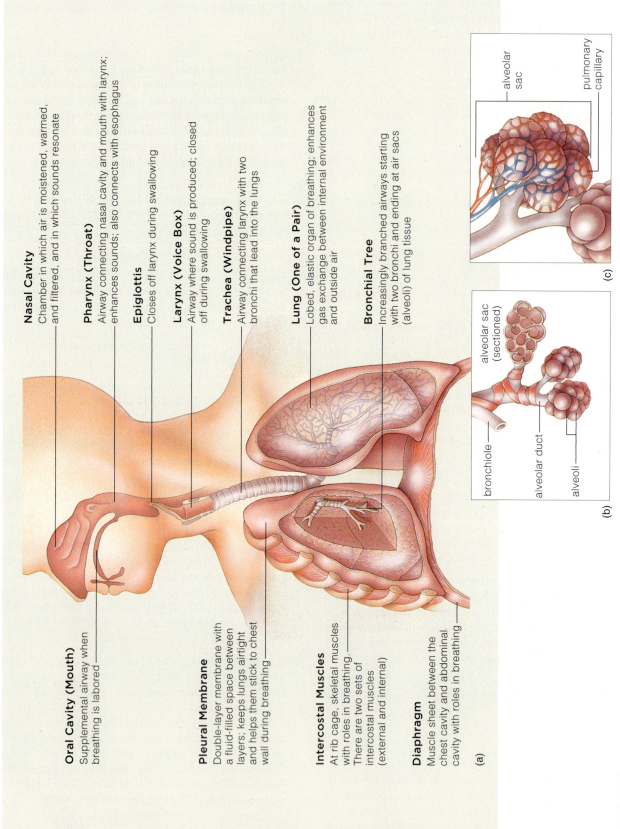

Oral Cavity (Mouth)
Supplemental airway when breathing is labored

Pleural Membrane
Double-layer membrane with a fluid-filled space between layers; keeps lungs airtight and helps them stick to chest wall during breathing

Intercostal Muscles
At rib cage, skeletal muscles with roles in breathing. There are two sets of intercostal muscles (external and internal)

Diaphragm
Muscle sheet between the chest cavity and abdominal cavity with roles in breathing

Nasal Cavity
Chamber in which air is moistened, warmed, and filtered, and in which sounds resonate

Pharynx (Throat)
Airway connecting nasal cavity and mouth with larynx; enhances sounds; also connects with esophagus

Epiglottis
Closes off larynx during swallowing

Larynx (Voice Box)
Airway where sound is produced; closed off during swallowing

Trachea (Windpipe)
Airway connecting larynx with two bronchi that lead into the lungs

Lung (One of a Pair)
Lobed, elastic organ of breathing; enhances gas exchange between internal environment and outside air

Bronchial Tree
Increasingly branched airways starting with two bronchi and ending at air sacs (alveoli) of lung tissue

bronchiole
alveolar sac (sectioned)
alveolar duct
alveoli

(b)

alveolar sac
pulmonary capillary

(c)

(a)

FIGURE 10-12 Human respiratory system. *(After Starr and McMillen, 2010.)*

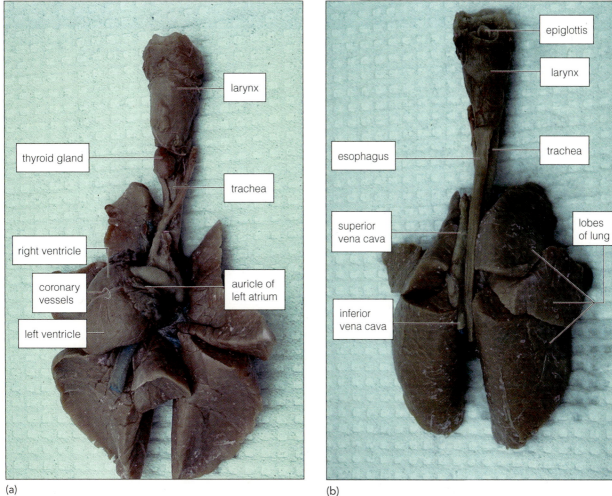

(a)

(b)

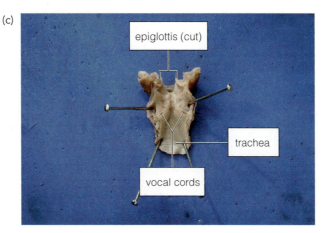

(c)

FIGURE 10-13 (a) Ventral and (b) dorsal views of the respiratory system of the fetal pig. In (c) the larynx has been slit open to show the vocal cords. *(Photos by D. Morton.)*

6. Now relocate the diaphragm and note its position in relation to the lungs. Contraction of this skeletal muscle in part results in inhalation.

7. Complete your study of the respiratory system of the fetal pig by tracing the pathway of a carbon dioxide molecule from an alveolus to the nostrils.

8. Place your pig and any excised organs back in the plastic bag. Tie it shut to prevent your pig from drying out. Dispose of any paper towels that contain preservative *as directed by your instructor.* Clean the tray, dissecting tools, and the laboratory table.

_____ **1.** A fetal pig
   (a) is a newborn pig
   (b) is an unborn pig
   (c) has structures quite similar to humans
   (d) b and c

_____ **2.** *To dissect* means primarily
   (a) to cut open
   (b) to remove all internal organs
   (c) to expose to view
   (d) all of the above

_____ **3.** In a fetal pig, *dorsal* and *ventral* refer respectively to
   (a) the head and tail regions of the body
   (b) the tail and the head regions of the body
   (c) the upper (back) portion and the lower (underside) portion of the body
   (d) the lower (underside) portion and the upper (back) portion of the body

_____ **4.** The umbilical cord functions to
   (a) carry waste products in the blood from the fetus to the mother
   (b) carry waste products in the blood from the mother to the fetus
   (c) carry oxygen in the blood from the mother to the fetus
   (d) do both a and c

_____ **5.** The female fetal pig is similar to the male fetal pig in that its body
   (a) has separate openings for the urinary system and the reproductive system
   (b) has a common opening for the urinary system and the reproductive system
   (c) has an opening for the urinary system but none for the reproductive system
   (d) has an opening for the reproductive system but none for the excretory system

_____ **6.** Pigs have digitigrade locomotion because they walk on
   (a) their ankles
   (b) the soles of their feet
   (c) the tips of their toes, which are modified as hooves
   (d) their hands and knees

_____ **7.** The diaphragm is a sheetlike skeletal muscle that separates the
   (a) thoracic and pleural cavities
   (b) thoracic and pericardial cavities
   (c) thoracic and abdominopelvic cavities
   (d) pleural and pericardial cavities

_____ **8.** The digestive system is concerned with
   (a) blood circulation
   (b) break down and absorption of nutrients
   (c) reproduction
   (d) excretion of urine

_____ **9.** The liver functions to
   (a) produce bile
   (b) pump blood
   (c) form urea
   (d) do both a and c

_____ **10.** The cardiac, pyloric, anal, and ileocecal sphincters are all part of the
   (a) digestive tract
   (b) respiratory tract
   (c) circulatory system
   (d) muscular system

EXERCISE 10 External Anatomy and Organs of the Digestive and Respiratory Systems

POST-LAB QUESTIONS

10.1 External Anatomy of the Fetal Pig

1. Identify the external features of a fetal pig noted in the drawing.

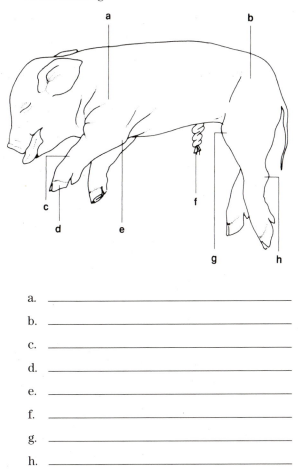

a. _____

b. _____

c. _____

d. _____

e. _____

f. _____

g. _____

h. _____

2. What is the function of the umbilical cord?

3. Briefly describe how human feet differ from those of a pig. What is the basic difference between digitigrade and plantigrade locomotion?

4. Using external features, briefly describe how you can determine the difference between a male and a female fetal pig.

10.2 Opening the Ventral Body Cavities

5. Describe the location of the two *major* ventral body cavities.

10.3 Digestive System

6. Describe the difference between the digestive system and the digestive tract.

7. List in order the organs through which food and other substances pass in their journey into, through, and out the digestive tract.

8. What are the main organs for digestion, and where are most of the nutrients absorbed?

10.4 Respiratory System

9. Describe the major similarities and differences in the location, structure, and function of the trachea and esophagus.

10. Describe the main function of the respiratory system.

Organs of the Circulatory, Urinary, and Reproductive Systems

OBJECTIVES

After completing this exercise, you will be able to

1. define *blood, heart, blood vessels (arteries, arterioles, capillaries, venules,* and *veins), pulmonary* and *systemic circuits* of *circulatory system, portal vein, urea, peritoneum, urine, urinary bladder, homologous, ovulation, semen, inguinal hernia, vasectomy,* and *nephron.*

2. locate the organs of the circulatory, urinary, and reproductive systems.

3. describe and give the functions of the organs of the circulatory, urinary, and reproductive systems in a living mammal.

4. explain the importance of the circulatory, urinary, and reproductive systems to a living mammal.

5. identify the major blood vessels.

6. locate, name, and describe the functions of the chambers of the heart.

7. name the internal structures of the heart.

8. locate, name, and describe the function of the internal structures of the kidney.

9. name the parts of the nephron and their basic functions.

Introduction

Along with the organs of the digestive system, the ventral body cavities (thoracic and abdominopelvic cavities) contain the heart and major blood vessels. Also, the abdominopelvic portion contains most of the internal organs of the urinary and reproductive systems. As in the previous exercise, you will dissect a fetal pig, observe a previously dissected fetal pig, or examine these organs in a human torso model.

MATERIALS

Per student pair:

- One preserved fetal pig injected with red and blue latex
- Plastic bag to store fetal pig
- Dissection pan
- Dissecting kit
- Dissecting pins
- Bone shears
- Four large rubber bands *or* two pieces of string, each 60 cm long
- Piece of string, 20 cm long
- Compound light microscope
- Prepared section of kidney

Per student group (4–6):

- Human torso model (optional)
- Fetal pigs prepared by instructor (optional)

Per lab room:

- liquid waste disposal bottle
- Boxes of different sizes of lab gloves
- Box of safety goggles

11.1 Blood Vessels and Surface Anatomy of the Heart

Mammals have a vast network of **blood vessels** that transport **blood**—a fluid containing *cells* and *plasma*—to and from every part of the body (Figure 11-1). Plasma consists of water, oxygen, carbon dioxide, nutrients, metabolic wastes, hormones, and other substances. In the pericardial cavity (part of the thoracic cavity) resides the muscular four-chambered **heart** that pumps blood into the **arteries**. Arteries branch and branch some more. The final branches are the **arterioles**, which deliver blood to the **capillaries**, where the exchange of substances between the blood and tissues takes place across their thin walls. After passing through capillaries, blood drains into **venules** and then into **veins**, which transport it back to the heart.

The circulatory system is divided into two circuits: the **pulmonary circuit**, which involves blood flow to and from the lungs, and the **systemic circuit**, which is concerned with the flow of blood to and from the rest of the body. In this section, you will study these two circuits and examine how the heart directs the flow of blood through them both in a fetus and in an adult.

PROCEDURE

A. Pulmonary Circuit and Surface Anatomy of the Heart

1. The fluid used to preserve fetal pigs may be irritating to the hands and eyes. Therefore, wear lab gloves during dissection and put on protective goggles. Place the pig, ventral side up, in the dissection pan and restrain it using rubber bands or string as you did in the first section of this exercise. Pin back the body flaps to the sides of the body.

> **CAUTION** *Preserved specimens are kept in preservative solutions. Use lab gloves whenever you handle a specimen. Wash any part of your body exposed to this solution with lots of water. If preservative solution is splashed into your eyes, wash them with the safety eyewash for 15 minutes. Even if you wear contact lenses, you should wear safety goggles during dissection and when observing or studying a dissection. Eyeglasses should suffice when goggles do not fit over them.*

2. If it is not already torn, open the pericardial sac. Similar to the situation in the pleural cavities, the inside of the pericardial sac and the outside of the heart are lined by the **pericardial membrane.**

3. Identify the four chambers of the heart (Figures 11-2 and 11-3)—the **right** and **left atria** (singular, *atrium*) and the larger **right** and **left ventricles.** On the surface of the heart, locate the **coronary vessels** lying in the diagonal groove between the two ventricles (Figure 11-2a and 11-3). The coronary arteries and their branches supply blood directly to the heart. (The heart is a muscle and as such has the same requirements as any other organ.) When these vessels become severely occluded, a heart attack may occur. The coronary arteries and their branches are replaced or "bypassed" in coronary bypass surgery.

4. Gently push the heart to the left and identify two relatively large blue veins (Figure 11-2b) called the **superior vena cava** and the **inferior vena cava** in humans and the *anterior vena cava* and the *posterior vena cava* in pigs.

   After birth, oxygen-poor (or carbon dioxide–rich) blood from all of the body except the lungs and heart returns from the systemic circuit to the right atrium of the heart through these large veins. Trace the inferior vena cava a short distance from the heart.
   Through what structure does the inferior vena cava pass? _____

5. The blood that enters the right atrium passes to the right ventricle and is then forced into the pulmonary circuit as the heart contracts. Identify the **pulmonary trunk,** which lies between the left and right atria and extends dorsally and to the left (Figure 11-3). It branches to form the **left** and **right pulmonary arteries** (Figure 11-2b).
   Do these arteries contain red or blue latex? _____
   After birth, do these arteries carry oxygen-rich or oxygen-poor blood? _____
   Carefully move the heart aside and follow the pulmonary arteries to the lungs.

6. After birth, after the blood is oxygenated (and the carbon dioxide removed) in the lungs, it returns to the left atrium of the heart via the **left** or **right pulmonary veins,** which complete the pulmonary circuit. Carefully move the lungs and heart aside and locate these large vessels (Figure 11-2b).

7. From the left atrium of the adult, the oxygenated blood passes to the left ventricle. Blood is forced into the **aorta** as the heart contracts, starting its trip through the systemic circuit. Locate the aorta (Figures 11-2), which leads dorsally out of the left ventricle of the heart. Note that its base is partially covered by the pulmonary trunk coming from the right ventricle.

8. Examine how blood circulation is different in fetal mammals (Figure 11-4). In the fetus, most of the pulmonary circuit is bypassed twice. First, most blood from the right ventricle enters the aorta directly from the pulmonary trunk through the **ductus arteriosus,** a large but short vessel connecting the pulmonary trunk directly to the aorta. The second bypass occurs when most of the blood delivered to the right atrium, mostly from the posterior portions of the body by the inferior vena cava, passes directly into the left atrium via a temporary opening in the

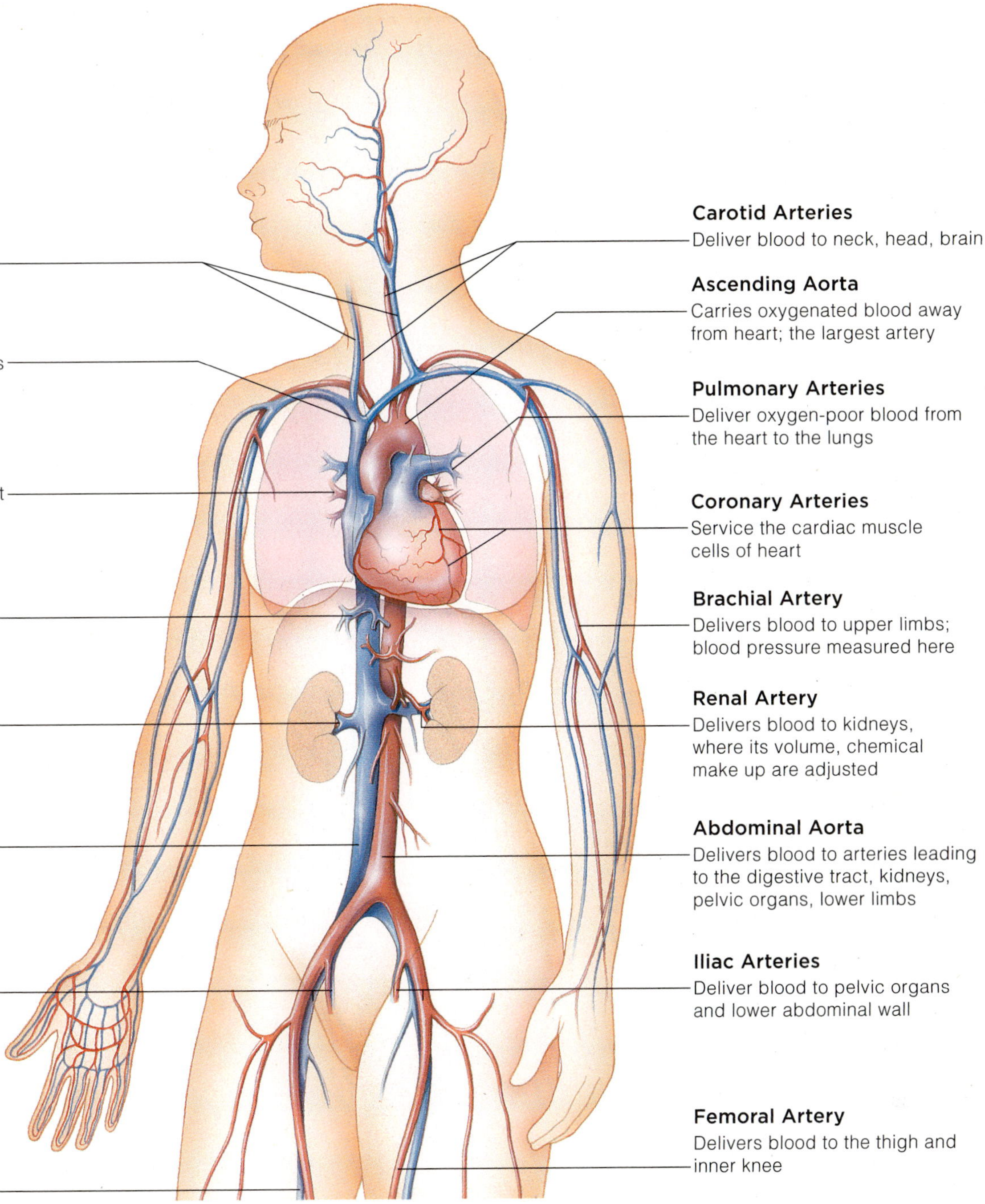

Jugular Veins
Receive blood from brain and from tissues of head

Superior Vena Cava
Receives blood from veins of upper body

Pulmonary Veins
Deliver oxygenated blood from the lungs to the heart

Hepatic Vein
Carries blood that has passed through small intestine and then liver

Renal Vein
Carries processed blood away from kidneys

Inferior Vena Cava
Receives blood from all veins below diaphragm

Iliac Veins
Carry blood away from the pelvic organs and lower abdominal wall

Femoral Vein
Carries blood away from the thigh and inner knee

Carotid Arteries
Deliver blood to neck, head, brain

Ascending Aorta
Carries oxygenated blood away from heart; the largest artery

Pulmonary Arteries
Deliver oxygen-poor blood from the heart to the lungs

Coronary Arteries
Service the cardiac muscle cells of heart

Brachial Artery
Delivers blood to upper limbs; blood pressure measured here

Renal Artery
Delivers blood to kidneys, where its volume, chemical make up are adjusted

Abdominal Aorta
Delivers blood to arteries leading to the digestive tract, kidneys, pelvic organs, lower limbs

Iliac Arteries
Deliver blood to pelvic organs and lower abdominal wall

Femoral Artery
Delivers blood to the thigh and inner knee

FIGURE 11-1 Major blood vessels and the parts of the body to which the arteries (red) deliver blood and from which the veins (blue) drain blood. (*Starr and McMillan, 2010.*)

wall separating the atria *(foramen ovale)*. Thus, this blood entirely bypasses the pulmonary circuit. Identify the ductus arteriosus (Figures 11-2 and 11-4).

Why is it not necessary for large quantities of blood to enter the pulmonary system of a fetus?

_____

_____

With the first breath of the newborn, the ductus arteriosus contracts, and the foramen ovale closes. Circulation through the pulmonary circuit is increased dramatically. Then, during the 8 weeks after birth, the ductus arteriosus forms a fibrous strand of

connective tissue, the *ligamentum arteriosum*, and the foramen ovale permanently fuses shut.

B. Systemic Circuit: Major Arteries and Veins Anterior to the Heart

1. The systemic circuit begins with the aorta. This large vessel leads anteriorly out of the left ventricle of the heart and makes a sharp turn to the left (the so-called **aortic arch,** Figure 11-5a) and proceeds posteriorly through the body as the **dorsal aorta** (Figure 11-5a). All of the major arteries of the body arise from the aortic arch and dorsal aorta).

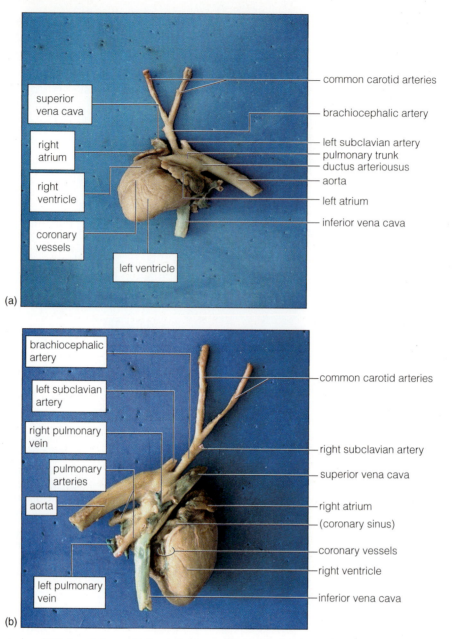

superior
vena cava

right
atrium

right
ventricle

coronary
vessels

left ventricle

common carotid arteries

brachiocephalic artery

left subclavian artery
pulmonary trunk
ductus arteriousus
aorta

left atrium

inferior vena cava

(a)

brachiocephalic
artery

left subclavian
artery

right pulmonary
vein

pulmonary
arteries

aorta

left pulmonary
vein

common carotid arteries

right subclavian artery

superior vena cava

right atrium

(coronary sinus)

coronary vessels

right ventricle

inferior vena cava

(b)

FIGURE 11-2 (a) Ventral and (b) dorsal views of a fetal pig heart. (*Photos by D. Morton.*)

2. Locate the first visible vessel, the **brachiocephalic artery** (Figure 11-5a), to branch from the aortic arch. The first vessels to branch from the aorta are the coronary arteries; the actual branching cannot be seen without dissecting the heart. The brachiocephalic artery branches to give rise to the **right subclavian artery** (Figure 11-5a), going to the right forelimb, and the **carotid trunk** (Figure 11-5a), whose branches course anteriorly through the neck and head. Trace the right subclavian artery and its branches through the shoulder region to the right forelimb. The name of the right subclavian artery changes to the *right axillary artery* in the shoulder and then to the *right brachial artery* when it enters the upper forelimb.

3. Return to the aorta and locate the second visible vessel to branch from the aortic arch, the **left subclavian**

artery (Figure 11-5a). The left subclavian artery and its branches pass through the shoulder and left forelimb or arm in the same manner as the right subclavian artery, described above. As you trace the course of the left subclavian, notice that some of its branches feed the muscles of the chest and back.

4. Return to the right forelimb and locate the venous system that passes through this appendage. Because the veins are relatively thin walled, this may be very difficult. Also, some of them may not be injected with blue latex and will appear a brownish color. If possible, follow the *brachial vein* to the *axillary* and the **subclavian vein** (Figure 11-5b) until the latter becomes the **brachiocephalic vein** (Figure 11-5b). It should be relatively easy to follow the brachiocephalic to its juncture with the previously identified superior vena

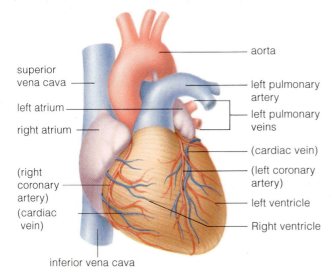

FIGURE 11-3 Anterior view of human heart. (*Starr and McMillan, 2010.*)

Labels in figure:
- aorta
- superior vena cava
- left atrium
- right atrium
- (right coronary artery)
- (cardiac vein)
- inferior vena cava
- left pulmonary artery
- left pulmonary veins
- (cardiac vein)
- (left coronary artery)
- left ventricle
- Right ventricle

cava (Figure 11-5a; it forms a prominent V), which returns blood to the right atrium of the heart.

5. To examine the arterial system that serves the throat and head, locate the carotid trunk (Figure 11-5a; a branch of the brachiocephalic artery; see above). This short branch of the brachiocephalic artery immediately splits into the **left** and **right common carotid arteries.** Each of these vessels divides into the *internal* and *external common carotid arteries.* Remove the thymus and thyroid glands and considerable muscle tissue in the throat to locate the anterior portions of the common carotid arteries.

As you locate and trace the carotid arteries, look for a white "fiber" that parallels them. This is the *vagus nerve.*

6. On either side of the neck are the major veins that drain the head and throat region. The **internal** and **external jugular veins** (Figure 11-5b) join the subclavian veins (from the forelimbs) to form the previously identified brachiocephalic vein, which leads into the superior vena cava.

C. Systemic Circuit: Major Arteries and Veins Posterior to the Heart

1. The posterior extension of the aortic arch is the dorsal aorta. As the name implies, the dorsal aorta lies near the dorsal body wall running parallel to the spine. From this large vessel arise all of the arterial branches that feed the organs, glands, and muscles of the abdominal region and the muscles of the hindlimbs and tail.

2. Follow the dorsal aorta posteriorly. Carefully move the liver and stomach of the pig and use a dissection needle to scrape away the sheet of tissue that connects the dorsal aorta to the pig's back. Locate the **celiac artery** (Figure 11-6a), the branches of which deliver blood to the stomach, spleen, and liver. Continue to follow the dorsal aorta posteriorly

and locate the **superior mesenteric artery** (Figure 11-6a). This vessel, located just posterior to the celiac artery, divides into branches to the pancreas and duodenum of the small intestine.

3. Posterior to the superior mesenteric artery are the **renal arteries** (Figure 11-6a), which are relatively short vessels that connect the dorsal aorta and the kidneys. At this time, it's easy to locate the **renal veins** (Figure 11-6b),which drain blood from the kidneys to the inferior vena cava.

4. As you follow the dorsal aorta posteriorly beyond the kidneys, the **external iliac arteries** (Figure 11-6a) branch, one into each hindlimb. Each leg is also drained with a major vein, the **common iliac vein** (Figure 11-6a), which joins the inferior vena cava.

5. Follow the branches of the dorsal aorta into the tail region, being careful not to cut the two intervening branches. The small extension toward the tail region is called the *sacral artery* as it leaves the dorsal aorta and the *caudal artery* when it enters the tail.

6. Just anterior to the sacral artery, the **internal iliac arteries** (Figure 11-4) branch from the dorsal aorta. These enlarge and form the two **umbilical arteries** (Figure 11-4), which run through the umbilical cord to the placenta. Cut the umbilical cord transversely and note the arrangement of the umbilical arteries within it. Consider the composition of the blood as it travels through the umbilical arteries to the placenta. Is it rich in oxygen or carbon dioxide?

_____

7. Locate the two pieces of vein that you tied with string in the previous exercise. This is the **umbilical vein** (Figure 11-4), through which blood rich in nutrients and oxygen flows from the placenta back to the fetus. Locate the umbilical vein in the umbilical cord and follow it anteriorly toward the liver. When the umbilical vein reaches the liver, it becomes the **ductus venosus** (Figure 11-4), which continues anteriorly within the substance of the liver and joins the inferior vena cava. The umbilical arteries, umbilical veins, and ductus venosus become modified into ligaments after the birth of the fetus.

What is the relationship between the navel and the umbilical cord?

_____

_____

_____

After birth, the *hepatic portal system*, which consists of a network of veins, collects all of the blood from the lower digestive tract and associated organs (stomach, small intestine, pancreas, and spleen). These veins drain into the *hepatic portal vein*, which carries the blood to the liver. In general, a **portal vein** is one that collects blood from the capillaries of one organ and transfers it to the capillaries of another organ.

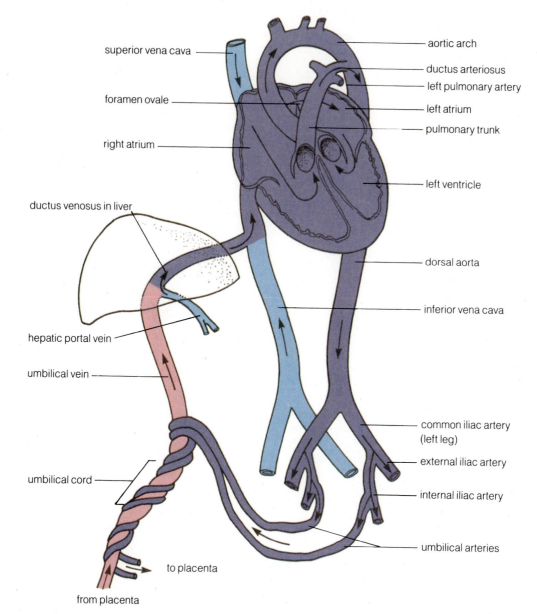

superior vena cava

foramen ovale

right atrium

ductus venosus in liver

hepatic portal vein

umbilical vein

umbilical cord

to placenta

from placenta

aortic arch

ductus arteriosus

left pulmonary artery

left atrium

pulmonary trunk

left ventricle

dorsal aorta

inferior vena cava

common iliac artery (left leg)

external iliac artery

internal iliac artery

umbilical arteries

FIGURE 11-4 Diagram of the circulatory system of a fetal mammal. Arrows indicate the flow of blood. Pink represents fully oxygenated blood. The blue indicates oxygen-depleted blood. *(After Weller and Wiley, 1985.)*

After birth, the *hepatic veins* drain the blood from the liver into the inferior vena cava (Figure 11-7).

11.2 Internal Structure of the Heart

The hearts of a fetal pig and a human are quite similar. For example, the heart is the primary pump of the circulatory system, it is four chambered, and it is primarily composed of cardiac muscle tissue.

PROCEDURE

1. Carefully free the heart from the fetal body by cutting through the superior and inferior venae cavae, the subclavians, the common carotids, and the dorsal aorta

just posterior to the heart. Cut through the left and right pulmonary veins and the pulmonary arteries at their juncture with the lungs. Remove the heart from the fetus and place it on paper towels with its ventral surface facing you, as it was in the thoracic cavity. If any remnants of the pericardial sac are still present, carefully remove them from around the heart.

2. Review the location of the four heart chambers: the left and right atria and the left and right ventricles. Locate the coronary artery and vein in the groove running between the left and right ventricles.

3. Place the heart dorsal side up and locate the inferior and superior venae cavae. Cut through these vessels with your scissors and expose the interior chamber of the right atrium (see Figure 11-8b, incision 1). Carefully remove any latex and coagulated blood in the right atrium. Between the atrium and the right ventricular cavity are three membranous cusps attached to the wall of the right atrium. This is the **tricuspid valve.** The open ends of its cusps face downward like open parachutes into the cavity of the right ventricle.

4. Continue working from the dorsal side and cut into the right ventricle (Figure 11-8a, incision 2). With your forceps and needle, remove any latex that obstructs your view. You should be able to see the three cusps of the **pulmonary semilunar valve** at the juncture of the pulmonary artery and the right ventricle. The open ends of the cusps face into the pulmonary trunk and thus prevent a backward flow of blood into the ventricle.

5. Examine the internal walls of the ventricle. If you wish, you may extend incision 2 to the ventral side of the right ventricle. Look for muscular ridges on the inside wall. These are the **papillary muscles,** and arising from them are relatively fine fibers, the **chordae tendinae.** The chordae tendinae are attached to the edges of the tricuspid valve and are commonly called the *heartstrings.*

6. Next, with the heart's ventral surface facing you, locate the ductus arteriosus and the aorta. (Remember that the ductus arteriosus is a shunt between the pulmonary trunk and the aorta.)

7. Cut open the left atrium (Figure 11-8a, incision 3) and the left ventricle (incision 4). Remove any latex. On

(a)

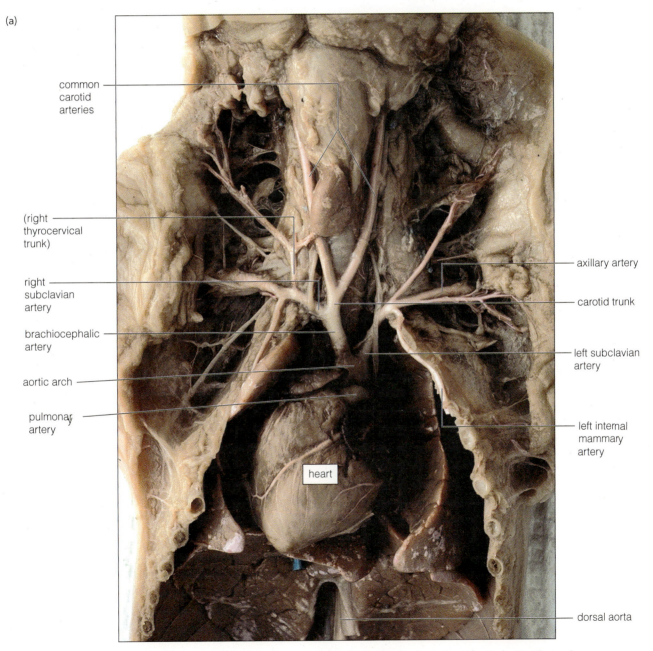

common carotid arteries

(right thyrocervical trunk)

right subclavian artery

brachiocephalic artery

aortic arch

pulmonary artery

axillary artery

carotid trunk

left subclavian artery

left internal mammary artery

heart

dorsal aorta

FIGURE 11-5 Ventral views of (a) arteries and (b) veins anterior to the heart of a fetal pig. *(Photos by D. Morton.)*

(b)

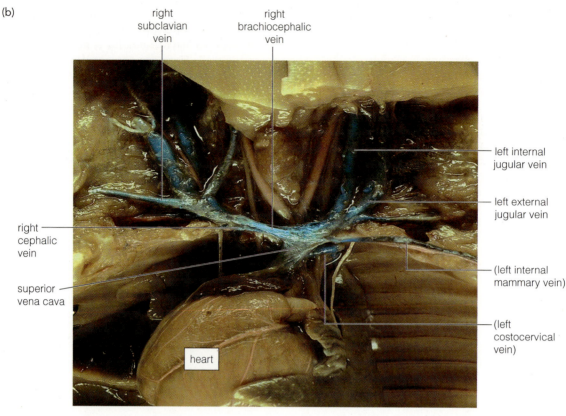

right
subclavian
vein

right
brachiocephalic
vein

right
cephalic
vein

superior
vena cava

left internal
jugular vein

left external
jugular vein

(left internal
mammary vein)

(left
costocervical
vein)

heart

FIGURE 11-5 *continued.*

the dorsal wall of the heart, find the pulmonary veins from the inside of the left atrium. Next, locate the *bicuspid valve* (consisting of two cusps) between the left atrium and left ventricle.

Do the cusps appear similar to the tricuspid valve? _____ (yes or no)

8. Turn the heart so that the ventral surface is facing you and examine the cavity of the left ventricle. Note the papillary muscles and the chordae tendinae in the left ventricle.

Do they appear similar to those in the right ventricle? _____ (yes or no)

9. Insert a probe into the aorta from the exterior of the heart and note where it enters the cavity of the left ventricle. At this point, there is another valve, the aortic semilunar valve, with three cusps.

Is the orientation of the aortic semilunar valve the same as that of the semilunar valve between the pulmonary trunk and the right ventricle? _____ (yes or no)

10. Recall that in the fetus a temporary opening, the foramen ovale (Figure 11-4) exists between the right and left atria. Look for it in your specimen.

11. Complete your study by looking for differences in the thickness of the walls of the atria and those of the ventricles.

Why is the wall of the left ventricle thicker than that of the right ventricle?

_____

_____

_____

11.3   Urinary and Reproductive Systems

The urinary and reproductive systems are often studied together (as the *urogenital system*) because they share several anatomical features.

PROCEDURE

A. Urinary System

Similar to humans, the pig is a terrestrial organism and, as such, must conserve water. At the same time, metabolic wastes must be continuously removed from the blood. Furthermore, the composition of the blood must be constantly monitored and adjusted so the cells of the body are bathed in fluids of constant composition.

Much of the potentially poisonous waste occurs in the form of **urea** and results from the metabolism of amino acids in the liver. Urea is filtered from the bloodstream in the kidneys, which also regulate water and salt balance. The main organs that carry on these functions are labeled in Figure 11-9.

1. Place your pig on its back in the dissection pan and use string or rubber bands to secure the legs. Pin the lateral body-wall flaps to the dorsal side of your specimen and pull the umbilical cord and surrounding tissue back between the hindlimbs.

2. Find the paired **kidneys** in the lumbar region of the body cavity pressed against the dorsal body wall (Figure 11-10). They are covered by the **peritoneum,** the smooth, rather shiny membrane that lines the abdominopelvic cavity. (You may have already removed much of this during the dissection of the circulatory system in the preceding exercise.) The main function of the kidneys is the production of **urine,** a fluid containing urea and other waste substances dissolved in water.

3. To expose the right kidney, carefully lift up the abdominal organs and move them anteriorly and to the left. Using a dissecting needle, carefully scrape away the peritoneum so that the kidney bean–shaped kidney and the **ureter**—the duct that connects it to the bladder—are easily seen. Note the central depression in the surface of the kidney. This is the *hilus,* the region where the ureter and *renal vein* leave and the *renal artery* enters the kidney.

(a)

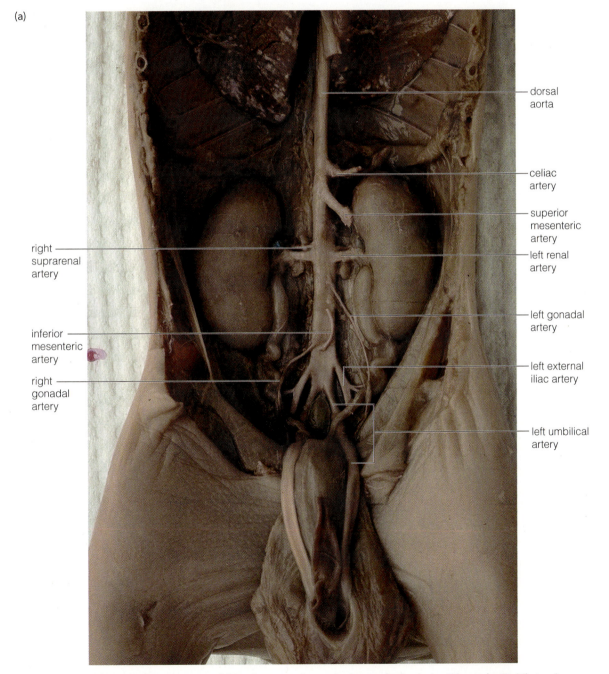

right suprarenal artery

inferior mesenteric artery

right gonadal artery

dorsal aorta

celiac artery

superior mesenteric artery

left renal artery

left gonadal artery

left external iliac artery

left umbilical artery

FIGURE 11-6 Ventral views of (a) arteries and (b) veins posterior to the heart of a fetal pig. *(Photos by D. Morton.)*

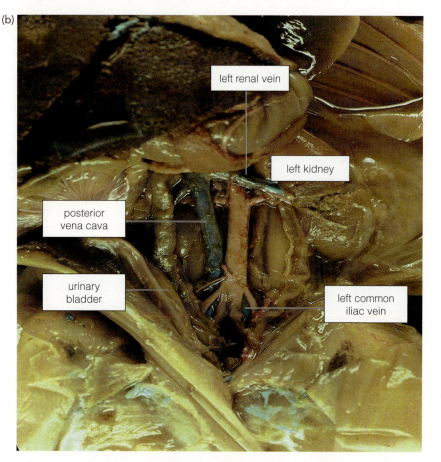

(b)

left renal vein

left kidney

posterior vena cava

urinary bladder

left common iliac vein

FIGURE 11-6 *continued.*

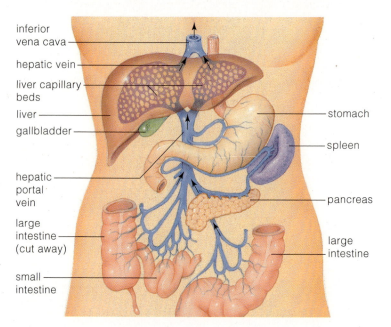

inferior vena cava

hepatic vein

liver capillary beds

liver

gallbladder

hepatic portal vein

large intestine (cut away)

small intestine

stomach

spleen

pancreas

large intestine

FIGURE 11-7 Anterior view of the hepatic portal system of an adult human. Blood from the capillaries of the organs of the lower digestive tract is delivered to those of the liver by the hepatic portal vein. *(After Starr and McMillan, 2010.)*

4. Carefully follow the ureter from the hilus to the *allantoic bladder*. Then lift the bladder and find the **urethra** (Figure 11-10). The latter is the structure through which urine passes from the bladder to the outside of the animal. In males, the urethra is very long and passes through the *penis* to the outside of the body. Notice that the urethra passes posteriorly for a distance of approximately 2 cm and then turns sharply anteriorly and ventrally before entering the penis. In female fetal pigs, the urethra is short and passes posteriorly to join with the *vagina* to form the *urogenital sinus* (Figure 11-11).

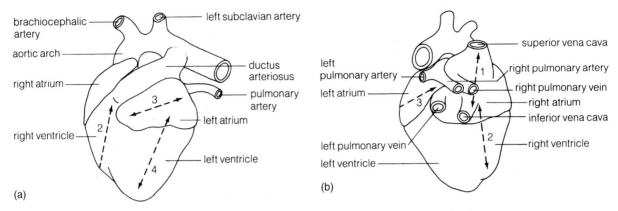

FIGURE 11-8 (a) Ventral and (b) dorsal views of fetal pig heart, showing numbered incisions for dissection.

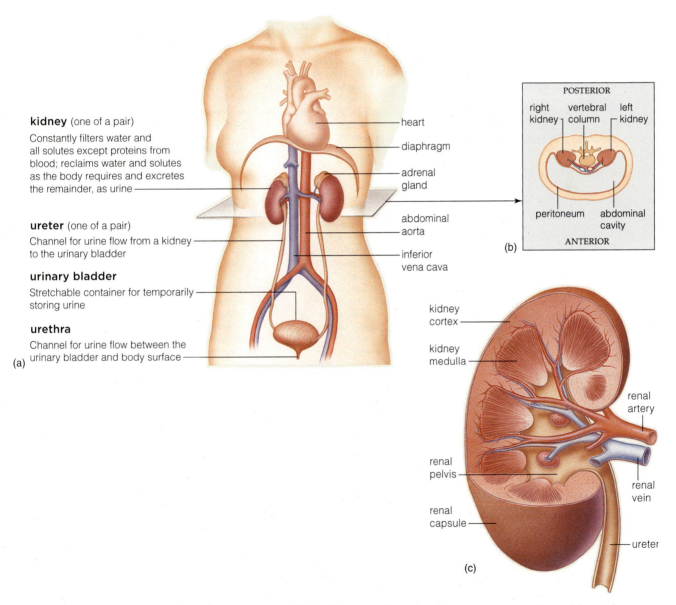

kidney (one of a pair)
Constantly filters water and all solutes except proteins from blood; reclaims water and solutes as the body requires and excretes the remainder, as urine

ureter (one of a pair)
Channel for urine flow from a kidney to the urinary bladder

urinary bladder
Stretchable container for temporarily storing urine

urethra
Channel for urine flow between the urinary bladder and body surface

heart
diaphragm
adrenal gland
abdominal aorta
inferior vena cava

POSTERIOR
right kidney — vertebral column — left kidney
peritoneum   abdominal cavity
ANTERIOR

kidney cortex
kidney medulla
renal artery
renal pelvis
renal vein
renal capsule
ureter

FIGURE 11-9 (a) Anterior view of human urinary system. (b) The kidneys and associated organs are retroperitoneal (located under the peritoneal membrane). (c) A longitudinal section of the kidney and ureter. The cortex consists mainly of the functional unit for the formation of urine, (the nephron). *(After Starr and McMillan, 2010.)*

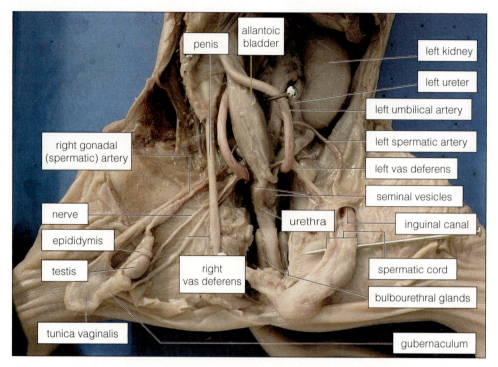

FIGURE 11-10 Ventral view of the male urogenital system of the fetal pig. *(Photo by D. Morton.)*

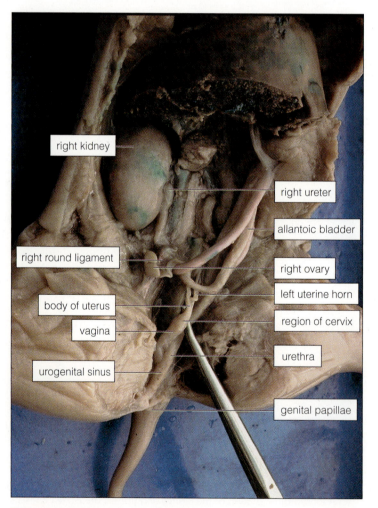

FIGURE 11-11 Ventral view of the female urogenital system of the fetal pig. *(Photo by D. Morton.)*

5. The *allantoic duct*, which leads from the allantoic bladder into the umbilical cord, is largely a vestigial structure because most of the wastes produced in the fetus are carried in the bloodstream to the placenta, where they are excreted by the body of the mother. After birth, the allantoic duct collapses, and the allantoic bladder is part of the **urinary bladder,** which functions to store urine.

B. Female Reproductive System

In terrestrial organisms, fertilization (the fusion of the nuclei of male and female gametes) occurs internally, in which a relatively stable aquatic-like environment is maintained. After fertilization has occurred, the zygote divides to form an embryo, implants in the uterine wall, and eventually develops into a fetus. A combined fetal and uterine structure, the placenta, nourishes, delivers oxygen to, and removes wastes from the developing fetus until it can pass through the birth canal and exist on its own in the outside world. The organs of the female reproductive system are labeled in Figure 11-12.

1. Examine the **vulva,** the collective term for the external genitalia in females. In fetal pigs, the vulva includes the *genital papilla* on the outside of the body; the *labia* or lips found on either side of the *urogenital sinus* (Figure 11-11), and the **clitoris**; a small body of erectile tissue on the ventral portion of the urogenital sinus. Also included in the vulva is the opening of the urogenital sinus itself.

2. In the female fetal pig, the **urogenital sinus** is the common passage for the urethra and the **vagina** (Figure 11-11). To locate these structures, carefully insert your scissors into the opening of the urogenital sinus and cut this structure from the side. Locate where the ducts of the vagina and urethra enter to form the urogenital sinus.

   The urogenital sinus is not present in adult female pigs. During the subsequent development of the fetus, the sinus is reduced in size until the vagina and the urethra each develop their own separate openings to the outside. Thus, in adult female pigs, urine exits through the urinary opening. This is the also the situation in human post partum (after birth) females.

   How does this compare with the structure of the reproductive system of most male mammals?

   _____

   _____

   _____

3. Again, locate the clitoris. This small, rounded region on the inner ventral surface of the urogenital sinus is **homologous** (i.e., similar in developmental origin) with part of the male penis. In males, the tissues of the penis develop around and enclose the urethra; in females, the urethra opens posteriorly to the clitoris.

4. Follow the urogenital sinus anteriorly and identify the thick-walled muscular vagina, which is continuous

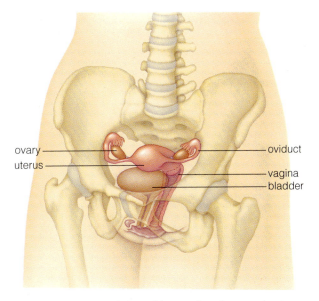

FIGURE 11-12 Ventral view of human female reproductive system. *(After Starr and McMillan, 2010.)*

with the **uterus.** In pigs, the uterus consists of three structures or regions: the **cervix** at the entrance to the uterus, the **uterine body,** and the two **uterine horns** (Figure 11-11). Note that the uterine horns unite to form the body of the uterus. The pig has a *bicornuate uterus*, in which the fetuses develop within the uterine horns. Human women have no uterine horns, and the fetus develops within the body of the *simplex uterus*.

5. From the uterine horns, follow the **oviducts** to the **ovaries** (Figure 11-11). The ovaries are small, yellowish kidney bean–shaped structures that lie just posterior to the kidneys. They are the sites of egg production and the source of the female sex hormones, estrogen and progesterone.

   All eggs that a female produces during her lifetime are present in the ovaries at birth. After puberty, batches of eggs mature, a few of which rupture from the surface of the ovaries (**ovulation**) and enter the oviducts.

   If viable sperm are present in the upper third of an oviduct when it contains eggs, fertilization can occur. In this case, the fertilized egg or zygote will develop into an embryo and pass down the oviduct to become implanted in the wall of the uterine horn. In human women, however, it becomes implanted in the wall of the uterus.

6. Identify the membranous broad and round ligaments. The **broad ligament,** which originates from the **dorsal body wall,** supports the ovaries, oviducts, and uterine horns. The **round ligament** (Figure 11-11), which also supports the ovaries, extends from the lateral wall and crosses the broad ligament diagonally.

C. Male Reproductive System

The male reproductive system of a mammal consists of organs and structures (Figure 11-13) that primarily function in the production of sperm, their transit to the base

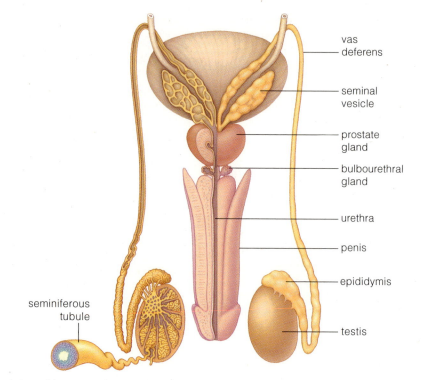

FIGURE 11-13 Ventral view of human male reproductive system. (*After Starr and McMillan, 2010.*)

of the penis during sexual excitement, and their subsequent ejaculation in **semen**—sperm plus the secretions of the *sex accessory glands*.

1. Locate the **testes** (male gonads; Figure 11-10), the site of **sperm** production and source of *testosterone*, the male sex hormone. In older fetuses, the testes are located in the sac-like **scrotum**, but in younger fetuses, they can be found anywhere between the kidneys and the scrotum.

   For viable sperm to be produced in adult males, the testes must be situated outside of the abdominopelvic cavity, where body temperatures are slightly lower than within. Thus, during normal development, the testes undergo a posterior migration into the scrotum.

2. Locate the scrotum. Make a midline incision through this structure, cutting through the muscle tissue. Pull out the two elongated bulbous structures covered with a transparent membrane. This membrane is the *tunica vaginalis* and is actually an outpocketing of the abdominal wall. Notice the tough white cord, the **gubernaculum**, that connects the posterior end of the testes to the inner face of the sac (Figure 11-10). This cord is homologous to the round ligament in the female reproductive system. It grows more slowly than the surrounding tissues and thus aids in pulling the testes posteriorly into the scrotal sacs.

3. Cut through the tunica vaginalis to expose a single testis and find the **epididymis** (Figure 11-10). This is a tightly coiled tube that lies along one side of the testis. Sperm produced in the testis are stored in the epididymis until ejaculation.

4. Note the slender, elongated **spermatic cord** (Figure 11-10) that emerges anteriorly from each testis. Gently pull the cord and note that it passes through an opening, the **inguinal canal,** which is actually an opening from the abdominopelvic cavity into the scrotum. It is through this opening that the testes descend during their migration into the scrotum.

   Some human males develop an **inguinal hernia,** a condition in which part of the intestine drops through the inguinal canal into the scrotum. Pigs and other four-legged mammals (hint) do not develop inguinal hernias. Why do you think this is so?

   _____

   _____

   _____

5. Using the tips of your scissors, slit open the spermatic cord attached to the dissected testis. Note that it contains the *sperm duct* or **vas deferens** (plural, *vasa deferentia*), spermatic vein, **spermatic artery,** and spermatic nerve (Figure 11-10). The vas deferens is severed when a human man has a **vasectomy.** Follow the vas deferens to the base of the bladder, where it loops up and over the ureter and then continues posteriorly to enter the urethra.

6. Expose the full length of the **penis** (Figure 11-10) and its juncture with the urethra. To find the latter, make an incision with your scalpel through the muscles in the midventral line between the hind legs. When the cut is deep enough, the muscles will lie flat. Carefully remove the muscle tissue and pubic bone on each

side of the cut until the urethra is exposed. With a blunt probe, tear the connective tissue connecting the urethra to the rectum, which lies dorsal to it.

7. Locate a pair of small glands, the **seminal vesicles** (Figure 11-10), on the dorsal surface of the urethra where the two vasa deferentia enter. Situated between the bases of the seminal vesicles is the **prostate gland.** The other sex accessory glands are the **bulbourethral glands,** two elongated structures lying on either side of the juncture of the penis and urethra.

The seminal vesicles, prostate gland, and bulbourethral glands all secrete fluids that, together with sperm, form semen, which is ejaculated during sexual intercourse. In addition to sperm, semen is mostly water, sugar, and other molecules that provide a supportive environment for the sperm.

11.3   Kidney

In the first part of this exercise, you located the kidneys, a pair of kidney bean–shaped structures lying on either side of the spine in the lumbar region of the body. Although the kidneys are situated below the diaphragm, they are actually located outside of the peritoneum, the membrane that lines the abdominal cavity. During this procedure, you will examine the internal anatomy of the kidney, including its functional unit, the nephron.

PROCEDURE

A. Anatomy

Each kidney contains numerous nephrons, which function to filter the blood and start the process of urine formation. A system of ducts completes urine formation and transports it to a space, the *renal pelvis*, which is drained by the ureter. The ureter transports the urine to the urinary bladder.

1. Return to the right kidney and attempt to identify the *adrenal gland* (Figure 11-9). Look for a tiny, cream colored, comma-shaped body located on the medial, anterior side of the kidney. This is another of the body's endocrine organs. The paired adrenal glands secrete several hormones, including adrenaline (epinephrine). Free the right kidney by severing the renal vein, renal artery, and ureter. Remove the kidney from the body cavity and place it on a paper towel with the central depression to the right.

2. With your scalpel, carefully cut the kidney in half lengthwise, as you would separate the two halves of a peanut. Examine the cut surface of one of the halves and locate the three major regions of the kidney—the outer **cortex,** inner **medulla,** and **renal pelvis** (Figure 11-9). The cortex and medulla contain different portions of the nephrons and their associated blood vessels.

B. Microscopic Structure of the Nephrons

Each **nephron** is a little tube or tubule that is closed at one end (Figure 11-14). The closed end is expanded and collapsed upon itself much like a deflated basketball pushed in by your fist, only the inside space is connected to the space within a tube attached to the other side. The expanded, collapsed end of the nephron is called **Bowman's capsule,** and the space within its inner wall contains a network of capillaries, the **glomerulus** (plural, glomeruli). The combination of Bowman's capsule and the glomerulus is called a *renal corpuscle.*

Urine formation begins with the filtration of blood across the capillary and inner capsule walls into the space within Bowman's capsule. After filtration, the filtrate travels through the rest of the nephron—the **proximal convoluted tubule** or just proximal tubule, **loop of Henle,** and **distal convoluted tubule** or proximal tubule. Much of the water, ions, sugars, and other useful substances are reabsorbed into the blood in the *peritubular capillaries.* Meanwhile other molecules, substances such as ammonia, potassium, and hydrogen ions are secreted to join the urea in the forming urine. Thus, the nephron carries out its excretory and water and salt balance functions in three steps: filtration, tubular reabsorption, and tubular secretion.

Although all of the activities of the nephron are extremely vital to the health of a mammal, the importance of the reabsorption function is easily illustrated with a few numbers. The renal corpuscles of human kidneys produce approximately 180 L (~180 quarts) of filtrate each day. About 99% of this filtrate is reabsorbed as

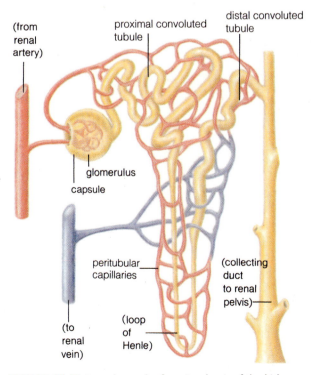

FIGURE 11-14 A nephron, the functional unit of the kidney. *(After Starr and Taggart, 2001.)*

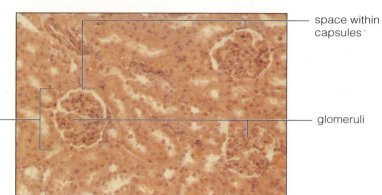

FIGURE 11-15 Section of the cortex of a kidney (186x). *(Photo by D. Morton.)*

water, primarily by the remainder of the nephrons. If they were not so efficient, we would have to drink constantly just to replenish the fluid lost.

   Look at Figure 11-14, which shows a nephron, its named portions, and the associated blood vessels.

1. Examine a prepared section of the kidney with your compound microscope. Identify the *cortex* and *medulla*. In the cortex, locate *renal corpuscles, glomeruli,* and *Bowman's capsules* (Figure 11-15).

2. Identify cross-sections of the tubular portion of nephrons.

3. Place your pig and any excised organs back in the plastic bag. Tie it shut to prevent your pig from drying out. Dispose of any paper towels that contain preservative *as directed by your instructor.* Clean the tray, dissecting tools, and laboratory table.

_____ 1. A vein is a blood vessel that always carries
   (a) blood toward the heart
   (b) blood away from the heart
   (c) oxygen-rich blood
   (d) oxygen-poor blood

_____ 2. The hearts of a fetal pig and a human are similar in that they are
   (a) the primary pump of the circulatory system of the body
   (b) both four chambered
   (c) composed of cardiac muscle tissue
   (d) all of the above

_____ 3. The urogenital system refers to the
   (a) urinary and reproductive systems
   (b) urinary and excretory systems
   (c) reproductive system
   (d) external genitalia

_____ 4. The ureters drain urine into the
   (a) renal pelvis
   (b) cecum
   (c) urinary bladder
   (d) small intestine

_____ 5. The clitoris of the female and a portion of the penis of the male are homologous structures. This means they have a similar
   (a) function
   (b) structure
   (c) developmental origin
   (d) none of the above

_____ 6. The testes of a male differ from the ovaries of a female in that the testes
   (a) develop in the body cavity and migrate to a position outside of the body cavity
   (b) require a slightly higher temperature than that of the body to produce viable gametes
   (c) produce zygotes
   (d) do both a and b

_____ 7. When a human man has a vasectomy, the operation involves
   (a) removal of the male gonads or testes
   (b) removal of the urethra
   (c) severing the vas deferens
   (d) removal of the prostate gland

_____ 8. Semen contains
   (a) sperm
   (b) the secretions of sex accessory glands
   (c) eggs
   (d) both a and b

_____ 9. The functional unit of the kidney is the
   (a) renal pelvis
   (b) ureter
   (c) cortex
   (d) nephron

_____ 10. The urinary system is concerned with
   (a) blood circulation
   (b) digestion and the absorption of nutrients
   (c) reproduction
   (d) excretion of urine

EXERCISE **11** **Organs of the Circulatory, Urinary, and Reproductive Systems**

POST-LAB QUESTIONS

11.1 Blood Vessels and the Surface Anatomy of the Heart

1. Identify the structures in the photo below.

2. What is the main difference between the pulmonary and the systemic circuits of the circulatory system?

11.2 Internal Structure of the Heart

3. What is the foramen ovale? What is its fate after birth?

4. What structures are found between the right atrium and the right ventricle, the right ventricle and the pulmonary trunk, the left atrium and the left ventricle, and the left ventricle and the aortic trunk? What are their functions?

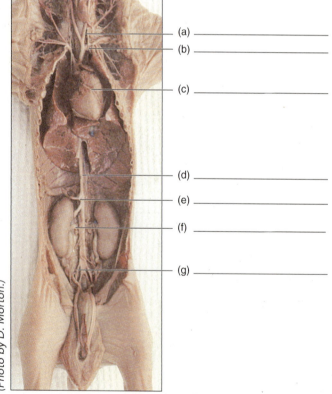

(a) _____

(b) _____

(c) _____

(d) _____

(e) _____

(f) _____

(g) _____

(Photo by D. Morton.)

11.3 Urinary and Reproductive Systems

5. Briefly describe the functions of the kidney, ureters, bladder, and urethra in adult pigs.

6. What is the vulva?

7. How does the uterus of female pigs and humans differ? Include in your discussion the site of embryo implantation.

8. Identify the structures in the photo below.

11.4 Kidney

9. Name the functional unit of the kidney. Briefly describe its parts and how they operate.

Food for Thought

10. Search the Internet for websites that describe artificial hearts, heart and kidney transplants, and dialysis. List a website for each topic and briefly summarize their contents.

http://

http://

http://

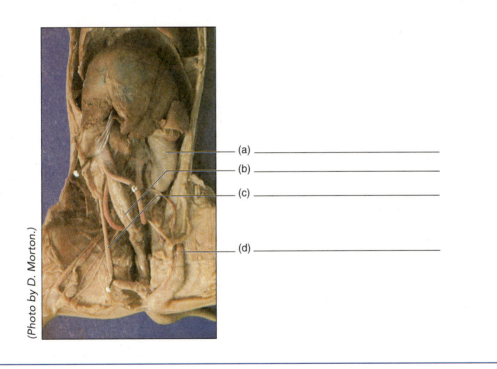

(Photo by D. Morton.)

(a) _____

(b) _____

(c) _____

(d) _____

Human Blood and Circulation

OBJECTIVES

After completing this exercise, you will be able to

1. define *blood vessels* (different types), *blood, heart, pulmonary circuit, systemic circuit, homeostasis, plasma, hematocrit, agglutination, blood pressure, elastic membranes, valves, sinoatrial node, acetylcholine,* and *epinephrine.*

2. identify and give the characteristics and functions of the different types of blood cells.

3. explain the ABO and Rh blood group systems.

4. describe how to perform a hematocrit and blood typing.

5. give the structure and function of the different types of blood vessels.

6. describe how blood flows through the circulatory system.

7. name the four chambers and four valves of the heart and describe the route blood takes through them.

8. describe how the heart contracts.

9. explain how the heart is controlled.

Introduction

Circulation—the bulk transport of fluid around the body—joins together the specialized cells of multicellular organisms, even though they are separated physically. The **heart** pumps the fluid **blood** around the circulatory system in pipelike **blood vessels.**

There are two completely separate routes leading to and from the heart: the pulmonary and systemic circuits (Figure 12-1). In each circuit, branching **arteries** convey blood to smaller and more numerous **arterioles,** which in turn deliver the blood to beds of capillaries. Across the capillary walls, the exchange of dissolved gases, nutrients, wastes, and so on takes place. Within the beds the capillaries branch and merge, finally merging into **venules.** Venules merge into a smaller number of **veins,** which continue merging and drain blood back toward the heart.

The **pulmonary circuit** carries oxygen-depleted blood to the capillary beds of the lungs, where oxygen is loaded and where excess carbon dioxide is unloaded. Pulmonary veins drain the oxygen-rich blood back to the heart. The **systemic circuit** takes the oxygen-rich blood from the heart and conveys it to the rest of the body's capillary beds, where oxygen is unloaded and excess carbon dioxide is picked up. Systemic veins drain the oxygen-depleted blood back to the heart.

In a few cases this pattern of blood flow—heart → arteries → arterioles → capillaries → venules → veins → heart—is interrupted by a **portal vein,** which connects two capillary beds. The most prominent example is the *hepatic portal vein,* which transports blood from capillary beds in the intestines, stomach, and spleen to beds of large capillaries in the liver.

12.1 Blood

Human blood is about 45% cells by volume, although it is only slightly thicker than water. Blood cells are suspended in a straw-colored fluid called **plasma** (~55% of total blood volume). Plasma is mostly water but contains many dissolved substances, including gases, nutrients, wastes, ions, chemical signals (hormones), enzymes, antibodies, and other proteins.

MATERIALS

Per student:

- Compound microscope, lens paper, bottle of lens-cleaning solution (optional), lint-free cloth (optional), dropper bottle of immersion oil (optional)

- Prepared slide of a Wright- or Giemsa-stained smear of human blood

Per lab room:

- Eosinophil on demonstration (compound microscope)

- Basophil on demonstration (compound microscope)

- Dropper bottle(s) of sheep blood

- Blood typing kit with
    - of simulated blood
    - Anti-A, anti-B, anti-D (Rh) simulated sera,
    - Stirrers color coded to the sera
    - Plastic blood typing trays

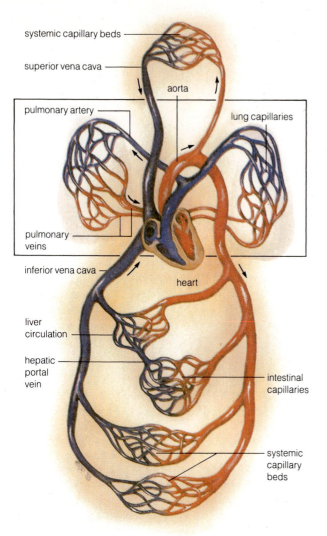

systemic capillary beds

superior vena cava

aorta

pulmonary artery

lung capillaries

pulmonary veins

inferior vena cava

heart

liver circulation

hepatic portal vein

intestinal capillaries

systemic capillary beds

FIGURE 12-1 The human circulatory system. The pulmonary circuit is enclosed by the box. Red indicates oxygenated blood; blue indicates oxygen-depleted blood. *(After Starr and Taggart, 1989.)*

Per student group (4):

- Plain capillary tubes
- Microhematocrit centrifuge

- Ruler or hematocrit reader
- Synthetic blood typing kit with simulated blood and sera

Per lab section:

- Boxes of different sizes of lab gloves
- Box of safety goggles
- Blood waste disposal jar

PROCEDURE

A. Formed Elements (Cells and Platelets) of Blood

The abundance and size of the various blood cells are summarized in Table 12-1. Table 12-2 lists their major functions.

1. Use the medium power of your compound microscope to examine a prepared slide of a stained smear of blood. Note the numerous, small pink-stained red blood cells, or **erythrocytes** (Figure 12-2a). Center a group of them and rotate the nosepiece to the high-dry objective. Erythrocytes have a biconcave disk shape and do not have nuclei. After examining them, explain in your own words their biconcave shape.

_____

_____

Scattered among the erythrocytes are a much smaller number of blue/purple-stained cells. These are white blood cells, or **leukocytes.** Center a leukocyte and rotate the nosepiece to the high-dry objective.

What part of the cell, nucleus or cytoplasm, is stained blue/purple? _____

2. Using high power, preferably, if your microscope has one *and with your instructor's permission*, the oil-immersion objective or move the slide slowly and look for cell fragments between the erythrocytes and leukocytes. They usually have one small blue-stained granule in them and are often clumped together. These cell fragments are **platelets,** or thrombocytes (Figure 12-2b).

TABLE 12-1	Characteristics of Formed Elements of Blood		
Number/mm³ in Cell or Fragment	**Peripheral Blood**	**Percent of Leukocytes**	**Size (μm)**
Erythrocytes	4.5–5.5 million	—	7 by 2
Platelets	250,000–300,000	—	2–5
Neutrophils	3000–6750	65	10–12
Eosinophils	100–360	3	10–12
Basophils	25–90	1	8–10
Lymphocytes	1000–2700	25	5–8
Monocytes	150–750	6	9–15

TABLE 12-2   Functions of Formed Elements of Blood

Cell or Fragment	Functions
Erythrocytes	Contain hemoglobin, which transports oxygen, and carbonic anhydrase, which promotes transport of carbon dioxide by the blood
Platelets	Source of substances that aid in blood clotting
Neutrophils	Leave the blood early in an inflammation to become phagocytes (cells that eat bacteria and debris)
Eosinophils	Phagocytosis of antigen–antibody complexes; numbers are elevated during allergic reactions
Basophils	Granules contain a substance (histamine) that makes blood vessels leaky and a substance (heparin) that inhibits blood clotting
Lymphocytes	Perform many functions central to immunity
Monocytes	Leave the blood to form phagocytic cells called macrophages

3. Locate at least three of the five leukocytes—**neutrophils, lymphocytes,** and **monocytes** (Figures 12-2c to e). Search for them with the high-dry objective. When you find one, center and examine it. If your microscope has an oil-immersion objective *(and with the permission of the instructor)*, use it to look at each white blood cell. You may also find **eosinophils** and **basophils** (Figures 12-2f and 12-2g), which are normally the rarest leukocyte types. If you have not found an eosinophil or a basophil by the time you have identified the three common leukocyte types, look at one or both of the demonstration microscopes set up by your instructor. The three types of leukocytes with the suffix *-phil* (for *philic,* meaning "to like") have large *specific granules.* The prefix in their names refers to the staining characteristics of the specific granules: *neutro-* for neutral (i.e., little staining by either of the two dyes in typical blood stains—eosin and methylene blue), *eosino-* because the specific granules stain with the pink dye eosin, and *baso-* because the specific granules stain with the *basic* dye methylene blue.

B. Hematocrit

If you have donated blood, you have probably had your **hematocrit** or percent packed red blood cell volume taken. A lower than normal hematocrit is one indicator of anemia (a lower than normal hemoglobin concentration).

1. Work in groups of four. Fill a plain capillary tube to the mark with sheep blood and seal it with clay. Fresh blood requires a capillary tube with the inside surface coated with the anticoagulant heparin to prevent clotting.

2. Put the sealed capillary tube in a microhematocrit centrifuge along with those from other groups. Place the clay end against the cushion that lines the inside of the outer rim of the centrifuge. Balance the centrifuge by placing the tubes opposite each other. If there is an odd number of tubes, use an empty one. Write down the number of your slot below._____

3. After your instructor has secured and spun the tubes and opened the centrifuge, recover your tube. Determine the percent packed red blood cell volume using a ruler (Figure 12-3) or a hematocrit reader provided and demonstrated by your instructor and record it here. _____%

4. Dispose of your capillary tube in the blood waste disposal jar.

C. Blood Typing

On the surfaces of your red blood cells are one or more *antigens* that will cause their agglutination if exposed to the complementary *antibodies.* **Agglutination** is the clumping of erythrocytes. This could theoretically occur during a blood transfusion. The transfusion of incompatible blood causes the destruction of donor erythrocytes and perhaps the death of the patient when the clumped cells block blood vessels. For these reasons, blood used for transfusion is very carefully matched for compatibility with the patient's blood.

In the **ABO blood typing system,** erythrocytes can have the A antigen alone or the B antigen alone, both, or neither. If one or both antigens are not present, the plasma contains the antibody or antibodies for the missing antigen (Table 12-3). For example, if an individual's blood is type A, then his or her plasma contains anti-B antibodies (blue in Figure 12-4a). Therefore, a type A individual can not safely receive blood from type B and AB donors because the anti-A antibodies in the host's plasma will agglutinate the donor erythrocytes (Figure 12-4b).

Individuals with blood type O are sometimes called *universal donors* because they can theoretically give blood to all other blood types. Explain why this is so.

_____

_____

_____

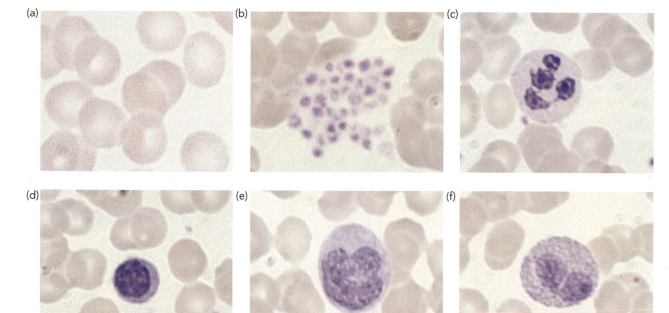

(a)

(b)

(c)

(d)

(e)

(f)

(g)

FIGURE 12-2 Formed elements of blood (a–g, 1439×).
(Photos by D. Morton.)

Individuals with blood type AB are sometimes called *universal recipients* and can theoretically receive blood from any other type. Explain why this is so.

_____

_____

_____

Usually a standard blood typing procedure includes a test for the Rh factor or D antigen. For example, people with A$^+$ blood have both the A and D antigens on the surface of their erythrocytes. An A$^-$ individual has only the A antigen. About 86% of the population in the United States is Rh$^+$. However, Rh$^-$ individuals do not have the anti-D antibody unless they have been exposed to the D antigen. This could happen during a transfusion of Rh$^+$ blood or during the birth of an Rh$^+$ child. The latter situation is usually blocked by injecting a solution containing antibodies (RhoGAM) during the pregnancy and shortly after birth. The injected antibody ties up any D antigen and prevents the mother's body from making anti-D antibody, which would otherwise attack a subsequent Rh$^+$ fetus.

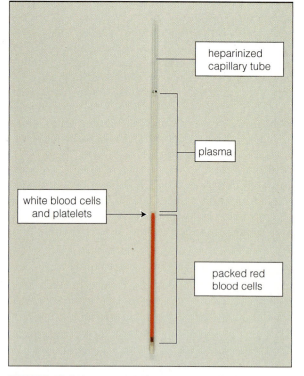

heparinized capillary tube

plasma

white blood cells and platelets

packed red blood cells

FIGURE 12-3 Hematocrit. The percent packed volume of red blood cells is calculated by dividing the height of the column of packed blood cells by the length of the total column and multiplying by 100. In this case, it is 53%.
(Photo by D. Morton.)

TABLE 12-3 · ABO Blood Types

Blood Type	Antigens Present on Erythrocytes	Antibodies Present in Plasma	Plasma Agglutinates
A	A	Anti-B	B and AB
B	B	Anti-A	A and B
AB	A and B	None	None
O	None	Anti-A and B	A, B, and AB

1. Work in groups of four. Gather enough materials to test four blood samples—blood typing slides or trays, mixing sticks or toothpicks.

2. Place one drop of simulated blood where indicated on the slide or in each depression in the tray. Then place one drop of anti-A serum next to the blood cells to be tested for A antigens, one drop of anti-B serum next to the blood cells to be tested for B antigens, and one drop of anti-D serum next to the blood cells to be tested for D antigens. To mix the drops, stir each pair of blood and sera with an unused end of a mixing stick or rock the tray back and forth.

3. If a specific antigen is present, the erythrocytes in that mixture will clump. If it is not present, the mixture will not change. Record the blood types in Table 12-4. The eight possible blood types are A^+, A^-, B^+, B^-, AB^+, AB^-, O^+ and O^-.

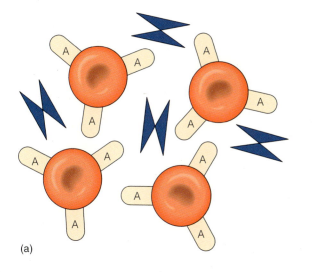

(a)

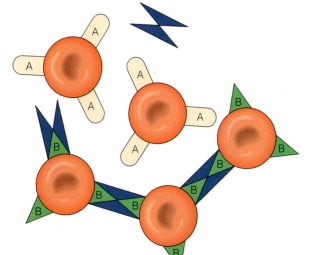

(b)

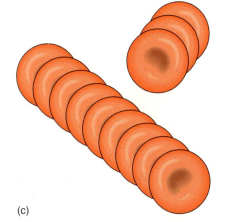

(c)

FIGURE 12-4 **(a)** Red blood cells (RBCs) from a type A individual. The RBCs have A antigen on their surfaces and anti-B antibody (blue) in the plasma. **(b)** If type B RBCs are introduced into the circulation, agglutination of these cells will occur. **(c)** The round surfaces of the RBCs have more antigen, so they tend to stack up in a "slipped stack of poker chips" pattern.

TABLE 12-4 Blood Typing Results

Sample	A Antigen Present	B Antigen Present	D Antigen Present	Blood Type

12.2 Blood Vessels

The basic structure and function of blood vessels is shown in Figure 12-5 and Table 12-5.

MATERIALS

Per student:

- Compound microscope
- Prepared slide of a companion artery and vein, cross-section
- Prepared slide of a whole mount of mesentery

Per student pair:

- Fish net
- Small freshwater fish (3–4 cm long) in an aquarium
- 3×7 cm piece of absorbent cotton
- Petri dish
- Coverslip
- Dissecting needle

Per student group (4):

- Container of anesthetic dissolved in dechlorinated water
- Squeeze bottle of dechlorinated water

Per lab room:

- Safe area to run in place
- Several meter sticks taped vertically to the walls with the lower numbers at the bottom
- Clock with a second hand

PROCEDURE

A. Arteries

Each contraction of the heart pumps blood into the space within the arteries. The rate of flow of blood out of the heart and into the arteries per minute is called *cardiac output* (CO). The arterial space is fairly constant, and it is somewhat difficult for blood to flow through the blood vessels, especially out of the arterioles. This resistance to the flow of blood is called *peripheral resistance* (PR). As CO, PR, or both increase, more blood has to fit into the arterial space, which increases the force that the blood exerts on the walls of the arteries. This force is called the **blood pressure** (BP). The BP is directly proportional to the product of CO and PR (BP $\propto$ CO $\times$ PR).

Pressure in blood vessels is highest in the arteries leaving the heart, gradually decreasing the farther away artery or arterioles are from the heart (Table 12-6). Blood or any other liquid or gas always flows from high to low

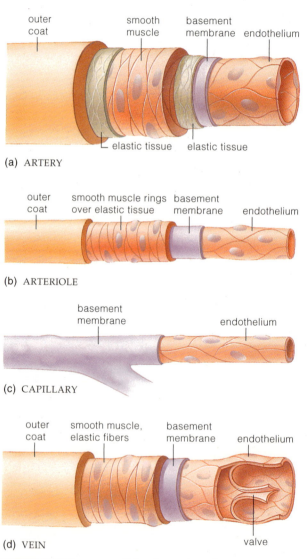

(a) ARTERY

outer coat — smooth muscle — basement membrane — endothelium — elastic tissue — elastic tissue

(b) ARTERIOLE

outer coat — smooth muscle rings over elastic tissue — basement membrane — endothelium

(c) CAPILLARY

basement membrane — endothelium

(d) VEIN

outer coat — smooth muscle, elastic fibers — basement membrane — endothelium — valve

FIGURE 12-5 Structure of blood vessels. (*After Starr and Taggart, 2001.*)

TABLE 12-5 Types of Blood Vessels and Their Major Functions

Type	Functions
Elastic arteries	1. Receive blood from heart 2. Deliver blood to more numerous muscular arteries 3. Maintain blood pressure between contractions of the heart
Muscular arteries	1. Deliver blood to more numerous and smaller arterioles 2. Regulate blood flow to organs
Arterioles	1. Deliver blood to more numerous and smaller capillaries 2. Regulate peripheral resistance 3. Precapillary sphincters of smooth muscle regulate blood flow into particular capillary beds
Capillaries	1. Exchange dissolved gases, nutrients, wastes, and so on with fluid surrounding cells (interstitial fluid) 2. Form interstitial fluid
Venules	1. Drain blood into fewer and larger veins 2. Serve as a blood reservoir
Veins	1. Drain blood into fewer and larger veins and finally back to heart 2. Serve as a blood reservoir

TABLE 12-6 Blood Pressure in Different Parts of the Circulatory System of a Young Man at Rest

Location	Blood Pressures (mm Hg)
Right atrium of heart	5/0 (systolic/diastolic)
Right ventricle of heart	25/5
Pulmonary arteries	20
Arterioles and capillaries of lung	20–10
Pulmonary veins	10
Left atrium of heart	10/0
Left ventricle of heart	120/10
Brachial artery	120/80
Arterioles	100–50
Capillaries	50–20
Veins	20–0

the blood, maintaining BP and flow. Valves at the bases of the pulmonary arteries and the aorta stop blood backflow when the heart is between contractions.

When a physician takes your BP, it is usually of the brachial artery of the upper arm and with the body at rest. A BP of 120/80 means the systolic pressure is 120 mm of mercury (Hg) and the diastolic pressure is 80 mm Hg. The difference between the systolic and diastolic pressures (*pulse pressure*) produces a pulse that you can feel in arteries that pass close to the skin.

A person's BP changes with his or her health, emotional state, activity, and other factors.

1. Get a prepared slide of an artery and vein. With your compound microscope, locate and examine the cross-section of an artery (Figure 12-6).

   Arteries have thick walls compared with other blood vessels. They have an outer coat of connective tissue, a middle coat of smooth muscle tissue, and an inner coat of simple squamous epithelium (*endothelium*). Elastic membranes separate the three coats. The middle coat is the thickest.

2. Find your *radial pulse* in the radial artery (Figure 12-7). Use the index and middle fingers of your other hand. A pulse occurs every time the heart contracts. The strength of the pulse is an estimate of the difference between the systolic and diastolic BP.

3. Sit down and count the number of pulses in 15 seconds. To calculate your heart rate, multiply by 4. Record these numbers in Table 12-7.

pressure. So blood flows from the arteries near the heart to the arterioles and into the capillaries.

The walls of the largest arteries contain many **elastic membranes,** which are stretched during contraction (*systole*) of the heart. When the heart is relaxing (*diastole*), these membranes rebound and the arterial walls squeeze

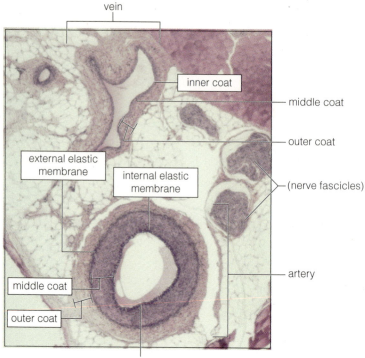

vein

inner coat

middle coat

outer coat

external elastic membrane

internal elastic membrane

(nerve fascicles)

artery

middle coat

outer coat

inner coat

FIGURE 12-6 Photomicrograph of cross section of an artery and vein (35×). *(Photo by D. Morton.)*

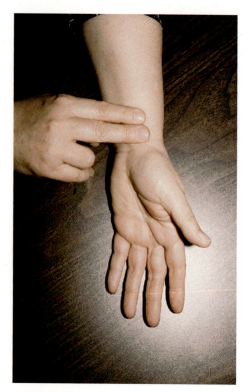

FIGURE 12-7 Feeling the radial pulse. *(Photo by D. Morton.)*

TABLE 12-7	Heart Rate Under Different Conditions	
Condition	**15-Second Counts**	**Heart Rate (beats/min)**
Sitting at rest	_____ × 4 =	
Holding breath	_____ × 4 =	
Running in place	_____ × 4 =	

 Do not do the following procedures if you have any medical problems with your lungs or heart. All subjects should be seated except when otherwise indicated and should stop immediately if they feel faint.

4. Hold your breath. After 10 seconds have passed, count the number of pulses and then calculate your heart rate as in step 3. Record these numbers in Table 12-7.

   Compared with your resting heart rate, does your heart rate increase, decrease, or remain the same?

   _____

   When you hold your breath, you decrease the return of blood to your heart. This reduces your pulse pressure. Homeostatic mechanisms increase your heart rate to compensate for the reduced BP.

5. Now run in place for 2 minutes in the area designated by your lab instructor. Immediately after sitting down, count the number of pulses, calculate your heart rate, and record these numbers.

   After running in place, does your heart rate increase, decrease, or remain the same? _____ Explain these results.

   _____

   _____

   _____

6. *Optional.* Instructions for measuring arterial BP are provided in Exercise 7.

B. Capillaries

Capillaries have a very thin wall that is only one cell thick.

1. Obtain a prepared slide of a whole mount of mesentery and examine it with your compound microscope. Look for groups of blood vessels running through the connective tissue (Figure 12-8). The smallest vessels, which branch and join with each other, are capillaries.
   Can you see red blood cells inside the capillaries? _____ (yes or no)

2. Use the net to catch a small fish from the aquarium and place it in the anesthetic fluid. Treat the fish gently, and it will not be harmed by this procedure.

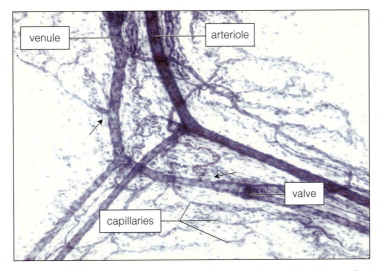

FIGURE 12-8 Arteriole, venule, and capillary bed in a whole mount of mesentery. Arrows indicate capillaries joining venule (100×). *(Photo by D. Morton.)*

3. Immediately after the fish turns belly up, wrap its body in cotton made soaking wet with dechlorinated water. Place the fish in half a petri dish so the tail is in the center.

4. Using dechlorinated water, make a wet mount of the posterior two-thirds of the fish's tail and examine it with the low-power, medium-power, and high-dry objectives of the compound microscope. Use the lowest illumination that still allows you to see the blood flowing in the vessels. If necessary, you can temporarily close the condenser iris diaphragm to create more contrast. The smallest blood vessels are capillaries.

Can you see red blood cells moving through them? _____ (yes or no)

What other vessels can you identify?

_____

_____

Is the blood flowing at the same speed in all of the capillaries? _____ (yes or no)

Describe blood flow in the fish's tail.

_____

_____

_____

_____

5. Immediately return the fish to the aquarium, wash the half Petri dish, and squeeze out the cotton in the sink before dropping it in the trash can.

C. Veins

For the blood to return to the heart after passing through capillary beds below the heart, it must overcome the force of gravity. Veins have **valves** to prevent the backflow of blood away from the heart. Primarily muscular and breathing movements move blood from one segment between valves to another.

1. Again look at a prepared slide of an artery and vein. Find and examine the cross-section of a vein (Figure 12-6). The vein has thinner walls and a larger lumen compared with the artery in the same section. Veins have an outer coat of connective tissue, a middle coat of smooth muscle tissue, and an inner coat of endothelium. Elastic membranes may be present. The outer coat is the thickest layer of the three coats. Compared with arteries, the walls of veins are more disorganized.

2. Work in pairs. Notice the veins as the subject's arm hangs down at the side of the body. You can easily see the veins because they are full of blood. This is usually best seen on the back of the hand and forearm. Now raise the arm above the head. Describe and explain any changes to the veins that take place.

_____

_____

_____

3. To estimate the venous pressure in the veins of the hand, have the subject stand to the left of one of the meter sticks vertically taped to the wall. Position the subject so the right lower forearm crosses the meter stick. Hold the subject's arm straight and laterally against the wall at heart level. Find a recognizable spot (or make a pen mark) on the forearm where it crosses the meter stick and read the corresponding number. Record the number in millimeters in the second row of the second column in Table 12-8 (measurement 1).

4. Raise the subject's arm slowly until the veins in the hand collapse (be sure that most of the muscles in the arm are relaxed). Read the number where the same

TABLE 12-8	Measurement of Venous Blood Pressure	
Measurement 1	_____	mm
Measurement 2	_____	mm
Measurement 2 − Measurement 1	_____	mm H₂O
_____ mm H₂O × 0.074 mm Hg/mm H₂O	_____	mm Hg

point on the forearm now crosses the meter stick. Record this number in millimeters in the third row of the second column in Table 12-8 (measurement 2).

5. Subtract measurement 1 from measurement 2 and record the result in the third row of Table 12-8. This is an estimate of the venous pressure in millimeters of water. Calculate the venous pressure in mm Hg by multiplying this number by 0.074 mm Hg/mm H₂O.

How does the venous pressure compare with typical arterial pressures?

_____

_____

6. Look at the veins of a subject's forearm and hand. The swellings that occur at various intervals are valves. Figure 12-8 shows a smaller valve in a venule. Choose a section between two swellings that does not have any side branches. Place one finger on the swelling away from the heart and with another finger press the blood forward (toward the heart) beyond the next swelling.

Does the vein fill up with blood again? _____ (yes or no)

Now remove your finger and observe what happens.

This time, does the vein fill up with blood? _____ (yes or no)

Try this again but press blood in the opposite direction away from the heart. Discuss your observations.

_____

_____

_____

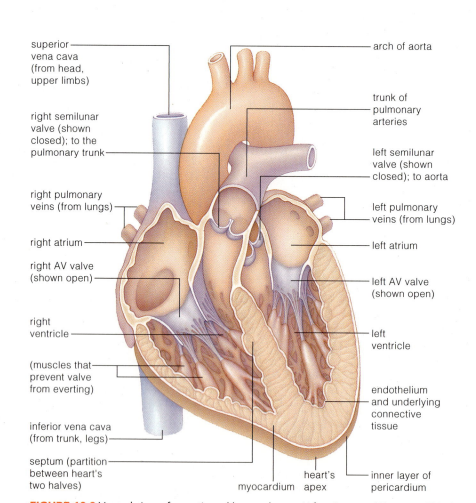

superior vena cava (from head, upper limbs)

right semilunar valve (shown closed); to the pulmonary trunk

right pulmonary veins (from lungs)

right atrium

right AV valve (shown open)

right ventricle

(muscles that prevent valve from everting)

inferior vena cava (from trunk, legs)

septum (partition between heart's two halves)

arch of aorta

trunk of pulmonary arteries

left semilunar valve (shown closed); to aorta

left pulmonary veins (from lungs)

left atrium

left AV valve (shown open)

left ventricle

endothelium and underlying connective tissue

inner layer of pericardium

myocardium

heart's apex

FIGURE 12-9 Ventral view of a sectioned human heart. (After Starr and Taggart, 2001.)

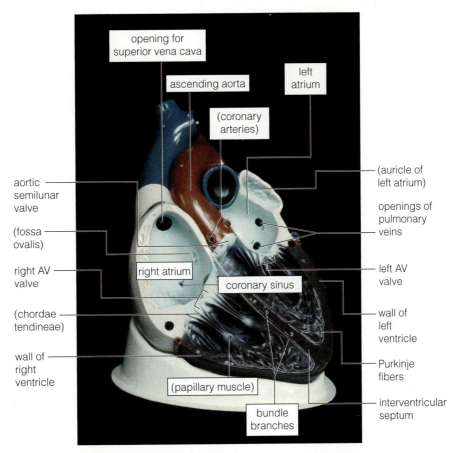

FIGURE 12-10 Model of a human heart. (*Photo by D. Morton.*)

12.3 The Heart

Normally, your heart beats about 100,000 times a day, pumping the blood around the circulatory system (Figure 12-9).

MATERIALS

Per lab group bench:

- Human heart model

Per lab section:

- Demonstration of the effects of acetylcholine and epinephrine on the heart of a double-pithed frog (no brain or spinal cord) kept moist with amphibian Ringer's solution (balanced salts solution) or a computer simulation

PROCEDURE

A. Human Heart Model

Examine a model of the human heart and identify and trace the flow through its four chambers and four valves (Figure 12-10).

1. The **right atrium** (Figures 12-9 and 12-10) receives blood from the systemic circuit via the *superior vena cava*, *inferior vena cava*, and *coronary sinus* (which drains blood from capillary beds in the heart itself).

2. When the right atrium contracts, blood is pushed through the **right atrioventricular (AV) valve** (*right AV* or *tricuspid valve*) into the **right ventricle**.

3. Contraction of the right ventricle forces this blood into the trunk of the *pulmonary arteries* to start its journey through the pulmonary circuit. The *left semilunar valve* prevents the backflow of blood back into the left ventricle.

4. The **left atrium** receives blood from the pulmonary circuit via the *pulmonary veins*.

5. Blood is forced through the **left AV valve** (*left AV* or *bicuspid valve*) into the **left ventricle** by the contraction of the left atrium.

6. Contraction of the left ventricle pushes blood into the *aorta* to begin its travel through the systemic circuit. The *aortic semilunar valve* prevents the backflow of blood into the left ventricle.

B. Demonstration of Frog Heart Function

1. Your instructor has set up a demonstration of a frog heart in the opened thorax of a frog. Alternatively, a computer program will be available in place of this procedure. Although the frog has a three-chambered heart—two atria and one ventricle—the heart's function and control are essentially the same as those of

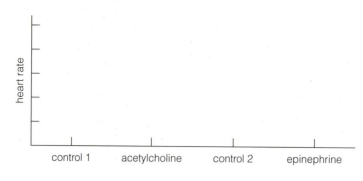

FIGURE 12-11 Bar graph for plotting heart rate data.

humans. The nervous system of the frog has been destroyed, so it does not feel pain or control heart action.

Is the heart beating? _____ (yes or no)

2. Carefully observe the beating heart. Does the entire heart muscle contract simultaneously or do some parts contract before others? _____

_____

Is there a pattern in the way each contraction sweeps across the heart or is it totally random as to when a particular part contracts?

_____

_____

If you observe carefully, you can see the order in which the chambers contract. Record your observations.

_____

_____

_____

The primary pacemaker of the heart, the **sinoatrial node**, is located in the right atrium. It functions independently of the nervous system, firing rhythmically. Each time the sinoatrial node fires, it initiates a message to contract. This message spreads over the atria. Then special heart cells amplify and conduct the message throughout the ventricle.

3. Count how many times the heart contracts in 15 seconds, record it in Table 12-9 as the control 1 rate, and then calculate the heart rate.

4. Note when your instructor places several drops of an acetylcholine solution on the heart. After 1 minute passes, repeat step 3, recording the count in row 2.

5. Watch your instructor thoroughly flush the thoracic cavity with balanced salt solution, wait 3 minutes, and then determine the control 2 heart rate. Now

TABLE 12-9	Effect of Acetylcholine and Epinephrine on the Heart Rate of a Frog	
Condition	**15-sec Counts**	**Heart Rate (beats/min)**
Control 1	_____ × 4 =	
After acetylcholine	_____ × 4 =	
Control 2	_____ × 4 =	
After epinephrine	_____ × 4 =	

observe as your instructor places several drops of an epinephrine solution on the heart. After 1 minute, repeat step 3, recording the count in row 4.

6. Plot your results on Figure 12-11.

7. Describe the effects of acetylcholine and epinephrine on the heart rate.

_____

_____

_____

_____

In an intact frog, the heart rate is modified by input from the central nervous system. The heart rate is affected by the amount of **acetylcholine** and **norepinephrine** secreted by neurons around the sinoatrial node. During restful activities, acetylcholine slows the heart rate and thus acts as a brake on the sinoatrial node. Pain, strong emotions, the anticipation of exercise, and the fight-or-flight response can all increase the secretion of norepinephrine. Norepinephrine, which has a molecular structure and action similar to epinephrine, speeds the heart rate and thus acts as an accelerator on the sinoatrial node and can override the brake.

_____ 1. The number of circuits in the circulatory systems of humans is
(a) one
(b) two
(c) three
(d) four

_____ 2. Blood contains
(a) dissolved gases
(b) dissolved nutrients
(c) dissolved hormones
(d) all of the above

_____ 3. Red blood cells are
(a) erythrocytes
(b) leukocytes
(c) platelets
(d) all of the above

_____ 4. The most common leukocyte in the blood is
(a) a lymphocyte
(b) an eosinophil
(c) a basophil
(d) a neutrophil

_____ 5. The cellular fragments in the blood that function in blood clotting are
(a) erythrocytes
(b) leukocytes
(c) platelets
(d) none of the above

_____ 6. Blood vessels that return blood from capillaries back to the heart are
(a) arteries
(b) veins
(c) portal veins
(d) arterioles

_____ 7. Blood vessels that connect networks of capillary beds are
(a) arteries
(b) veins
(c) portal veins
(d) both b and c

_____ 8. From which chamber of the heart does the right ventricle receive blood?
(a) right atrium
(b) left atrium
(c) left ventricle
(d) none of the above

_____ 9. How many chambers does the frog heart have?
(a) one
(b) two
(c) three
(d) four

_____ 10. The primary pacemaker of the heart is the
(a) bicuspid valve
(b) tricuspid valve
(c) aorta
(d) sinoatrial node

EXERCISE 12 Human Blood and Circulation

POST-LAB QUESTIONS

12.1 Blood

1. Name and give the staining characteristics and functions of the three leukocytes with specific granules.

   a.

   b.

   c.

2. Describe the shape, content, and function of an erythrocyte.

3. Why can't you give type A blood to a type B patient?

12.2 Blood Vessels

4. Describe how to take someone's hematocrit.

5. Identify the indicated blood vessel at the bottom of the left column of this page.

6. Pretend you are an erythrocyte in the right atrium of the heart. Describe one trip through the human circulatory system, ending back where you started.

12.3 The Heart

7. Explain how the heart of a double-pithed frog can continue to contract in an organized manner after the nervous system is destroyed.

8. How is the heart rate controlled by the nervous system in an intact organism?

Food for Thought

9. Considering its relationship with the other systems of the body, what is the importance of the circulatory system?

10. Search the Internet for sites about arteriosclerosis. List two web sites and briefly summarize their contents.

    http://

    http://

    _____

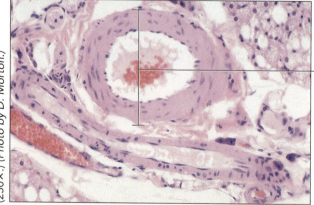

(250×.) (Photo by D. Morton.)

Human Respiration

OBJECTIVES

After completing this exercise, you will be able to

1. define *ventilation*, breathing, *inhalation*, *exhalation*, *cohesion*, *tidal volume*, *inspiratory reserve volume*, *expiratory reserve volume*, *residual volume*, *vital capacity*, and *chemoreceptor*.

2. list the skeletal muscles used in breathing and give the specific function of each.

3. trace the flow of air through the organs and structures of the respiratory system.

4. explain how air moves in and out of the lungs during human breathing.

5. distinguish between negative-pressure inhalation and positive-pressure exhalation.

6. describe the relationship between vital capacity and lung volumes and the interrelationships among lung volumes.

7. explain the importance of carbon dioxide concentration in the blood and other body fluids to the control of respiration.

Introduction

Cellular respiration breaks down glucose, consumes oxygen (O_2) and produces carbon dioxide (CO_2) and water (H_2O) to fuel cellular activities. Assuming glucose supplies are adequate, for cellular respiration to continue, O_2 must be replenished and CO_2 removed from cells by the process of diffusion.

The efficiency of diffusion to transport substances is great over short distances but decreases rapidly as distance increases. However, there are organisms of different sizes—from one-celled species to the largest living animal, the blue whale. Because diffusion works well only over short distances, animals about the size of earthworms and larger have circulatory systems that move dissolved gases around the body. Animals a little larger also have respiratory systems.

In humans and other mammals, O_2 uptake and the removal of excess CO_2 occur by diffusion across the moistened thin membranes lining millions of alveoli (singular, *alveolus*) and the thin walls of surrounding capillaries located in the lungs (Figure 13-1). These animals are protected from excessive water loss via evaporation from the very large, moist respiratory surface by having the lungs positioned inside the body. The main function of the rest of the respiratory system is **ventilation**—the exchange of gases between the lungs and the atmosphere. **Breathing,** the movement of gases in (**inhalation**) and out (**exhalation**) of the respiratory system, requires the rhythmical contraction of skeletal muscles.

13.1   Breathing

Before continuing with this exercise, let us review some anatomical terms. The trunk of the body is divided into an upper *thorax*, which is supported by the rib cage and contains the *thoracic cavity*, and the lower *abdomen*. The thoracic and abdominopelvic cavities are separated by a partition of skeletal muscle called the *diaphragm*. The thoracic cavity contains two *pleural cavities*, which contain the lungs, and the *pericardial cavity* surrounding the heart.

The muscles of breathing and their roles in inhalation and exhalation are listed in Table 13-1.

The external and internal intercostal muscles are located between the ribs. Contraction of the diaphragm increases the size of the thoracic cavity by lowering its floor. Relaxation of the diaphragm allows it to spring back to its original position. Contraction of the abdominal muscles squeezes the internal organs, pushing up the diaphragm to further decrease the size of the thoracic cavity.

MATERIALS

Per student:

- Two pieces of paper (each 14 × 21.5 cm—half of a sheet of notebook paper)

Per group (2):

- Metric tape measure
- Large caliper with linear scale (for example, Collyer pelvimeter)

Per group (4):

- Functional model of lung

NASAL CAVITY
Chamber in which air is moistened, warmed, and filtered and in which sounds resonate

PHARYNX (THROAT)
Airway connecting nasal cavity and mouth with larynx; enhances sounds; also connects with esophagus

EPIGLOTTIS
Closes off larynx during swallowing

LARYNX (VOICE BOX)
Airway where sound is produced; closed off during swallowing

TRACHEA (WINDPIPE)
Airway connecting larynx with two bronchi that lead into the lungs

LUNG (ONE OF A PAIR)
Lobed, elastic organ of breathing; enhances gas exchange between internal environment and outside air

BRONCHIAL TREE
Increasingly branched airways starting with two bronchi and ending at air sacs (alveoli) of lung tissue

ORAL CAVITY (MOUTH)
Supplemental airway when breathing is labored

PLEURAL MEMBRANE
Double-layer membrane that separates lungs from other organs; the narrow, fluid-filled space between its two layers has roles in breathing

INTERCOSTAL MUSCLES
At rib cage, skeletal muscles with roles in breathing

DIAPHRAGM
Muscle sheet between the chest cavity and abdominal cavity with roles in breathing

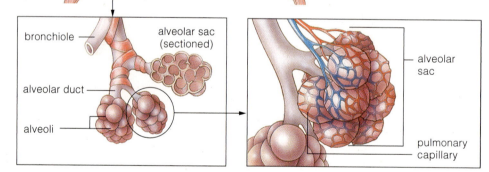

bronchiole

alveolar sac (sectioned)

alveolar duct

alveoli

alveolar sac

pulmonary capillary

FIGURE 13-1 Human respiratory system. *(After Starr and Taggart, 2001.)*

TABLE 13-1	Muscles Used to Breathe				

	Muscle				
Stage of Respiratory Cycle	**External Intercostals**	**Internal Intercostals**	**Diaphragm**	**Neck and Shoulder**	**Abdominal**
Restful inhalation	Relaxed	Relaxed	Contracted	Relaxed	Relaxed
Forced inhalation	Contracted	Relaxed	Contracted	Contracted	Relaxed
Restful exhalation	Relaxed	Relaxed	Relaxed	Relaxed	Relaxed
Forced exhalation	Relaxed	Contracted	Relaxed	Relaxed	Contracted

- Models of the organs and structures of the respiratory system
- Prepared slide of a section of mammalian lung

PROCEDURE

A. Ventilatory Ducts and Lungs

The respiratory system consists of the lungs and the ducts that shuttle air between the atmosphere and the lungs. Its organs and structures from the outside in are the two **external nares** (nostrils); the **nasal cavity** (partitioned into right and left sides by a **nasal septum**); the **pharynx** (shared with the digestive tract); the **glottis** (and its cover the **epiglottis**); the **larynx** (voice box); the **trachea** (windpipe); and two **bronchi** (singular, bronchus) and its branches, which terminally open into myriad **alveoli** in the **lungs.** Along with respiration, sound production is a major function of the respiratory tract.

1. Look at the various models of respiratory tract organs and structures and identify as many of the preceding boldfaced terms as possible.

2. Get your compound light microscope and examine a lung section at high power; identify the alveoli (Figure 13-2).

B. Ventilation

All flow occurs down a pressure gradient. When you let go of an untied inflated balloon, it flies away, propelled by the jet of air flowing out of it. The air flows out because the pressure is higher inside the balloon. The high pressure inside the balloon is maintained by the energy stored in its stretched elastic wall.

When the thoracic cavity expands during inhalation, first the pressure in the pleural sacs decreases, and then the pressure within the lungs decreases. Because the pressure outside the body is now higher than that in the lungs, and assuming the connecting *ventilatory ducts* (trachea and so on) are not blocked, air flows into the lungs (Figure 13-3). This is called **negative-pressure inhalation.**

The opposite occurs during exhalation. The size of the thorax and pleural sacs decreases, the pressure in the lungs increases, and air flows out of the body down its pressure gradient. This is called **positive-pressure exhalation.**

The pressure in the pleural sacs is actually always below atmospheric pressure, which means the lungs are always partially inflated after birth. Thus, a hole in a pleural sac or lung will result in a collapsed lung. **Cohesion** (sticking together) of the wet pleural membranes lining the outsides of the lungs and insides of the body walls of the pleural cavities aids inhalation. Exhalation depends in part on the *elastic recoil* (similar to letting go of a stretched rubber band) of lung tissue.

1. Work in groups of four. Look at the functional lung model. The "Y" tube is analogous to the ventilatory ducts. The balloons represent the lungs. The space within the transparent chamber represents the thoracic spaces and its rubber floor, the muscular diaphragm.

2. Pull down the rubber diaphragm. Describe what happens to the balloons.

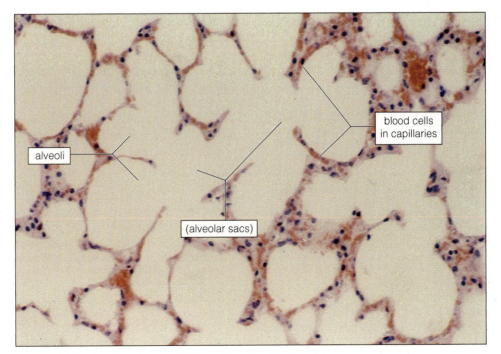

FIGURE 13-2 Alveoli in the lung (500×). *(Photo by D. Morton.)*

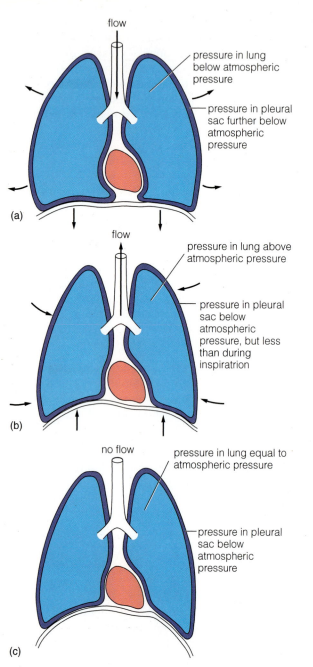

(a)
flow

pressure in lung below atmospheric pressure

pressure in pleural sac further below atmospheric pressure

(b)
flow

pressure in lung above atmospheric pressure

pressure in pleural sac below atmospheric pressure, but less than during inspiratrion

(c)
no flow

pressure in lung equal to atmospheric pressure

pressure in pleural sac below atmospheric pressure

FIGURE 13-3 Changes in the thoracic cavity during (a) negative pressure inhalation and (b) positive pressure exhalation, and corresponding movements of air. (c) The thoracic cavity at the end of an expiration. *(After Weller and Wiley, 1985.)*

As you pull down the rubber diaphragm, does the volume of the space in the container increase or decrease?

_____

As the volume changes, does the pressure in the container increase or decrease?

_____

As the balloons inflate, does the volume of air in the balloons increase or decrease?

_____

Why do the balloons inflate?

_____

_____

_____

_____

_____

_____

3. Push up on the rubber diaphragm. Describe what happens to the balloons and why it takes place.

_____

_____

_____

_____

_____

4. Pull the rubber diaphragm down and push it up several times in succession to simulate breathing.

5. Pucker up your lips and inhale. As you inhale, place one piece of paper directly over your lips. What occurs?

_____

_____

The negative pressure created in your lungs by the contraction of the muscles of inhalation causes this suction.

6. Fold the narrow ends of the two pieces of paper to produce 2- to 3-cm flaps. Open the flaps and use them as handles. Hold a piece of paper with each hand and touch the papers' flat surfaces together in front of you (Figure 13-4). Pull them apart.

Thoroughly wet both pieces of paper with water and again touch their flat surfaces together in front of you. Pull them apart. What difference did the water make?

_____

_____

C. Breathing Movements

1. Place your hands on your abdomen and make a deep inhalation followed by a deep exhalation followed by another deep inhalation and so on for three complete breaths. What occurs to the abdomen, if anything, during each inhalation?

_____

_____

What occurs to the abdomen during each exhalation?

FIGURE 13-4 Use of two pieces of paper to demonstrate cohesion. *(Photo by D. Morton.)*

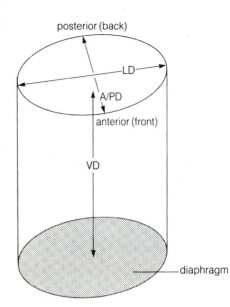

FIGURE 13-5 Thoracic diameters: LD, lateral diameter; A/PD, anterioposterior diameter; and VD, vertical diameter.

---

2. Place your hands on your chest and repeat step 1. What occurs to the size of the thorax during each inhalation?

What occurs to the size of the thorax during each exhalation?

D. Measurements of the Thorax

The size of the thorax can be described by three so-called diameters: the lateral diameter (LD), the anteroposterior diameter (A/PD), and the vertical diameter (VD; Figure 13-5). The VD is the only one that can not be measured easily.

Make the following observations and record them in Table 13-2.

1. Work in pairs. Take turns measuring the circumference of each other's chest with a tape measure at two levels, under the armpits (axillae, C_{AX}) and at the lower tip of the sternum (xiphoid process, C_{XP}) for the conditions listed in Table 13-2. While the measurements are being taken, it's extremely important not to tense any muscles other than those used for respiration. For example, do not raise the arms.

TABLE 13-2	Chest Measurements (cm)				
Subject	**Condition**	**CAX**	**CXP**	**A/PD**	**LD**
You	At the end of a restful inhalation				
	At the end of a restful (passive) exhalation				
	At the end of a forced (maximum) inhalation				
	At the end of a forced (maximum) exhalation				
Your lab partner	At the end of a restful inhalation				
	At the end of a restful (passive) exhalation				
	At the end of a forced (maximum) inhalation				
	At the end of a forced (maximum) exhalation				

2. With calipers, also measure the A/PD from the center of the back to the center of the chest at the nipple line for these same conditions. The distance between the tips of the calipers is read off the scale in centimeters. Likewise, measure the LD from one side to the other at the nipple line.

3. About two-thirds of the air inhaled during a restful inhalation is attributable to contraction of the diaphragm. Interpret the data in Table 13-2 and in your own words describe changes in the size of the thorax during:

    (a) A restful inhalation:

CAX _____

CXP _____

A/PD _____

LD _____

    (b) A passive exhalation:

CAX _____

CXP _____

A/PD _____

LD _____

    (c) A forced inhalation:

CAX _____

CXP _____

A/PD _____

LD _____

    (d) A forced exhalation:

CAX _____

CXP _____

A/PD _____

LD _____

Does the size of the thorax change significantly during a restful inhalation? _____ (yes or no)
Does the size of the thorax change significantly during a passive exhalation? _____ (yes or no)
How does the shape of the thorax change during a forced inhalation?

_____

_____

_____

_____

How does the shape of the thorax change during a subsequent forced exhalation?

_____

_____

_____

_____

13.2 Spirometry

Air in the lungs is divided into four separate volumes (Figure 13-6): tidal volume (TV), inspiratory reserve volume (IRV), expiratory reserve volume (ERV), and residual volume (RV).

Tidal volume is the volume of air inhaled or exhaled during breathing. It normally varies from a minimum at rest to a maximum during strenuous exercise.

Inspiratory reserve volume is the volume of air you can voluntarily inhale after inhalation of the TV. **Expiratory reserve volume** is the volume of air you can voluntarily exhale after an exhalation of the TV. IRV and ERV both decrease as TV increases.

Residual volume is the volume of air that cannot be exhaled from the lungs. That is, normal lungs are always partially inflated.

There are four capacities derived from the four volumes:

Inspiratory capacity (IC) = TV + IRV

Functional residual capacity (FRC) = ERV + RV

Vital capacity (VC) = TV + IRV + ERV

Total lung capacity (TLC) = Total of all four lung volumes

All of the lung volumes except the RV can be measured or calculated from measurements obtained using a simple spirometer or lung volume bag. **Vital capacity** is simply the sum of TV, IRV and ERV. A more sophisticated recording spirometer plots respiration over time (a spirogram). Figure 13-6 shows a spirogram and the relationships of the lung volumes and some of the capacities.

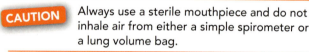
CAUTION   Always use a sterile mouthpiece and do not inhale air from either a simple spirometer or a lung volume bag.

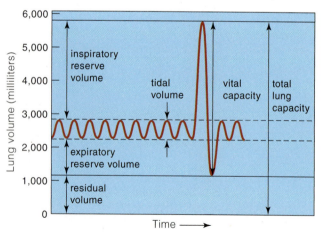

FIGURE 13-6 Spirogram. There are nine passive breaths followed by a forced inhalation and a forced exhalation, and then two more passive breaths. *(Starr and McMillan, 2010.)*

MATERIALS

Per group (4):

- Noseclip (optional)
- Simple spirometer or lung volume bags

PROCEDURE

1. Work in groups of four. Sit quietly and breathe restfully. Use a noseclip or hold your nose. After you feel comfortable, start counting as you inhale. After the fourth inhalation, exhale normally into the spirometer or lung volume bag. Read the volume indicated by the spirometer or squeeze the air to the end of the lung volume bag and read the volume from the wall of the bag. Record the volume below (trial 1). Reset the spirometer or squeeze the air out of the lung volume bag. Repeat this procedure two more times (trials 2 and 3) and calculate the total and average TV at rest.

   Trial 1 _____ mL Trial 3 _____ mL

   Trial 2 _____ mL Total = _____ mL

   Divide the total by 3 = _____ mL to calculate the average TV at rest. Record the TV in Table 13-3.

2. Determine the volume of air you can forcibly exhale after a restful inhalation (average of three trials).

   Trial 1 _____ mL Trial 3 _____ mL

   Trial 2 _____ mL Total = _____ mL

   Divide the total by 3 = _____ mL. This is the average sum of ERV and TV at rest.

3. Determine the volume of air you can forcibly exhale after a forceful inhalation (average of three trials).

   Trial 1 _____ mL Trial 3 _____ mL
   Trial 2 _____ mL Total = _____ mL

   Divide the total by 3 = _____ mL to calculate the average VC. Record VC in Table 13-3.

4. Calculate the ERV at rest by subtracting step 1's result from step 2's result.

   _____ mL (step 2) − _____ mL (step 1) = _____ mL
   Record the ERV in Table 13-3.

TABLE 13-3	Vital Capacity and Lung Volumes at Rest
Measure	**Volume (mL)**
Tidal volume	_____
Inspiratory reserve volume	_____
Expiratory reserve volume	_____
Vital capacity	_____

5. Calculate the IRV at rest by subtracting step 2's result from step 3's result.

   _____ mL (step 3) − _____ mL (step 2) = _____ mL
   Record the IRV in Table 13-3.

6. Does VC change as TV increases or decreases? _____ (yes or no)

7. Measure and record your height in centimeters. _____ cm

   Write your VC/height on the board (your name is not necessary).

   Plot the vital capacity versus weight of each student in your lab section on the following graph.

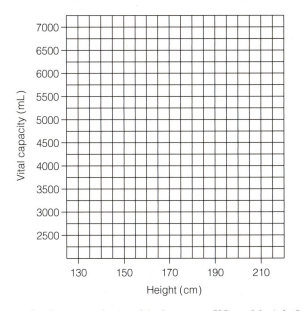

Is there a relationship between VC and height? If so, describe it mathematically using a graphing calculator or with words.

_____

_____

_____

13.3 Control of Respiration

The control of respiration, both the rate and depth of breathing, is very complex. Simply stated, **chemoreceptors** (receptors for chemicals such as O_2, CO_2, and hydrogenions, or H^+), stretch receptors in the ventilatory ducts, and centers in the brain stem (part of the brain that connects to the spinal cord) control respiration.

Your own experience has taught you that respiration is to some extent under the control of the conscious mind. We can decide to stop breathing or to breathe more rapidly and deeply. However, the unconscious mind can override voluntary control. The classic example of this is the inability to hold one's breath for more than a few minutes. In this section, you will make further pertinent observations related to the hypothesis that CO_2 concentration in the blood and other body fluids is the most important stimulus for the control of respiration.

CAUTION Do not do the following activities if you have any medical problems with your lungs or heart. All subjects should stop immediately if they feel faint.

MATERIALS

Per student pair:

- Small mirror

Per lab room:

- A safe place to exercise
- Clock with a second hand

PROCEDURE

1. Work in pairs. Write two predictions in Table 13-4. In prediction 1, forecast the effects on respiratory rate and the ability to hold one's breath after *hyperventilation* (overventilating of the lungs by forced, rapid and deep breathing). (For example, if I hyperventilate, then. . .). In prediction 2, forecast the effects on respiratory rate and the ability to hold one's breath after exercise.

2. Sit down, and after you feel comfortable, have your lab partner count the number of times you breathe in 3 minutes. If it is difficult to see you breathe, have your partner place a small mirror under your nose. Record the number of breaths: _____ breaths

   Divide this number by 3 to calculate the average respiratory rate *at rest* and write it in the third column of Table 13-4.

3. Determine how many minutes you can hold your breath after a restful inhalation and record the time in the fourth column of Table 13-4.

4. Now hyperventilate--breathe as rapidly and deeply as possible. Try to take at least 10 breaths, but stop as soon as you can answer the following question. (In any case, do not continue for more than 20 breaths.) As times goes on, does it become easier or more difficult to continue rapid deep breathing?

   _____

   Immediately have your lab partner count and record the number of breaths you take in the next 3 minutes: _____ breaths

5. Now determine how many minutes you can hold your breath *immediately* after hyperventilating. Record this time in the fourth column of Table 13-4.

6. Divide the number in step 4 by 3 to calculate the average respiratory rate *after hyperventilation* and write it in the third column of Table 13-4. Does hyperventilation increase, decrease, or have no effect on the CO_2 concentration of the blood?

   _____

7. When fully recovered, carefully run in place for 2 minutes in the area designated by your lab instructor. Immediately after sitting down, again have your lab partner count how many breaths you take in 3 minutes and record it: _____ breaths

   Now determine how long you can hold your breath *immediately* after 2 minutes of exercise and record it in the fourth column of Table 13-4.

   As you did earlier, calculate the average respiratory rate *after exercise* and record it in the third column of Table 13-4. Does running in place increase or decrease the CO_2 concentration of the blood? _____

TABLE 13-4	Respiratory Data		
Prediction 1:			
Prediction 2:			
	Condition	**Respiratory Rate (breaths/min)**	**Breath Holding (min)**
You	At rest		
	After hyperventilation		
	After exercise		
Your lab partner	At rest		
	After hyperventilation		
	After exercise		
Conclusions:			

What causes the CO_2 concentration to change while you are running in place?

_____

_____

_____

8. Switch roles with your lab partner and repeat steps 1 to 7, recording these data in Table 13-4.

9. Look over Table 13-4 and summarize your results:

_____

_____

_____

_____

Write a conclusion in Table 13-4 as to whether your results supported your predictions.

13.4 Experiment: Physiology of Exercise

When an animal increases its activity, its muscles need more O_2 to support their higher level of cellular metabolism. Increasing both breathing and the circulation of blood through the lungs, heart, and skeletal muscles enhances O_2 delivery.

From observing our own bodies, we know that exercise is accompanied by increases in the rate of breathing, the depth of breathing, and the heart rate. Intuitively, we also expect a similar elevation in arterial blood pressure because this would increase the blood pressure difference between the elastic arteries near the heart and the arterioles near the capillaries. This steeper pressure gradient would result in a higher blood velocity—the speed at which blood flows—just as a boulder rolls faster down a steeper hill. Higher blood velocity along with increased activity of the respiratory system means more O_2 is transported to the heart and skeletal muscles.

So, how do heart rate, systolic blood pressure, respiratory rate, and TV change as the intensity of exercise increases? This experiment addresses the hypothesis that *all of these factors will increase as the intensity of exercise increases to contribute to delivering more and more O_2 to the skeletal muscles.*

MATERIALS

Per student:

- Calculator

Per lab room:

- Television or monitor
- Video cassette or DVD player
- Video cassette or DVD: "Experiment: Biology; The Physiology of Exercise"

PROCEDURE

As a class, you will watch a video of a young man on a stationary bicycle. After recording his heart rate, systolic blood pressure, respiratory rate, and TV at rest, you will record data for these same observations after four periods of work—pedaling against a constant resistance at 10, 15, 20, and 25 km/hr.

1. State a prediction for this experiment and write it in Table 13-5.

2. Watch the video and record the data in Table 13-5. Certain numbers need to be multiplied by 10 or 100 to convert them to the units used in the table. However, do not convert any of the values read off the various meters in the video because the conversion factors are included in the table.

3. List the experimental variables.

Independent variable _____

Dependent variables _____

TABLE 13-5	Physiology of Exercise			
Prediction:				
Intensity of Exercise	**Heat rate (beats/min)**	**Systolic Blood Pressure (mm Hg)**	**Respiratory Rate (breaths/min)**	**Tidal Volume (L)**
0 km/hr (rest)	____ × 10	____	____ × 10	____
10 km/hr	____ × 10	____	____ × 10	____
15 km/hr	____ × 100	____	____ × 10	____
20 km/hr	____ × 100	____	____ × 10	____
25 km/hr	____ × 100	____	____ × 10	____
Conclusion:				

Controlled variables (as many as you can identify)

_____

4. Plot these results in the following graphs:

5. Let's analyze the respiratory data further.

At 10 km/hr, did respiratory rate increase? _____
(yes or no)

Did TV increase? _____ (yes or no)

6. During which period of work did the respiratory rate first increase? _____ km/hr

7. Calculate the amount of air exhaled per minute during the most strenuous exercise period (25 km/hr).

$$\frac{Breaths}{min} \times \frac{L}{Breath} = \frac{L}{min}$$

This is the respiratory minute volume. This value is very close to this subject's maximum respiratory minute volume.

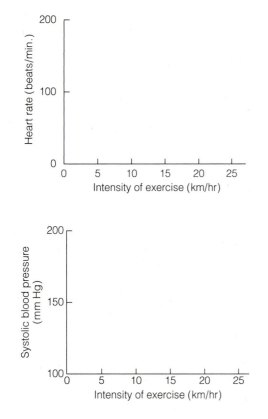

_____ **1.** Which muscles may contract during inhalation?
(a) external intercostals
(b) internal intercostals
(c) abdominal
(d) both b and c

_____ **2.** Which muscles contract during a restful exhalation?
(a) external intercostals
(b) internal intercostals
(c) diaphragm
(d) none of the above

_____ **3.** Which muscles may contract during a forceful exhalation?
(a) external intercostals
(b) diaphragm
(c) abdominal
(d) both b and c

_____ **4.** An untied inflated balloon flies because
(a) the pressure is higher inside than outside the balloon
(b) the pressure is lower inside than outside the balloon
(c) air flows down its pressure gradient
(d) both a and c occur

_____ **5.** Human ventilation is
(a) negative-pressure inhalation
(b) positive-pressure inhalation
(c) negative-pressure exhalation
(d) both b and c

_____ **6.** The amount of air breathed in and out of the lungs at any given time is the
(a) tidal volume
(b) inspiratory reserve volume
(c) expiratory reserve volume
(d) residual volume

_____ **7.** Vital capacity is always equal to
(a) tidal volume
(b) inspiratory reserve volume
(c) expiratory reserve volume
(d) a + b + c

_____ **8.** An instrument that measures lung volumes is a
(a) caliper
(b) spirometer
(c) barometer
(d) stethoscope

_____ **9.** Respiration is controlled by
(a) chemoreceptors
(b) stretch receptors
(c) centers in the brain stem
(d) all of the above

_____ **10.** The most important stimulus in the control of respiration is the concentration in the blood and other body fluids of
(a) oxygen (O_2)
(b) carbon dioxide (CO_2)
(c) hydrogen ions (H^+)
(d) nitrogen (N_2)

EXERCISE 13 Human Respiration

POST-LAB QUESTIONS

13.1 Breathing

1. Which skeletal muscles are contracted during

   a. restful inhalation?

   b. forced inhalation?

   c. restful exhalation?

   d. forced exhalation?

2. How does the size of the thorax change during

   a. inhalation?

   b. exhalation?

3. How does the potential volume of the pleural sacs change during

   a. inhalation?

   b. exhalation?

4. Define these terms:

   a. negative-pressure inhalation

   b. positive-pressure exhalation

5. Define *residual volume*. What prevents all of the gas in the lungs from being expelled during a forced exhalation?

13.2 Spirometry

6. Explain the relationship among vital capacity, tidal volume, inspiratory reserve volume, and expiratory reserve volume.

13.3 Control of Respiration

7. What substance is the most important stimulus in the control of respiration? How is its production linked to changes in metabolic rate, such as occur during exercise?

13.4 Experiment: Physiology of Exercise

8. Give an explanation for the fact that athletes' resting heart rates are usually slower than the average rate for healthy humans.

Food for Thought

9. Explain why hyperventilation can prolong the time you can hold your breath. Can this be dangerous (e.g., hyperventilation followed by swimming under water)?

10. Search the Internet for sites about emphysema and chronic obstructive pulmonary disease (COPD). List two websites below and briefly summarize their contents.

   http://

   http://

Human Sensations, Reflexes, and Reactions

OBJECTIVES

After completing this exercise, you will be able to

1. define *sensory receptors, stimuli, sensory neurons, motor neurons, effectors, interneurons, integration, sensations, perception, proprioception, modality, free neuron endings, encapsulated neuron endings (Meissner's and Pacinian corpuscles), sensory projection, phantom pain, sensory adaptation, reflex, reflex arc, stretch reflexes, patella reflex, muscle spindle, pupillary reflex, swallowing reflex,* and *reaction.*

2. describe the flow of information through the nervous system.

3. state the nature and function of sensations.

4. describe a stretch reflex.

5. describe the pupillary reflex.

6. distinguish between a reflex and a reaction.

7. measure visual reaction time.

Introduction

Interactions between parts of your body and between your body and the external environment depend on the uninterrupted flow of information through the nervous system (Figure 14-1).

Sensory receptors detect changes in the energy of variables **(stimuli)** such as an increase in blood pressure or a decrease in skin temperature. The sensory receptors stimulated are typically those most sensitive to the particular stimulus type. **Sensory neurons** transmit this information to the brain and spinal cord, from which **motor neurons** transmit information directing an appropriate response to the **effectors,** which actually respond. The brain and spinal cord together make up the central nervous system (CNS).

Effectors are muscles or glands. *Somatic motor neurons* control skeletal muscles that usually respond to the external environment. *Autonomic motor neurons* control smooth muscles, cardiac muscle, and glands that usually respond within the body. The processes of sensory and motor neurons along with connective tissue are located in the *nerves* of the peripheral nervous system (PNS).

Within the CNS, sensory neurons can directly stimulate motor neurons; more frequently, though, the sensory neurons stimulate one or more **interneurons,** which lie entirely within the CNS. In turn, these interneurons stimulate the motor neurons, other interneurons, or both. The sum of all of these interconnections allows for **integration,** a process during which sensory and other information is processed and appropriate actions taken.

14.1 Sensations

The *conscious mind*, which is located in the forebrain, feels some of the sensory information as **sensations**. The brain understands this information (**perception**) and uses it to initiate responses (e.g., avoid a source of pain or find water and drink when thirsty).

Our bodies have sensory receptors for light, sound, chemicals, temperature, tissue damage, and mechanical displacement. Sensations we feel include sight, hearing, taste, smell, hot, cold, pain, touch, pressure, vibration, equilibrium, and **proprioception** (knowledge of the position and movement of the various body parts). There are also a number of complex sensations such as thirst, hunger, and nausea.

Sensory receptors and the sensations they produce have three characteristics: *modality, sensory projection,* and *sensory adaptation.* These characteristics can be easily demonstrated by investigating the skin's sensory receptors.

MATERIALS

Per student pair:

- Compound microscope
- Prepared slide of mammalian skin stained with hematoxylin and eosin (containing Meissner's corpuscles)
- Felt-tip, nonpermanent pen

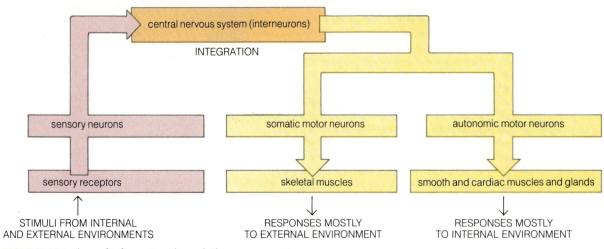

FIGURE 14-1 Flow of information through the nervous system.

- Bristle (those from a moderately sized house paint brush will do)
- Dissecting needle
- Scientific calculator
- Two blunt probes in a 250-mL beaker of ice water
- Two blunt probes in a 250-mL beaker of hot tap water (the hot water will have to be replenished every 5 minutes)
- Ice bag
- Camel-hair brush
- Reflex hammer
- Three 1000-mL beakers containing
  - Ice water
  - 45°C water
  - Room-temperature water
- Tissue paper

Per lab room:

- Demonstration slide of a Pacinian corpuscle

PROCEDURE

A. Modality

Modality is the particular sensation that results from the stimulation of a particular sensory receptor. For example, the modalities of taste—bitter, sour, salty, sweet, and umami (savory)—are associated with five different types of taste buds. However, although every sensory receptor is most sensitive to one type of stimulus, the modality actually depends on where in the brain the sensory neurons from the sensory receptor (or the interneurons to which they connect) terminate in the brain. Modality cannot be encoded in the messages carried by sensory neurons, because every impulse (action potential) in that message is identical. The only information neurons transmit is the absence or presence of a stimulus and (if one is present) its intensity—low-stimulus intensities

produce a low frequency of impulses, and high-stimulus intensities produce a high frequency of impulses.

1. Examine a prepared section of skin (Figure 14-2). Two categories of sensory receptors are present: **free neuron endings and encapsulated neuron endings**.

   Free neuron endings are almost impossible to see in typically stained sections, but note their distribution in Figure 14-2. Stimulating different free neuron endings produces sensations of pain, crude touch, and perhaps cold and hot. Encapsulated neuron endings consist of neuron endings surrounded by a connective tissue capsule.

   Find **Meissner's corpuscles** in the *dermal papillae* (Figure 14-2). Meissner's corpuscles are sensory receptors for fine touch and low-frequency vibration. Now look for **Pacinian corpuscles** between the dermis and hypodermis. Pacinian corpuscles look like a cut onion and are sensory receptors for pressure and high-frequency vibration. Not all skin sections contain a Pacinian corpuscle. If you cannot locate one, look at the demonstration slide.

2. With a felt-tip, nonpermanent ink pen, draw a 25-cell, 0.5-cm grid (Figure 14-3) on the inside of your lab partner's forearm, just above the wrist.

3. You are now the investigator, and your lab partner is the subject. At this point, ask your lab partner to close his or her eyes. Using a bristle, touch the center of each box in the grid, if the bristle bends, you are pressing too hard. Ask your lab partner to announce when a touch is felt. Do not count responses given when you remove the bristle. Just count those that coincide with the initial touch. Mark each positive response with a T in the upper left-hand corner of the corresponding box in Figure 14-3.

CAUTION Do not press; simply let the tip of the dissecting needle rest on the surface of the skin.

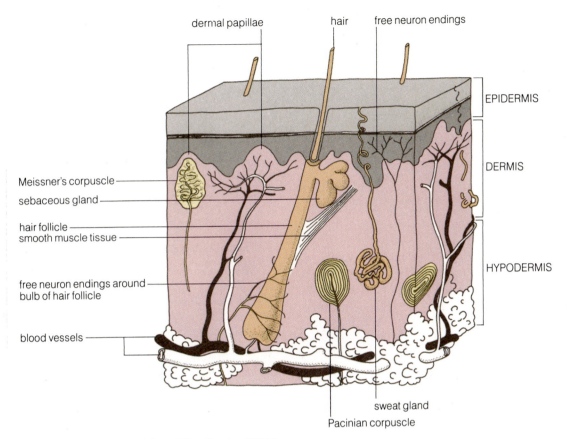

dermal papillae   hair   free neuron endings

EPIDERMIS

DERMIS

HYPODERMIS

Meissner's corpuscle

sebaceous gland

hair follicle
smooth muscle tissue

free neuron endings around
bulb of hair follicle

blood vessels

sweat gland

Pacinian corpuscle

FIGURE 14-2 Diagram of skin. *(After Fowler, 1984.)*

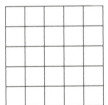

FIGURE 14-3 Grid for testing skin stimuli and recording modality data.

4. Repeat the above with a clean dissecting needle. This time, if you feel a prick, mark P for "pain" in the upper right-hand corner of the corresponding box in Figure 14-3.

5. Repeat the above with a chilled blunt probe. Before using the blunt probe, dry it with tissue paper. The blunt probe will warm up over time, so switch it with the second chilled blunt probe every five trials. This time, mark each positive response with a C for "cold" in the lower left-hand corner of the corresponding box in Figure 14-3.

6. Repeat the above with a heated blunt probe. Before using the blunt probe, dry it with tissue paper. Use the two blunt probes alternately every five trials. This time, mark each positive response with an H for "hot" in the lower right-hand corner of the corresponding box in Figure 14-3.

7. Record the total number of positive responses for each stimulus in Table 14-1. Calculate the density of sensory receptors for each modality (multiply the number of positive responses for each stimulus by 4) and record them in the third column of Table 14-1.

8. Repeat this procedure for your lab partner.

9. Does each cell in the grid contain a sensory receptor for all four modalities studied? _____ (yes or no)

10. Can you see a pattern or patterns in the distribution of positive responses marked in Figure 14-3? _____ If yes describe the pattern(s):

_____

_____

11. Are the densities for the sensory receptors for each modality the same? _____ (yes or no)

B. Sensory Projection

All sensations are felt in the brain. However, before the conscious mind receives a sensation, it is assigned back to its source, the sensory receptor. This phenomenon is called **sensory projection.** This is a very important characteristic of sensations because it allows the conscious mind to perceive the body as part of the world around it. You have probably experienced sensory projection. A common example is the "pins and needles" you feel in your hand and forearm when you accidentally jar

TABLE 14-1	Positive Identifications to Stimuli Applied to 25 Cells in a 0.25-cm² Patch of Skin	
Stimulus	**Number of Responses**	**Density (Responses/cm²)**
Touch		
Pain		
Cold		
Hot		

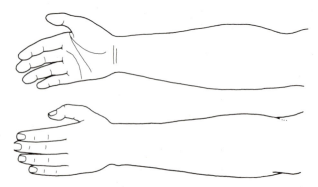

FIGURE 14-4 Front and back views of forearm and hand for recording projection data.

the nerve that passes over the inside of the elbow (so-called funny bone). The sensory neurons in the nerve are stimulated, and your brain projects the sensation back to the sensory receptors. Another example is the **phantom pain** and other sensations that recent amputees sometimes "feel" in missing limbs. This occurs because the sensory neurons that once served the missing body part are activated by the trauma of the amputation.

1. Obtain an ice bag from the freezer.

2. Hold the ice bag against the inside of your elbow for 2 to 5 minutes.

3. Describe any sensations felt in the hand or forearm for your lab partner to record them in Figure 14-4.

4. Continue to hold the ice bag on your elbow and check for any loss of sensation by your lab partner gently stroking your forearm and hand with a camel-hair brush. Sensations may also be felt after the ice bag is removed.

5. If no results are obtained, try tapping the inside of the elbow just above the funny bone with the reflex hammer.

6. Similarly test your lab partner.

7. What can you conclude about sensory projection and the sensory receptors on the surface of the hand and forearm?

_____

_____

_____

C. Sensory Adaptation

The intensity of the signal produced by a sensory receptor depends in part on the strength of the stimulus and sometimes on the degree to which the sensory receptor was stimulated before the current stimulus. Most sensory receptors undergo **sensory adaptation** to a constant stimulus over time. For example, when you first enter a dark room after being in bright light, you cannot see. After a while, your photoreceptors adapt to the new light conditions, and your vision improves.

1. Partially fill each of three 1000-mL beakers with ice water, water at room temperature, and water at about 45°C.

2. Place one hand in the ice water and one in the warm water. After 1 minute, place both hands *simultaneously* in the water at room temperature.

3. Describe the sensation of temperature in each hand to your lab partner, who should record these descriptions in Table 14-2.

4. What can you conclude about the skin sensory receptors for temperature and their capacity for sensory adaptation?

_____

_____

_____

What about the ability of other kinds of sensory receptors to adapt? Use your own experiences to answer question 4 above for touch, smell, and pain. (Hint for touch: Compare how your clothes feel after a morning shower with how they feel later.)

_____

_____

_____

_____

TABLE 14-2	Sensations Felt When Preadapted Hands Are Placed in Room-Temperature Water
Relative Temperature of Preadaption	**Result**
Cold	
Warm	

14.2 Reflexes

A **reflex** is an involuntary response to the reception of a stimulus. The simplest **reflex arc** consists of a sensory receptor, sensory neuron, motor neuron, and effector. *Involuntary* means that your conscious mind does not decide the response to the stimulus. However, the conscious mind may be aware after the reflex has taken place. Reflexes of which we are unaware occur most often in the internal environment (e.g., reflexes involved in adjustments of blood pressure).

MATERIALS

Per student pair:

- Reflex hammer
- Penlight

PROCEDURE

A. Stretch Reflexes

Stretch reflexes are an example of the simplest type because the interneurons are not directly involved (Figure 14-5). The sensory neuron connects directly with the motor neuron in the spinal cord. Stretch reflexes are important in controlling balance and complex skeletal muscular movements such as walking. Physicians often test these reflexes during physical examinations to check for spinal nerve damage. You have probably experienced one of these tests, the **patella reflex.** In this test, the sensory receptor is the **muscle spindle** in the quadriceps

femoris muscle group located on the front of the thigh, which is attached through its tendon and the patellar ligament to the top of the front surface of the tibia. The tibia is the larger of the two lower leg bones. The patella (kneecap) is embedded in the middle of the combined tendon/ligament. The muscle spindle detects any stretching of the muscle. The effector is the muscle itself.

1. Sit on a clean lab bench and shut your eyes.

2. When you are not expecting it, your lab partner's role is to tap the patella ligament with a reflex hammer (Figure 14-6). Describe the response.

_____

_____

_____

If you have trouble producing a response, repeat steps 1 and 2 but this time distract yourself by counting backward from 10.

3. Even with your eyes shut, are you aware of the stimulus and the response? _____ (yes or no) This is because of pressure sensory receptors that sense the tap and because of proprioceptors that sense movement of the leg.

4. Stretch reflexes are *somatic reflexes* because they involve somatic motor neurons and skeletal muscles. Can you willfully inhibit a stretch reflex ? _____ (yes or no)

5. Similarly test your lab partner.

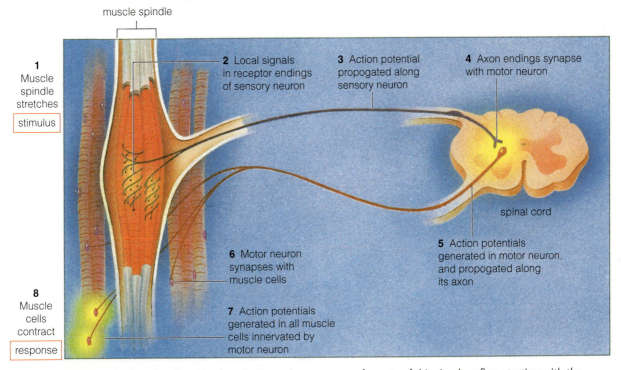

FIGURE 14-5 Parts of a stretch reflex. Numbers indicate the sequence of events of this simple reflex, starting with the stimulus—the stretching of the muscle spindle.

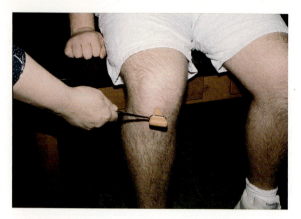

FIGURE 14-6 Area to tap to produce patella reflex. *(Photo by D. Morton.)*

B. Pupillary Reflex

1. Shine the penlight into your lab partner's eyes. Does the size of the pupil (the diameter of the opening into the eye that is surrounded by the pigmented iris) get larger or smaller? _____

2. Now turn off the penlight. Does the size of the pupil get larger or smaller? _____

3. Repeat steps 1 and 2. Which is faster, constriction of the iris (which makes the pupil smaller) or dilation of the iris (which makes the pupil larger)? _____

4. Are you aware of the pupil's changing diameter? _____ (yes or no)

5. The **pupillary reflex** is an autonomic reflex because it involves an autonomic motor neuron and in this case, smooth muscle. Can you willfully inhibit the pupillary reflex? _____ (yes or no)

6. Similarly test your lab partner.

C. Complex Reflexes

Complex reflexes involve many reflex arcs and interneurons. A good example is swallowing. The stimulus in the **swallowing reflex** is the movement of saliva, food, or drink into the posterior oral cavity. The response is swallowing.

1. Cup your hand around your neck and swallow. Feel the complex skeletal muscular movements involved in swallowing. Do you consciously control all these muscles? _____ (yes or no)

2. Test whether it is possible to swallow several times in quick succession. Can you do this? _____ (yes or no)

3. Explain this result. (Hint: It has something to do with the stimulus.)

_____

_____

_____

4. What part of swallowing does your conscious mind control, and what part is a reflex?

_____

_____

_____

14.3   Reactions

A **reaction** is a voluntary response to the reception of a stimulus. *Voluntary* means that your conscious mind initiates the reaction. An example is swatting a fly when it has landed in an accessible spot. Because neurons must carry the sensory message to the cerebral cortex and the message to react back to the motor neuron, a reaction takes more time than a reflex. *Reaction time* is the sum of the time it takes for

- The stimulus to reach the sensory receptor

- The sensory receptor to process the message

- A sensory neuron to carry the message to the integration center

- The integration center to process the information

- A motor neuron to carry the response to the effector

- The effector to respond

Visual reaction time can easily be measured with a reaction-time ruler. This device makes use of the principle of progressive acceleration of a falling object.

MATERIALS

Per student pair:

- Reaction-time ruler (Reaction Time Kit available from Carolina Biological Supply Company)

- Chair or stool

- Scientific calculator

PROCEDURE

1. Sit on a chair or stool (Figure 14-7).

2. Your lab partner stands facing you and holds the *release end* of the reaction-time ruler with the thumb and forefinger of the dominant hand at eye level or higher.

3. Position the thumb and forefinger of your dominant hand around the *thumb line* on the ruler. The space between the thumb and forefinger should be about 1 inch.

4. Tell your lab partner when you are ready to be tested.

FIGURE 14-7 Two students measuring visual reaction time.

TABLE 14-3 Reaction-Time Data		
Trial	Subject 1	Subject 2
1		
2		
3		
4		
5		
6		
7		
8		
9		
10		
Total		
Average (Total/10)		

5. Any time during the next 10 seconds, your lab partner releases the ruler.

6. Catch the ruler between the thumb and forefinger as soon as it starts to fall. The line under your thumb represents visual reaction time in milliseconds.

7. Your lab partner reads your reaction time from the ruler and records it in Table 14-3.

8. Repeat steps 2 to 7 ten times and calculate the average reaction time from the 10 trials.

9. Similarly test your lab partner.

10. The reaction times of most of the 10 trials should be similar, but perhaps the first few or one at random may be relatively different from the others. If this is true for your own or your lab partner's data, suggest some reasons for this variability.

_____

_____

_____

_____

_____

11. If opportunity and interest allow, the *Reaction Time Kit Instructions* booklet has a number of suggestions for other experiments you can easily do with the reaction-time ruler.

_____ **1.** Neurons that carry messages from sensory receptors to the CNS are
(a) sensory
(b) motor
(c) interneurons
(d) both a and b

_____ **2.** Neurons that carry messages from the CNS to effectors are
(a) sensory
(b) motor
(c) interneurons
(d) both a and b

_____ **3.** Neurons that carry messages within the CNS are
(a) sensory
(b) motor
(c) interneurons
(d) autonomic

_____ **4.** Knowledge of the position and movement of the various body parts is
(a) modality
(b) sensory projection
(c) sensory adaptation
(d) proprioception

_____ **5.** Skin contains
(a) free neuron endings
(b) encapsulated neuron endings
(c) no nervous tissue
(d) both a and b

_____ **6.** Which characteristic of sensory receptors does phantom pain illustrate?
(a) modality
(b) sensory projection
(c) sensory adaptation
(d) proprioception

_____ **7.** A simple reflex arc is made up of a sensory receptor and
(a) a sensory neuron
(b) a motor neuron
(c) an effector
(d) all of the above

_____ **8.** A stretch reflex is
(a) somatic
(b) autonomic
(c) both a and b
(d) none of the above

_____ **9.** A pupillary reflex is
(a) somatic
(b) autonomic
(c) both a and b
(d) none of the above

_____ **10.** A reaction is
(a) a reflex
(b) involuntary
(c) voluntary
(d) both a and b

EXERCISE **14** **Human Sensations, Reflexes, and Reactions**

POST-LAB QUESTIONS

Introduction

1. Where in the brain does your consciousness reside?

14.1 Sensations

2. In your own words, define the following terms:

a. modality

b. sensory projection

c. sensory adaptation

3. Identify the structure indicated in these photos.

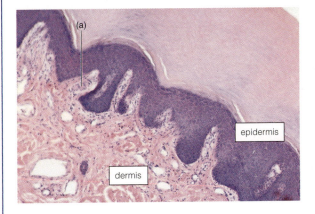

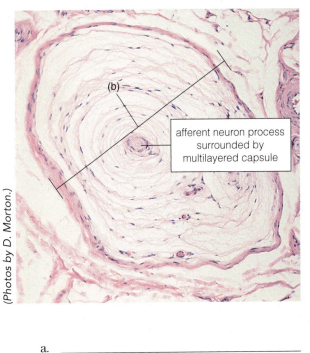

(b)

afferent neuron process surrounded by multilayered capsule

(Photos by D. Morton.)

a. _____

b. _____

14.2 Reflexes

4. Diagram the basic steps of a stretch reflex arc.

5. How does the patella reflex differ from the pupillary reflex?

14.3 Reactions

6. Indicate whether the following actions are caused by reactions or reflexes by putting a check in the correct column in the table.

Action	Reaction	Reflex
A baby wetting a diaper		
Braking a car to avoid an accident		
Withdrawing your hand from a hot stove surface		
Sneezing		
Waving to a friend across the street		

Food for Thought

7. What are the advantages and disadvantages to an organism of sensory receptor adaptation?

8. All animals do not perceive the external environment in exactly the same way. List some examples from your own knowledge and readings in the textbook.

9. To survive, an animal needs all its sensory receptors working, and even then, it cannot fully sense the external environment. What extra sensory receptors do you think would be advantageous to the survival of humans in this modern world?

10. Why is it advantageous for organisms *not* to be consciously aware of all the activity in their nervous systems?

15

Structure and Function of the Sensory Organs

OBJECTIVES

After completing this exercise, you will be able to

1. define *visual acuity, myopia, hyperopia, astigmatism, near point of eye,* and *presbyopia.*

2. explain the differences among photoreceptors, mechanoreceptors, and chemoreceptors.

3. recognize and describe the structures of the external eye, eyeball, and ear.

4. determine visual acuity and the presence or absence of astigmatism.

5. describe the various reflexes important to proper vision.

6. recognize the retina, organ of Corti, taste buds, and olfactory epithelium.

7. explain briefly the function of rods and cones.

8. describe afterimages.

9. explain how we sense the source of a sound.

10. discuss the similarities and differences between the senses of taste and smell.

Introduction

What we know about the world around us is the direct result of the activity of our sensory receptors in sensory organs and structures on the surface of our bodies, especially our heads. These sensory organs and structures contain sensory receptors that receive information about the form and amount of light, atmospheric sound waves, movements of our heads, chemicals that enter our mouths and nostrils, and other stimuli. They contain specialized neurons and sometimes accessory cells that process this information and transform it into bioelectrical energy. This bioelectrical energy produces impulses in sensory neurons that extend from the sensory receptors along neuron pathways to particular parts of our brains, where the impulses are received and perceived as particular sensations. Typically, the frequency of the impulses is interpreted as intensity. Being aware of the nature and limitations of our senses is crucial to understanding ourselves.

15.1 The Eyes and Vision

Of all our senses, vision is the most developed. The retina contains **photoreceptors** (sensory receptors sensitive not only to light intensity but also to particular groups of wavelengths of light, which the brain interprets as color). You will discover that the eye is similar to a digital camera in structure. For example, our eyes use a lens to focus an image on an array of light-sensitive receptors.

MATERIALS

Per student:

- Compound light microscope
- Section of eye with retina and optic nerve

Per student pair:

- Two sheets of white paper, one blank and one with dime-sized black dot; black and colored construction paper; a white paper dot cut-out and cut-outs of different colored paper shapes (heart, star, and so on)

- Mirror
- Preserved sheep eye
- Dissecting scissors
- Blunt probe or dissecting needle
- Forceps
- Scalpel
- Measuring tape
- Penlight

Per student group (4):

- Eye model

Per lab room:

- Liquid waste disposal bottle
- Boxes of different sizes of lab gloves
- Safety goggles
- Snellen chart
- Astigmatism chart
- Place to stand with a taped line 10 ft away and a red taped line 20 ft away from charts

PROCEDURE

A. External Eye Structures

The surface of the eyes is kept moist by *lacrimal fluid* (tears) secreted by the *lacrimal gland,* which is located under the skin above and just lateral to the center of the visible portion of the eyeball (Figure 15-1). Its ducts open under the lateral third of the upper *eyelid.* This fluid is swept across the visible surface of the eyeball

when we blink. Excess fluid drains into the openings (*lacrimal puncta*) of ducts located at the tip of small elevations (*lacrimal papillae*) situated on the medial rims of the eyelids. These canals join to the *nasolacrimal sac,* which connects via the *nasolacrimal duct* to the *nasal cavity.*

1. *Wash your hands thoroughly.* Consult Figure 15-1 and examine one of your eyes with a mirror. Identify the **eyelids, eyelashes, lacrimal papillae, lacrimal punta,** and **lacrimal caruncle** (a reddish body located at the inner corner of each eye that contains lubricating glands). From your own experience, suggest an additional function for eyelids and eyelashes other than moving tears over the eyeball.

   _____

   _____

   _____

2. Which eyelid moves when you open and close your eyes?

   _____

3. Why do your eyes water when you have a cold?

   _____

   _____

4. Open your eyes wide. Note that only a small part of the eyeball can be seen through the fissure in the skin located between the eyelids. The "white of the eye" is the **sclera.** It is actually covered by an unseen transparent epithelial membrane, the **conjunctiva,** which is continuous with the epithelium lining the insides of

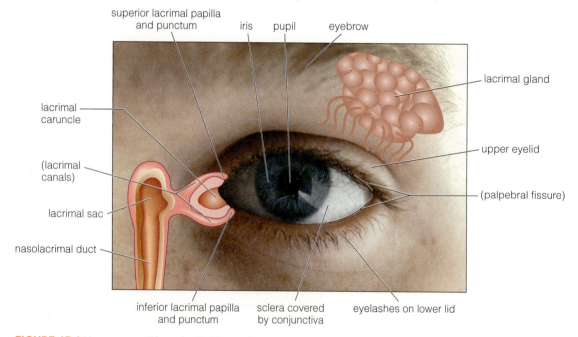

superior lacrimal papilla
and punctum   iris   pupil   eyebrow

lacrimal
caruncle

lacrimal gland

(lacrimal
canals)

upper eyelid

lacrimal sac

(palpebral fissure)

nasolacrimal duct

inferior lacrimal papilla
and punctum   sclera covered   eyelashes on lower lid
                     by conjunctiva

FIGURE 15-1 Human eye. *(Photo by D. Morton.)*

the eyelids. Identify the equally transparent cornea, a dome of sclera at the center of the visible eyeball. Through it you can see the pigmented **iris,** which functions as a diaphragm that opens (dilates) and closes (constricts) the hole or **pupil** through which light passes into the eyeball.

B. Structure and Function of the Eyeball

Many human eye models are available. Figure 15-2 shows two such models.

1. Examine a model of the eye and, consulting Figure 15-2, identify the following features:

   - **Outer eye muscles:** Skeletal muscles that move the eyeball

   - Sclera and cornea

   - **Choroid:** The black-pigmented middle layer of the eye. This layer is continuous with the blue, brown, or other-colored iris, which surrounds the pupil. Just behind the iris and also part of the choroid is

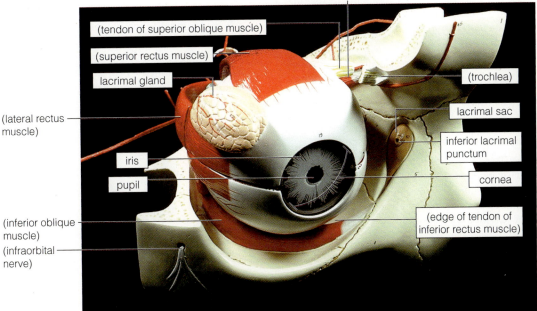

(superior oblique muscle)

(tendon of superior oblique muscle)

(superior rectus muscle)

lacrimal gland

(lateral rectus muscle)

iris

pupil

(inferior oblique muscle)

(infraorbital nerve)

(trochlea)

lacrimal sac

inferior lacrimal punctum

cornea

(edge of tendon of inferior rectus muscle)

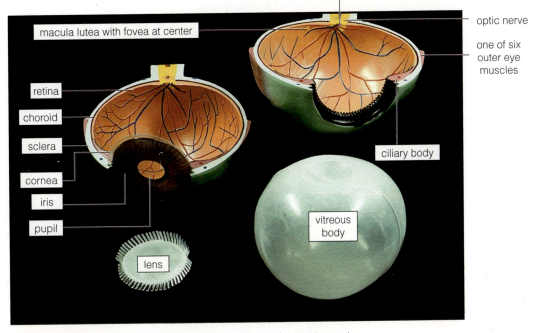

optic disk (blind spot)

macula lutea with fovea at center

optic nerve

one of six outer eye muscles

retina

choroid

sclera

cornea

iris

pupil

ciliary body

vitreous body

lens

FIGURE 15-2 Typical eye models, with labels. *(Photos by D. Morton.)*

the ringlike **ciliary body.** Both the iris and ciliary body have *inner eye muscles* composed of smooth muscle tissue.

- **Retina:** The inner layer of the eye, which lines the back two-thirds of the eyeball. The retina contains photoreceptors. At the optical center of the retina is a slight depression called the **fovea,** which contains the greatest concentration of photoreceptors for color and is surrounded by a yellowish spot, the **macula lutea.** Neuron pathways from the photoreceptors converge to form the root of the optic nerve (sometimes called the optic disk). Because there are no photoreceptors in this region, its presence creates a hole in the image perceived by the brain and therefore it is also called the **blind spot.**

- **Optic nerve:** Connects the retinal neuron pathways to the brain. Note that its outer layer is continuous with the sclera.

- **Lens:** A transparent, convex structure suspended by threadlike ligaments connected to the ciliary body and located just behind the pupil

- **Spaces:** The space behind the lens is called the **posterior cavity** (Figure 15-3a), which is filled with a thick, gooey fluid or humor, forming a transparent **vitreous body.** The space between the lens and the cornea, including the pupil, is called the **anterior cavity** and is filled by the clear, watery **aqueous humor.** The iris divides the anterior cavity into an anterior chamber in front of it and a posterior chamber behind.

2. Use your compound light microscope to look at a prepared slide (or slides) with a section (or sections) of the optic nerve and retina. Note the blind spot (Figure 15-3a). From the inside toward the outside of the eyeball, identify the three layers of neurons and the pigment epithelium that comprise the retina (Figure 15-3b). Photoreceptors comprise the layer of neurons closest to the pigment epithelium. There are two types of photoreceptors, **cones** and **rods.** Cones are concentrated in the center of the retina, in and around the fovea, and are stimulated only by bright light. Different levels of stimulation of the three subtypes of cones—red, green, and blue—allow for color vision. Color blindness results when one or two of the cone subtypes are impaired or missing. Rods are found away from the center of the retina and are more sensitive to light, functioning only in dim conditions.

C. Dissection of the Sheep Eye

Sheep eyes are a byproduct of the slaughtering of sheep for food. They are purchased from biological supplies companies.

(a)

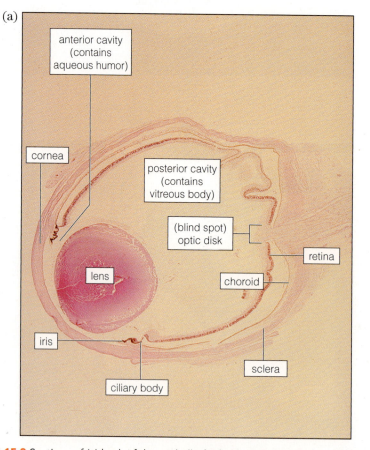

FIGURE 15-3 Sections of (a) back of the eyeball of a fetal eye (15′) and (b) a higher magnification view of the retina of a monkey (300′). The pigment epithelium and three layers of neurons comprise the retina. *(Photos by D. Morton.)*

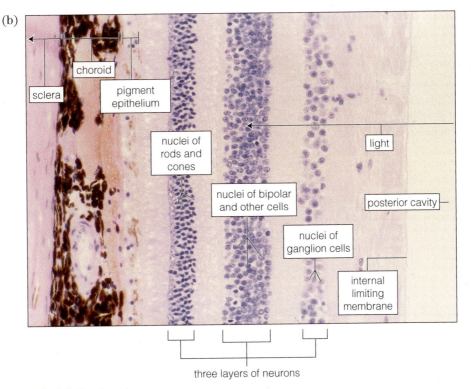

(b)

choroid

sclera

pigment epithelium

nuclei of rods and cones

nuclei of bipolar and other cells

light

posterior cavity

nuclei of ganglion cells

internal limiting membrane

three layers of neurons

FIGURE 15-3 *Continued.*

CAUTION Sheep eyes are kept in preservative solutions. Use lab gloves whenever you handle a specimen. Wash any part of your body exposed to this solution with lots of water. If preservative solution is splashed into your eyes, wash them with the safety eyewash for 15 minutes. Even if you wear contact lenses, you should wear safety goggles during dissection and when observing or studying a dissection. Eyeglasses should suffice in a situation when the goggles do not fit over them.

1. Work in pairs and wear lab gloves and safety goggles. Obtain a preserved sheep eye and place it on a stack of several paper towels. Identify its external features (Figure 15-4a).

2. Use dissecting scissors to dissect out the *outer eye muscles* and expose the tough, white, fibrous connective tissue of the sclera (Figure 15-4b). In life, these muscles move the eyeball.

3. With a scalpel, make an incision about 1/4 inch in back of and parallel to the edge of the *cornea* into the *anterior cavity*. Use scissors to extend the

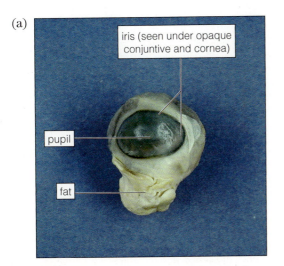

(a)

iris (seen under opaque conjuntive and cornea)

pupil

fat

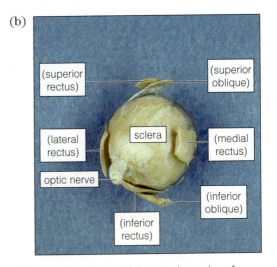

(b)

(superior rectus)

(superior oblique)

(lateral rectus)

sclera

(medial rectus)

optic nerve

(inferior oblique)

(inferior rectus)

FIGURE 15-4 (a) Anterior view of the external features and (b) posterior view with dissected muscles of a preserved sheep eye. *(Photos by D. Morton.)*

incision completely around the eyeball and gently pull apart the two pieces. Identify the internal features of the eyeball illustrated in Figures 15-5a and 15-5b.

4. Allow the *lens* and *vitreous body* of the posterior cavity to flow out of the posterior portion of the eyeball. Fill the posterior portion of the eyeball with water so as to flatten the delicate, whitish retina for examination. Identify the *blind spot* (Figure 15-5c) and, depending on the state of preservation, perhaps the macula lutea (Figure 15-2).

5. Examine the lens and vitreous body. One after the other, place them on an index card bearing a small letter *e*. If the lens is cloudy, do not use it. What is the effect of the lens and vitreous body on the image you see?

_____

_____

D. Vision

Now that you know the anatomy and function of the structures of the eye, let's investigate some of the important concepts related to the sense of vision—visual acuity, the lens and pupil accommodation reflexes, depth perception, the blind spot, and afterimages. **Visual acuity** refers to the "sharpness" of your vision, or how well you can see detail.

1. Determine your visual acuity using a Snellen chart as directed by your instructor. If you wear eyeglasses, perform this test both with and without them. Record your results in Table 15-1. A normal eye can read the line of letters marked 20 at 20 ft (red line) and is designated 20/20. If the eye can read only the letter marked 200, it is designated 20/200, and so on. Such an eye has **myopia** and is nearsighted. It is also possible for an eye to be farsighted (**hyperopia**). These conditions are usually the result of an elongated (myopia) or shortened (hyperopia) eyeball.

2. Visual acuity also can be reduced by **astigmatism**. Astigmatism is caused when one of the transparent surfaces (i.e., the cornea or lens) of the eye is not uniformly curved in all planes. Check for astigmatism by viewing a series of radiating lines on an astigmatism chart from a distance of 10 ft. To astigmatic individuals, some lines will look different, appearing sharper, thicker, or darker than other lines. Determine whether your right or left eye (or both) is

TABLE 15-1	Visual Acuity	
Eye	With Corrective Glasses or Contacts	Without Corrective Lenses
Right		
Left		

FIGURE 15-5 Internal structures of a dissected sheep eye showing (a) inside of front of eyeball, (b) inside of back of eyeball, and (c) retina. *(Photo by D. Morton.)*

astigmatic. If you wear eyeglasses, perform this test with them and then without them. Note the numbers of any lines that appear sharper, thicker, or darker in Table 15-2. If you wear glasses, detect the correction for astigmatism in your lenses by looking at the chart after rotating them 90°.

3. Your eyes accommodate focusing on objects at varying distances from the eye. Light rays are virtually parallel from an object 20 ft and farther away. Light rays from closer objects diverge, and these divergent rays must be bent at a sharper angle to focus them on the retina. **Lens accommodation** is accomplished by contracting the ciliary muscles to take tension off the lens, thus allowing it to bulge and become more convex.

   Have your lab partner measure the distance from your eyes to the nearest point at which you can sharply focus on an object such as the words in this lab manual. Record this value in Table 15-3. Repeat this procedure with and without glasses, if you wear them. This distance is called the **near point** and is a relative measure of the ability of the lens to accommodate for close vision. Lens accommodation requires an elastic lens. As an individual grows older, the lens becomes more inelastic, and to focus on an object, it must be held farther and farther from the eye. However, the object then appears smaller and smaller. This condition is called **presbyopia.** To get letters large enough to read, older people use bifocals (or similar eyeglasses) or extra big print.

4. To demonstrate pupil accommodation for changes in light intensity, shine a penlight intermittently into your lab partner's eye and note any changes in the diameter of the pupil. These responses are caused by the **photopupil reflex.** Describe these changes.

   _____

   _____

   _____

Which occurs more rapidly, constriction or dilation (opening)? /NL_Rule/_____

What advantage might this difference confer in changing light conditions?

_____

_____

_____

   Note that when your lab partner switches focus from a distant to a nearby object, the pupils also constrict (**pupil accommodation for distance reflex**). This is done to block out the most divergent (most difficult to focus) rays of light.

5. Observe that when your lab partner switches focus from a distant to a nearby object, the eyeballs converge (**eye convergence reflex**). Convergence allows for binocular vision (overlapped images), which is necessary for depth perception. Gently press one eyeball out of line by pressing on an eyelid while viewing an object.

How many objects do you see?

_____

6. Hold Figure 15-6 about 20 inches from your right eye. Close your left eye and place your right eye directly over the dot. Stare at the dot while bringing the page closer to the eye until the dot disappears. At this point, the image of the dot is falling on the blind spot of the retina.

7. Place a sheet of white paper with a black spot about the size of a dime onto another sheet of blank white paper. Under a bright light, stare at the black spot for 20 seconds, trying to move your head and eyes as little as possible. Slide off the top sheet, continuing to look at the blank sheet. Describe the **afterimage.**

   _____

   _____

Repeat this test, substituting a top sheet of black construction paper with a white dot on it. Describe the afterimage.

_____

_____

Bits of bright red, blue, green, and yellow paper can also be used in this activity. Just as the afterimage

TABLE 15-2	Astigmatism	
Eye	With Corrective Glasses or Contacts	Without Corrective Lenses
Right		
Left		

TABLE 15-3	Near Point	
Eye	With Corrective Glasses or Contacts	Without Corrective Lenses
Right		
Left		

FIGURE 15-6 Demonstration of blind spot.

of black is a more intense white and vice versa, color vision has afterimages that are essentially the "opposite" color of the original color. To demonstrate this, place a red heart on a green sheet of construction paper; then place this combination over the sheet of blank white paper. Under a bright light, stare at the red heart for 20 seconds, trying to move your head and eyes as little as possible. Slide off the top sheet, continuing to look at the blank white sheet. Describe the **afterimage.**

_____

_____

Repeat the experiment using a green heart and a piece of red construction paper and describe the results.

_____

_____

15.2   The Ears and Hearing

The ear contains several groups of mechanoreceptors (sensory receptors for mechanical energy). Depending on their specific location, their stimulation is interpreted by the brain as the sense of hearing or the sense of equilibrium.

MATERIALS

Per student:

- Compound light microscope
- Section of cochlea

Per student pair:

- Mug-sized container stuffed to overflowing with cotton wool

Per student group (4):

- Ear model

PROCEDURE

A. Structure and Function of the Ear

Except for the visible flaps (*auricles*) located on either side of the head, the paired ears are embedded in the temporal bones of the skull.

1. Examine a model of the ear and, consulting Figure 15-7, identify the following features:

    - **Outer ear:** Includes **auricles** and **outer ear canal,** which funnel sound waves in air to the **tympanic membrane.**

    - **Middle ear:** Contains inner ear **ossicles** (three small bones shaped like a hammer, anvil, and stirrup), which conduct vibrations in bone tissue from the tympanic membrane to the **oval window** of the cochlea. The middle ear space is connected to the upper pharynx (space behind the nasal and oral cavities) by the **auditory tube,** which functions to equalize air pressure on both sides of the tympanic membrane.

    - **Inner ear:** Composed of the snail shell-like cochlea, vestibule, and semicircular canals. The **cochlea** contains fluid-filled spaces that conduct vibrations from the end plate of the stirrup-shaped ossicle, which is lodged in the **oval window,** to mechanoreceptors for hearing in the **organ of Corti** (or *spiral organ*; Figure 15-8). A membrane-covered *round window* dissipates old vibrations from the cochlea into the air in the middle ear space. The **semicircular canals** and **vestibule** contain fluid-filled sacs with structures containing mechanoreceptors that aid in equilibrium.

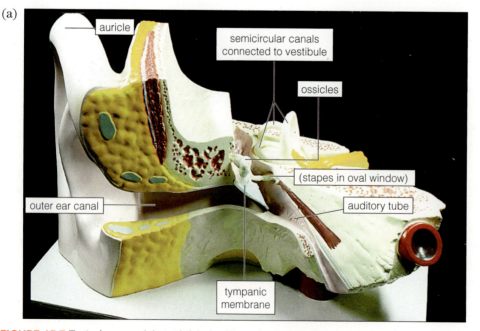

(a)

FIGURE 15-7 Typical ear models, with labels. *(Photo by D. Morton.)*

(b)

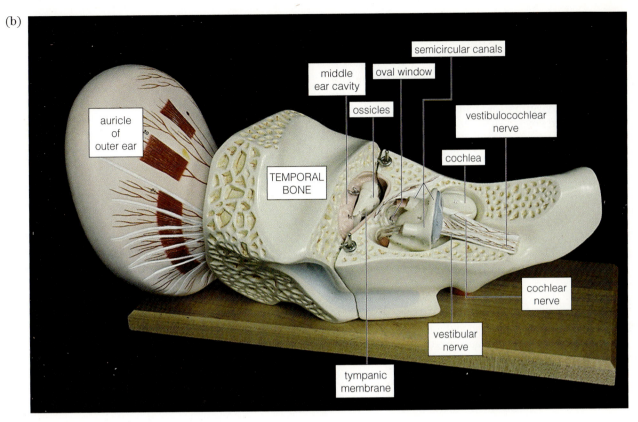

FIGURE 15-7 Continued.

(a)   (b)

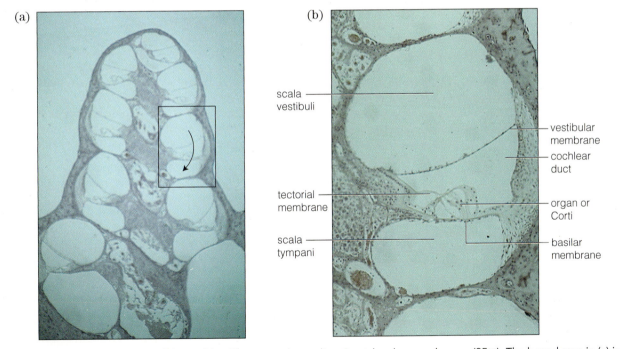

FIGURE 15-8 Cross section of the cochlea. The arrow shows direction taken by sound waves (25×). The boxed area in (a) is magnified in (b) 107×. (Photos by D. Morton.)

2. Observe a section of the cochlea with your compound light microscope. Consult Figure 15-8 and identify the *scala vestibuli, vestibular membrane, cochlear* *duct, organ of Corti, basilar membrane,* and *scala tympani*. The *arrow* indicates the direction followed by sound waves.

B. Sense of Direction of Sound

Why do we have two ears? Two ears enable our brains to perceive the direction of sound waves.

1. To demonstrate this, work in pairs. One of you sits on a stool with eyes closed. The other circles the stool quietly, stops, and calls the sitter's name in a high-pitched voice. (This might seem funny, but high-pitched sound does not penetrate bone as well as low-pitched sound.) Each time the sitter's name is called, the sitter tries to point to the caller. If a straight line from the sitter's finger "hits" the caller, place a check mark after trial 1 in the second column of Table 15-4. Repeat this test for a total of 10 trials.

2. Now cover one ear with a mug-sized container stuffed to overflowing with cotton wool and repeat step 1 ten more times. Use the third column of Table 15-4 to record "hits."

3. Total the number of check marks in both columns.

4. Consider the time of arrival and the relative intensity of the waves at each ear and suggest two ideas as to how the brain determines the direction of the sound source.

_____

_____

_____

_____

15.3 Taste and Smell

Chemoreceptors (sensory receptors for chemicals) are located in the taste buds of the tongue and in the olfactory epithelium of the nasal cavity.

TABLE 15-4	Sense of Sound Direction	
Trial	**Ears Uncovered**	**One Ear Covered**
1		
2		
3		
4		
5		
6		
7		
8		
9		
10		
Total		

MATERIALS

Per student:

- Compound light microscope
- Section of rabbit tongue
- Section of nasal cavity

Per student group (4):

- Box of tissues
- Permanent marker
- Four 250-mL Erlenmeyer flasks, one each with
  - /BLL/10% sucrose solution marked SW
  - 1% acetic acid solution marked SR
  - 5% NaCl solution marked SL
  - 0.5% quinine sulfate solution or tonic water marked BT
- Three dishes containing cubes of a different vegetable or fruit (e.g., apple, onion, and potato)

Per lab room:

- Box of applicator sticks
- Container of small disposable beakers
- Box of paper cups
- Source of drinking water

PROCEDURE

A. Location of Chemoreceptors

Five types of taste buds function to detect sweet, salty, savory, acidic, and bitter substances in food and drink. Savory taste buds are also called umami after the name of the scientist who discovered them. Taste buds are probably sensitive to all of these substances but are particularly sensitive to one. A much larger number of different types of chemoreceptors in the olfactory epithelium detect a larger number of more vaporizable substances.

1. Examine a section of rabbit tongue with your compound light microscope. Look at the outer edge of the organ between the projections (foliate papillae) and identify the **taste buds** (Figure 15-9a). They consist of neuron endings surrounded by _taste cells_ (chemoreceptors) with _microvilli_ projecting through a _taste pore_ and supportive epithelial cells.

2. Likewise, examine a section of the nasal cavity. Identify the **olfactory epithelium** lining the roof of the nasal cavity (Figure 15-10). Similar to taste buds, the olfactory epithelium contains neuron endings, chemoreceptors, and supportive epithelial cells.

B. Interaction Between Taste and Smell

On an everyday level, we tend to mix up the sense of taste and smell. For example, how many times have you heard a statement like, "This pizza tastes great"? How much of this "great taste" is actually taste, and how much of it is smell?

(a)

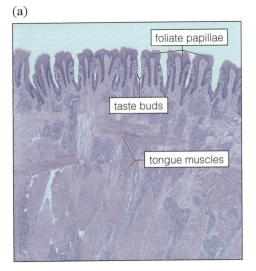

foliate papillae

taste buds

tongue muscles

(b)

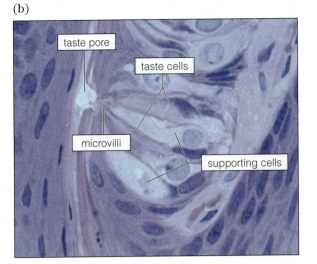

taste pore

taste cells

microvilli

supporting cells

FIGURE 15-9 (a) Foliate papillae in rabbit tongue (25×) and (b) magnified taste bud (1000×). *(Photos by D. Morton.)*

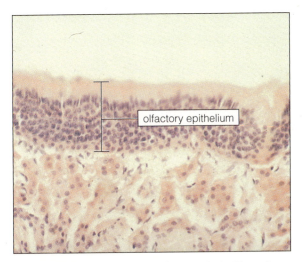

olfactory epithelium

FIGURE15-10 Olfactory epithelium (250×). *(Photo by D. Morton.)*

1. Rinse your mouth with drinking water. Shut your eyes, pinch your nostrils closed, and have your lab partner place a small piece of vegetable or fruit on your tongue. Attempt to identify it. Record its identity or, if you can not identify it, put a question mark in Table 15-5. With your nostrils still pinched closed, chew the piece thoroughly but do not swallow it. Again attempt to identify it. Record this and subsequent results in Table 15-5. Open your nostrils and identify the vegetable or fruit.

2. Repeat step 1 twice, using two more different vegetables or fruits.

3. In a strictly biological sense, when your nostrils are blocked by congestion, can you taste and smell as well as when they are open? Explain your answer.

_____

_____

_____

TABLE 15-5	Identity of Vegetable/Fruit		
Trial	**Placed on Tongue**	**After Chewing**	**Nostrils Open**
1			
2			
3			

_____ **1.** The cornea is part of the
   (a) sclera
   (b) choroid
   (c) retina
   (d) lens

_____ **2.** Parts of the choroid form
   (a) the lens
   (b) the ciliary body
   (c) the iris
   (d) both b and c

_____ **3.** Photoreceptors are found in the
   (a) lens
   (b) ciliary body
   (c) iris
   (d) retina

_____ **4.** The space between the lens and iris is
   (a) the posterior cavity
   (b) the posterior chamber
   (c) the anterior chamber
   (d) none of the above

_____ **5.** Nearsightedness is also called
   (a) myopia
   (b) hyperopia
   (c) presbyopia
   (d) none of the above

_____ **6.** Ossicles are found in
   (a) the outer ear
   (b) the middle ear
   (c) the inner ear
   (d) both a and c

_____ **7.** Sensory receptors found in the ear are
   (a) chemoreceptors
   (b) mechanoreceptors
   (c) photoreceptors
   (d) none of the above

_____ **8.** Taste buds detect
   (a) light
   (b) atmospheric sound waves
   (c) chemicals in food and drink
   (d) movements of the head

_____ **9.** The olfactory epithelium is found in the
   (a) ear
   (b) nasal cavity
   (c) eye
   (d) tongue

_____ **10.** The most developed sense in humans is
   (a) hearing
   (b) vision
   (c) taste
   (d) smell

EXERCISE 15 Structure and Function of the Sensory Organs

POST-LAB QUESTIONS

15.1 The Eyes and Vision

1. Define the following terms:

a. myopia

b. hyperopia

c. presbyopia

2. How does uncorrected myopia or hyperopia affect the near point?

3. What types and subtypes of photoreceptors are present in the retina? Briefly describe their function.

4. What is an afterimage? How is it formed?

5. List the eye structures that light passes through as it travels from outside the body to the retina.

15.2 The Ears and Hearing

6. List the ear structures that sound waves and resulting vibrations pass through as they travel from outside the body to the organ of Corti.

15.3 Taste and Smell

7. Why do humans and other animals have two nostrils?

8. When your nasal passages are blocked because of a heavy cold, food seems to lose its flavor. For example, pizza tastes like salt. Explain why this is so.

15.4 Food for Thought

9. According to your textbook and the Internet, in what part of the eye are cataracts located?

10. According to your textbook and the Internet what is the relationship between the fluid in the anterior cavity and the disease glaucoma?

Reproduction and Development

OBJECTIVES

After completing this exercise, you will be able to

1. define *sexual reproduction, gametes, ovum, sperm, zygote, fertilization, gonads, ovaries, testes, gametogenesis, oogenesis, spermatogenesis, meiosis, yolk, acrosome, blastodisc, vegetal pole, animal pole, seminiferous tubules, interstitial cells, testosterone, Sertoli cells, follicle, estrogen, progesterone, ovulation, corpus luteum, implantation, uterus, blastomere, morula, blastula, blastocoel, gray crescent, blastocyst, inner cell mass, trophoblast, placenta, archenteron, blastopore, primitive streak, notochord, neural tube, somites, amniotic cavity,* and *embryonic disk.*

2. draw a mammalian sperm and label its segments.

3. describe the structure of chicken and frog ova and how they differ from a typical mammalian ovum.

4. recognize interstitial cells, seminiferous tubules, Sertoli cells, and sperm in a prepared section of a mammalian testis.

5. recognize primary oocytes, primordial and secondary follicles, and a corpus luteum in a prepared slide of a mammalian ovary.

6. describe spermatogenesis and oogenesis in mammals.

7. describe the events and consequences of sperm penetration and fertilization.

8. compare and contrast cleavage in the sea star, frog, chicken, and human.

9. explain differences in cleavage according to (a) the amount and distribution of yolk in the ovum and (b) the evolution of mammals.

10. describe gastrulation, the formation of the primary germ layers (ectoderm, mesoderm, endoderm), and their derivatives.

Introduction

Sexual reproduction typically involves the fusion of the nuclei of two **gametes**, called the **ovum** (plural, ova) in females and **sperm** in males, to form the first cell of a new individual, the **zygote**. This fusion is referred to as **fertilization**.

Gametes are produced in reproductive organs called **gonads**, usually in individuals of two separate sexes—**ovaries** in females and **testes** in males. **Gametogenesis**, gamete formation, reduces the number of chromosomes by half. Gametes have 22 *autosomes*, one of each of the 22 pairs found in most other cells of the body plus one *sex chromosome*, an X or Y. Gametes with a half set of chromosomes are referred to as *haploid*. Cells with a full set of chromosomes are *diploid*. Fertilization produces a diploid zygote, whose combination of genes is unlike those of either parent. Other than cases of identical twins, triplets, and so on, it is also very unlikely that the genes of one zygote will be the same as those of any other zygote, even those derived from the same parents. This variation is an advantage to a species in a changing and unpredictable environment. Why?

_____

_____

_____

_____

NOTE Depending on the timing of your lab and to prepare for the observation of a live frog zygote, your instructor may demonstrate the fertilization of frog eggs before you start this exercise.

16.1 Gametes

Most sperm have at least one flagellum and are specialized for motility. They contribute little more than their chromosomes to the zygote.

Ova are specialized for storing nutrients, and they contain the molecules and organelles needed to fuel, direct, and maintain the early development of the embryo. Nutrients are stored as **yolk** in the cytoplasm. Consequently, mammalian ova are larger than body cells and in some species reach a diameter of 0.2 mm. In frogs, additional yolk increases the diameter of the ovum to 2 mm, and in chickens, it reaches about 3 cm.

MATERIALS

Per student:

- Compound light microscope
- Lens paper
- Bottle of lens-cleaning solution (optional)
- Lint-free cloth (optional)
- Glass microscope slide
- Coverslip
- Dissecting needle
- One-piece plastic dropping pipet

Per student pair:

- Unfertilized hen's egg
- Several paper towels
- Dissection pan
- Syracuse dish
- Two camel-hair brushes
- Dissection microscope

Per lab group (table or bench):

- Model of a frog ovum (optional)

Per lab section:

- Live frog sperm in pond water
- Live frog ova in pond water
- Pond water

Per lab room:

- Phase-contrast compound light microscope (optional)

PROCEDURE

A. Sperm

1. With the high-dry objective of your compound microscope, study a prepared slide of bull sperm. Draw several sperm in Figure 16-1. Each sperm has three major segments: the *head, midpiece,* and *tail.* Label the major segments of one sperm in your drawing. The tail is typically composed of a single *flagellum.*

FIGURE 16-1 Drawing of bull sperm (_____ X).

2. *Skip this step if your compound microscope does not have an oil-immersion objective.* Switch to the oil-immersion objective. Find the **acrosome** covering the *nucleus* in the head of a sperm and *mitochondria* in its midpiece. The acrosome contains enzymes that aid in the penetration of the egg. Why does a sperm have a lot of mitochondria relative to its cell size?

_____

_____

_____

3. Place 1 drop of frog sperm suspension on a glass microscope slide and make a wet mount. Observe the movements of sperm flagella using either the phase-contrast compound microscope or your compound microscope with the iris diaphragm partially closed to increase contrast. Describe what you see.

_____

_____

_____

B. Ova

1. *Chicken egg.*

   (a) Obtain an unfertilized chicken egg (Figure 16-2). Crack it open as you would in the kitchen and carefully spill the contents into a hollow made from paper towels placed in a dissection pan.

   (b) Only the yolk is the ovum. Look on its surface for the **blastodisc**, a small white spot just under the cell membrane. This area is free of yolk and contains the nucleus. This is where fertilization occurs if sperm are present in the hen's oviducts. The walls of the oviduct secrete the *albumin* (egg white).

   (c) Examine the *shell* and the two *shell membranes.* The shell membranes are fused except in the

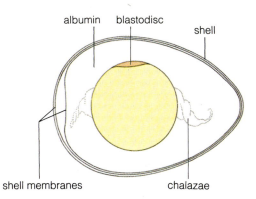

albumin   blastodisc
                      shell

shell membranes   chalazae

FIGURE 16-2 Unfertilized chicken egg.

region of the air space at the blunt end of the egg. Note the two shock absorber–like *chalazae* (singular, *chalaza*), which suspend the ovum between the ends of the egg. They are made of thickened albumin and may help rotate the ovum to keep the blastodisc always on top of the yolk.

2. *Frog egg.*

(a) Gently place a live frog egg in a Syracuse dish half filled with pond water and examine it with a dissection microscope. Use two clean camel-hair brushes to transfer the egg. *Do not let the egg dry out.* The ovum is enclosed in a protective jelly membrane.

Which surface (light side or dark side) of the ovum floats up in the pond water?

_____

(b) Look at a model of a frog ovum. Ova from different species of animals vary in the amount and distribution of yolk. Frog ova have a moderate amount of yolk that is concentrated in the lower half of the ovum (Figure 16-3). This half of the ovum is called the **vegetal pole**. The nucleus is located in the upper, yolk-free half, the **animal pole**. Note that the animal pole is black. This is because it contains pigment granules. Why does a frog ovum have less yolk than a bird ovum?

_____

_____

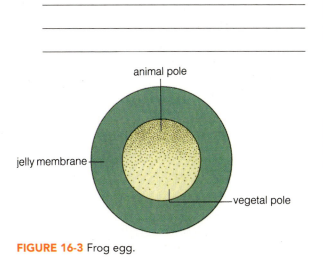

animal pole

jelly membrane

vegetal pole

FIGURE 16-3 Frog egg.

Suggest one or more possible functions for the black pigment in the animal pole. (*Hint:* One function is the same as that for the pigment in your skin that increases when exposed to sunlight.)

_____

_____

_____

3. *Human egg.* Figure 16-8 diagrams a human egg as it would appear in the upper oviduct. The egg is surrounded by a membrane, the *zona pellucida*, and a layer of follicle cells. Why do most mammalian ova have very little yolk?

_____

_____

_____

16.2   Gametogenesis

Gamete formation (**gametogenesis**) is called **oogenesis** in females and **spermatogenesis** in males. The general scheme and terminology of gametogenesis are summarized in Table 16-1. **Meiosis** is a process by which events in the nucleus of a diploid cell use two cytoplasmic divisions to form haploid gametes or cells that will mature into gametes. The nuclear events associated with the first division are referred to as meiosis I, and those of the second division are called meiosis II.

MATERIALS

Per student:

- Compound light microscope
- Prepared slides of a section of
  - Adult mammalian testis stained with iron hematoxylin
  - Adult mammalian ovary with follicles
  - Adult mammalian ovary with a corpus luteum

TABLE 16-1	Gametogenesis	
	Type of Cell	
Condition of Cell	**Male**	**Female**
Mitotically active	Spermatogonium	Oogonium
Before meiosis I	Primary spermatocyte	Primary oocyte
Before meiosis II	Secondary spermatocyte	Secondary oocyte and first polar body
After meiosis II	Spermatid	Ovum and three polar bodies
After differentiation	Sperm	—

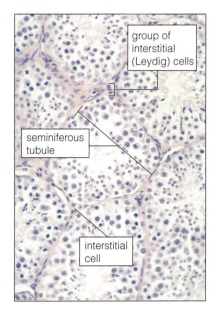

FIGURE 16-4 Transverse section of the testis (200X). *(Photo by D. Morton.)*

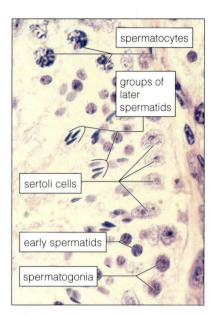

FIGURE 16-5 Spermatogenesis (800X). *(Photo by D. Morton.)*

PROCEDURE

A. Mammalian Spermatogenesis

1. With your compound microscope, look at a prepared slide of the testis (Figure 16-4). Most of the interior of a testis is filled with coiled **seminiferous tubules**. Transverse and oblique sections are present in your slide. Look for glandular **interstitial cells** between the seminiferous tubules. Interstitial cells secrete the male sex hormone **testosterone.**

2. Center a cross-section of a seminiferous tubule in the field of view and increase the magnification until you see only a portion of the tubule's wall. In the wall of the seminiferous tubule, identify as many stages of spermatogenesis as possible, as well as **Sertoli cells**, which function to nurture the developing sperm (Figure 16-5).

B. Mammalian Oogenesis

After birth, in humans and most other mammals, oogonia are not present because all of them have started oogenesis. There is an excess supply of primary oocytes present at birth (~2 million in a newborn girl).

In the ovaries, primary oocytes are located within cellular balls called **follicles.** In addition to supporting the developing oocytes, follicle cells also secrete the female sex hormones, **estrogen** and **progesterone.** A typical mammalian ovary contains a number of stages of follicular development (Figure 16-6).

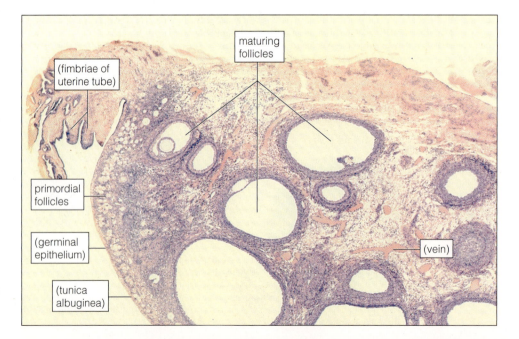

FIGURE 16-6 Mammalian ovary (25X). *(Photo by D. Morton.)*

A primordial follicle whose thin walls are one cell thick initially surrounds each primary oocyte. Most follicles degenerate, and in humans, only about 300,000 primary oocytes remain at puberty. The majority of these oocytes also degenerate. With each turn of the female cycle, several follicles begin to mature, and one (rarely two or more) of each batch of oocytes finally bursts from the surface of the ovary and is swept into the oviduct. The release of oocytes from the ovary is called **ovulation** (Figure 16-7).

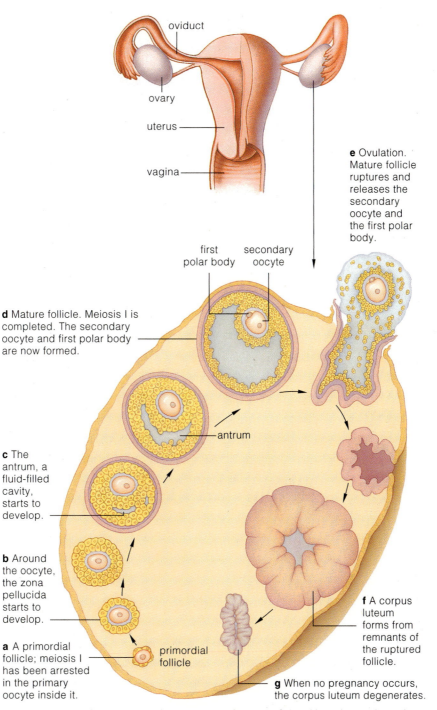

oviduct

ovary

uterus

vagina

e Ovulation. Mature follicle ruptures and releases the secondary oocyte and the first polar body.

first polar body

secondary oocyte

d Mature follicle. Meiosis I is completed. The secondary oocyte and first polar body are now formed.

antrum

c The antrum, a fluid-filled cavity, starts to develop.

b Around the oocyte, the zona pellucida starts to develop.

a A primordial follicle; meiosis I has been arrested in the primary oocyte inside it.

primordial follicle

f A corpus luteum forms from remnants of the ruptured follicle.

g When no pregnancy occurs, the corpus luteum degenerates.

FIGURE 16-7 Cyclic events in a human ovary, drawn as if sliced lengthwise through its midsection. Events in the ovarian cycle proceed from the growth and maturation of follicles, through ovulation (rupturing of a mature follicle with a concurrent release of a secondary oocyte). For illustrative purposes, the formation of these structures is drawn as if they move clockwise around the ovary. In reality, their maturation occurs at the same site. *(After Starr, 1999.)*

Around the time of ovulation, the primary oocyte divides to form a secondary oocyte and the first polar body. The zona pellucida and a capsule of follicle cells (Figure 16-8) surround an ovulated secondary oocyte. A sperm must first penetrate this barrier before penetration of the egg membrane and fertilization can occur. Upon fertilization, the secondary oocyte divides again to form an ovum and another polar body. The first polar body may also divide, resulting in a total of four polar bodies.

Follicle cells left behind in the ovary after ovulation develop collectively into a large roundish structure called the **corpus luteum** (the name means "yellow body" and refers to its color in a live ovary). The corpus luteum continues to secrete female sex hormones, especially progesterone.

1. With the compound microscope, examine the prepared slide of an ovary. Look for primordial follicles (Figure 16-9a).

Both the follicles and their primary oocytes are small compared with maturing follicles and their primary oocytes. Groups of primordial follicles tend to occur between maturing follicles or between maturing follicles and the ovarian wall. Note that the wall of the primordial follicle is composed of a single layer of smaller cells.

2. Now look for maturing follicles. In maturing follicles, the size of the cells in the follicle wall and of the primary oocyte itself increases. Also, the follicle cells divide, causing the wall to become first two cells thick and then multilayered. Then a space appears between the follicle cells (Figure 16-9b). This fluid-filled space increases in size until the primary oocyte and the follicle cells immediately around the primary oocyte are suspended in it (Figure 16-9c). The mass of cells is connected to the rest of the wall by a narrow stalk of follicle cells. Just before ovulation, the follicle reaches its maximum size and bulges from the surface of the ovary.

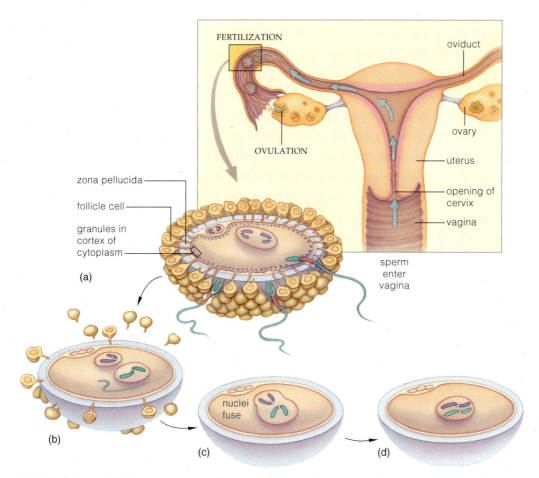

FIGURE 16-8 Fertilization. (a) Many sperm surround a secondary oocyte. Acrosomal enzymes clear a path through the zona pellucida. (b) When a sperm does penetrate the secondary oocyte, granules in the egg cortex release substances that make the zona pellucida impenetrable to other sperm. Penetration also stimulates meiosis II of the oocyte's nucleus. (c) The sperm's tail degenerates and its nucleus enlarges and fuses with the oocyte nucleus. (d) With fusion, fertilization is over. The zygote has formed. (*After Starr, 2000.*)

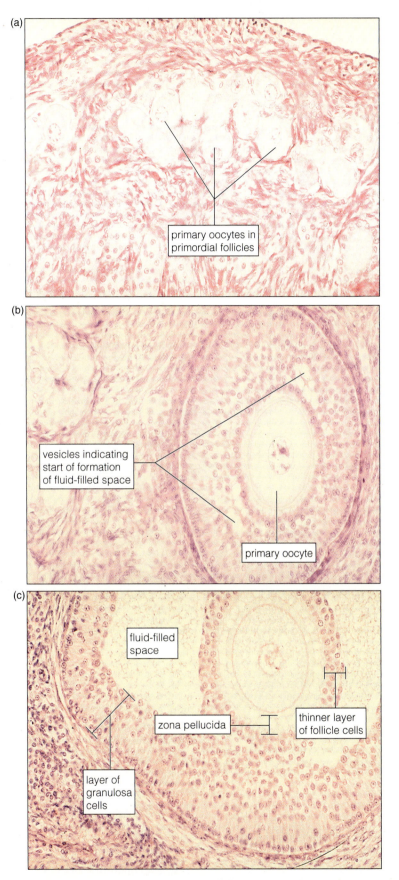

FIGURE 16-9 (a) Group of primordial follicles (450X), (b) maturing follicles (450X); and (c) portion of a mature secondary follicle (40X). (*Photos by D. Morton.*)

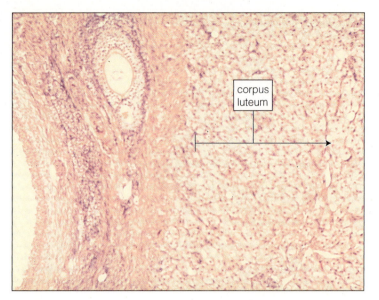

FIGURE 16-10 Corpus luteum (100X). *(Photo by D. Morton.)*

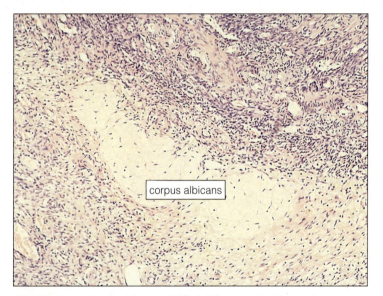

FIGURE 16-11 Corpus albicans (100X). *(Photo by D. Morton.)*

3. Replace the slide on the microscope with one of a mammalian ovary with a corpus luteum. Locate a corpus luteum (Figure 16-10).

If an embryo does not successfully implant in the **uterus**, the corpus luteum degenerates and is replaced by scar tissue. This scar is now called the *corpus albicans* (Figure 16-11).

16.3 Sperm Penetration and Fertilization in the Frog

In humans and most other mammals, both fertilization (in the upper oviduct) and **implantation** and subsequent development to a late stage (in the uterine wall) take place in the mother. Other animals, such as frogs, have external fertilization and development. At the beginning of the lab, or just before, your instructor induced ovulation in a female frog. Sperm obtained from the shredded testes of a male frog were then added to fertilize these ova.

MATERIALS

Per student pair:

- Syracuse dish
- Two camel-hair brushes
- Dissection microscope

Per lab group (table or bench):

- Model of a frog zygote (optional)

Per lab section:

- Live frog zygotes in pond water
- Pond water
- Video or link(s) to Internet site(s) with video of frog development (optional)

PROCEDURE

1. Using two clean camel-hair brushes, gently transfer to a Syracuse dish half-filled with pond water a live frog zygote that has not yet begun to cleave and examine it with a dissecting microscope. *Do not let the zygote dry out.*

   Which surface of the zygote (light or dark side) floats up in the pond water?

   _____

   Compared with ova, do frog zygotes show a change in their pattern of coloration after fertilization? (*Hint:* Look between the darkly pigmented and unpigmented areas.) _____ (yes, no)

   If so, describe what you have observed.

   _____

2. If you can not observe fertilization directly, study a model of a frog zygote, watch a video, or follow links provided by your instructor on early frog development.

16.4   Cleavage

Cleavage is a special type of cell division that occurs first in the zygote and then in the cells formed by successive cleavages, the **blastomeres.** Unlike typical cell division, there is no period of cytoplasmic growth between the cleavage divisions, causing the blastomeres to become smaller and smaller. After a number of cleavages, the blastomeres form a solid cluster of cells called the **morula.** The formation of a hollow ball of cells, called the **blastula**, marks the end of cleavage. Because there is no cytoplasmic growth, the size of the blastula is only slightly larger than that of the zygote.

In many organisms whose ova have little yolk (e.g., sea stars and humans), cleavage is *complete* and nearly *equal*, resulting in separate blastomeres that are all about the same size. In frogs and other animals with moderate amounts of yolk, cleavage is complete but unequal. There is so much yolk in the ova of many animals (e.g., most fish, reptiles, birds, and the two mammals that lay eggs—platypuses and spiny anteaters) that complete cleavage of the entire zygote is impossible, so it forms a disklike embryo that sits on the surface of the yolk. This type of cleavage is incomplete and is called *discoidal.*

 CAUTION Preserved specimens are kept in preservative solutions. Use lab gloves whenever you handle a specimen. Wash any part of your body exposed to this solution with lots of water. If preservative solution is splashed into your eyes, wash them with the safety eyewash for 15 minutes. Even if you wear contact lenses, you should wear safety goggles during dissection and when observing or studying a dissection. Eyeglasses should suffice when goggles do not fit over them.

MATERIALS

Per student:

- Compound light microscope
- Lens paper
- Bottle of lens-cleaning solution (optional)
- Lint-free cloth (optional)
- Prepared slide of a whole mount of sea star development through gastrulation

Per student pair:

- Two Syracuse dishes
- Two blue camel-hair brushes (optional)
- Two red camel-hair brushes
- Dissection microscope

Per lab section:

- Frog embryos in pond water from eggs fertilized 1 hour before the lab (optional)

Per lab room:

- Preserved specimens of two-, four-, and eight-cell cleavage stages; morulae (32-cell cleavage stage) and blastulae of frog in easily accessible screw-top containers
- Source of dH_2O
- Pond water (optional)
- Models of early frog development
- Models of human development
- Boxes of different sizes of lab gloves
- Safety goggles

PROCEDURE

A. Cleavage in Sea Stars

1. With your compound microscope, observe a prepared slide with whole mounts of early sea star embryos. Find a zygote; two-, four-, and eight-cell cleavage stages; a morula; and a blastula and draw them in Figure 16-12. Be sure to adjust the fine-focus knob to see the three-dimensional aspects of these stages. The blastula is a hollow ball of flagellated blastomeres

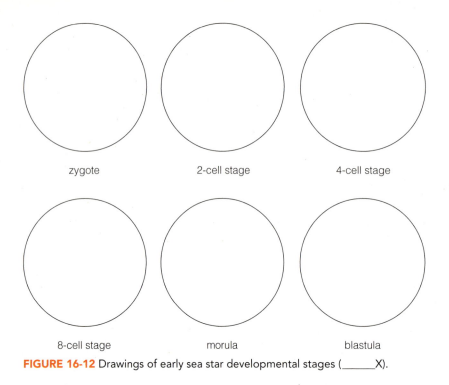

zygote

2-cell stage

4-cell stage

8-cell stage

morula

blastula

FIGURE 16-12 Drawings of early sea star developmental stages (_____X).

surrounding a cavity called the **blastocoel**. Label the blastomeres and blastocoel in your drawing of the sea star blastula.

2. What is the orientation (parallel or perpendicular) relative to each other of the first two cleavage planes?

_____

What is the orientation (parallel or perpendicular) of the third cleavage plane compared with the first cleavage plane?

_____

B. Cleavage in Frogs

1. *Optional.* If your instructor prepared fertilized frog eggs 1 hour before the lab, you will be able to watch the first two cleavage divisions during this laboratory. At room temperature about 1 hour after addition of the sperm to the eggs, a region of less pigmented cytoplasm, the **gray crescent,** appears opposite the site of sperm penetration (Figure 16-13). The first cleavage division occurs about 2 hours after fertilization, the second a half hour later, and the third after another 2 hours.

pigmented cortex

gray crescent

yolk-rich cytoplasm

sperm penetrating frog egg

FIGURE 16-13 Formation of the gray crescent. *(After Starr and Taggart, 2001.)*

Gently place several live frog zygotes in a Syracuse dish half filled with pond water and examine them with a dissection microscope. Use two blue camel-hair brushes to transfer the zygote. *Do not let the zygote dry out.*

2. Wear lab gloves and safety goggles. If living embryos are unavailable, look at preserved specimens under the dissecting microscope. Use two red camel-hair brushes to transfer each stage in turn to a Syracuse dish half filled with distilled water. When you are done, return the specimen to its container. *The embryo specimens are in a preservative solution, so make sure you do not use the red brushes to manipulate the live embryos because the residual preservative will harm them.*

3. Examine preserved specimens of eight-cell, morula, and blastula stages.

4. Study the models of the early stages of frog development.

5. Draw the stages of frog development in Figure 16-14.

6. Note that cleavage in the vegetal pole lags behind that of the animal pole. Why? (*Hint:* What is present in the vegetal pole that would hinder cleavage?)

_____

_____

_____

C. Cleavage in Chickens

In chickens, cleavage of the blastodisc forms a layer of cells called the *blastoderm*, which in time becomes

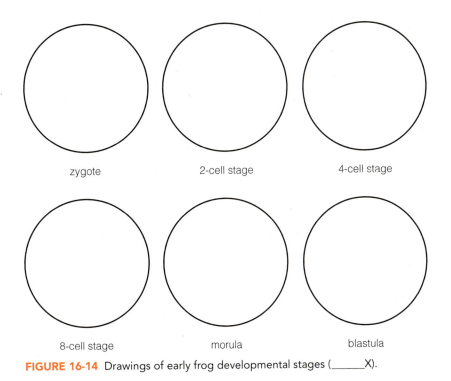

zygote | 2-cell stage | 4-cell stage

8-cell stage | morula | blastula

FIGURE 16-14 Drawings of early frog developmental stages (_____X).

separated from the yolk by a cavity, the blastocoel. Further development of the embryo occurs only in the blastoderm (Figure 16-15).

D. Cleavage in Humans

Examine models of the early stages of human development. Because the ovum has little yolk, cleavage is complete and nearly equal. At the end of cleavage, a hollow ball of cells is formed. However, this **blastocyst** differs from a blastula in that a group of cells aggregate at one pole of the inner surface of the blastocoel (Figure 16-16). These cells are called the **inner cell mass**, and similar to the blastoderm of chickens, further development of the embryo proceeds there only. The remaining cells that surround the blastocoel are called the **trophoblast.** The blastula stage coincides with implantation of the embryo in the uterus.

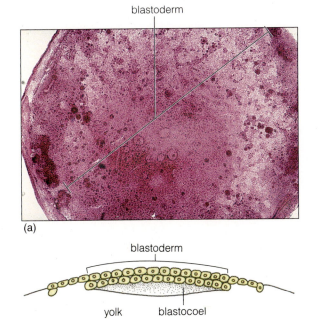

FIGURE 16-15 Cleavage in the chicken. (a) Surface view (42X). (*Photo courtesy Biodisc, Inc.*) (b) Transverse section along diameter indicated in (a).

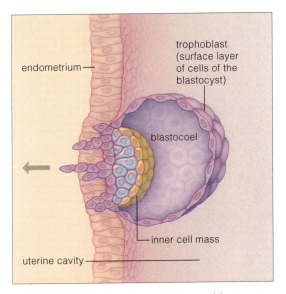

FIGURE 16-16 Section through a human blastocyst as it attaches to the inner lining (endometrium) of the uterus and starts implantation by burrowing in to it (day 6–7 after fertilization). (*After Starr, 1999.*)

Current evolutionary thought is that modern reptiles, birds, and mammals evolved from earlier reptilian ancestors that had external development and ova rich in yolk. Internal development in mammals linked nourishment of the embryo directly to the mother and removed the need for large, yolky eggs. The cost of producing excessive yolk would have been a biological liability; in this case, natural selection (an evolutionary process) resulted in mammalian ova with little yolk. However, gastrulation and subsequent developmental events in reptiles, birds, and mammals are remarkably similar.

16.5 Gastrulation

Gastrulation is the first time embryonic cells undergo typical cell division with periods of growth between divisions. These cells start the process of specialization and undertake migrations from one embryonic location to another. Groups of some cells even undergo programmed cell death (*apoptosis*) as part of the developmental process. The basic body plan is established. This body is typically made up of three **primary germ layers** or tissues and a front and back end, which can be divided along its longitudinal axis into near identical right and left sides. In animals having at least the organ level of organization, the primary germ layers are called

Layer	Derivatives
Ectoderm	Nervous tissues, epidermis (keratinized stratified squamous epithelium) and the structures it forms
Mesoderm	Muscle tissues, connective tissues, and epithelia of the urinary and reproductive systems
Endoderm	Epithelial lining of most of the digestive tract and respiratory tract, and associated glands (e.g., the liver)

TABLE 16-2 Tissue Derivatives of the Primary Germ Layers

the **ectoderm, mesoderm, and endoderm** and give rise to organs and, after continued cell specializations, subtypes of the four adult basic tissues (Table 16-2).

As development proceeds, four extraembryonic membranes (also called *fetal membranes* in mammals) form in support developing bird embryos in shelled eggs (Figure 16-17) and mammalian embryos in the uterine wall. The exchange of gasses, nutrients, wastes, and so on takes place across the **placenta** (Figure 16-18), which is partly maternal and partly embryonic in origin.

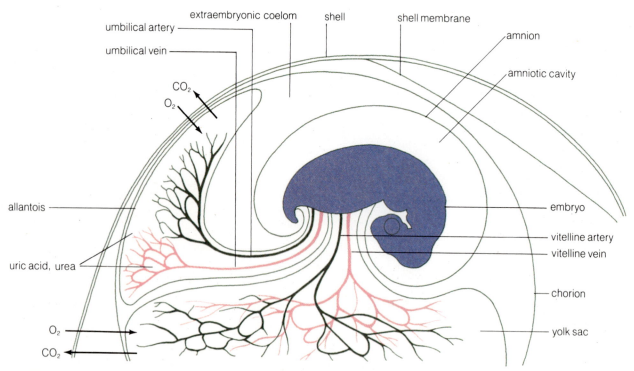

FIGURE 16-17 Extraembryonic membranes of the chicken. (*After B. M. Patten, National Sigma Xi Lecture, reprinted in American Scientist, 39:225–243, 1951. Used by permission.*)

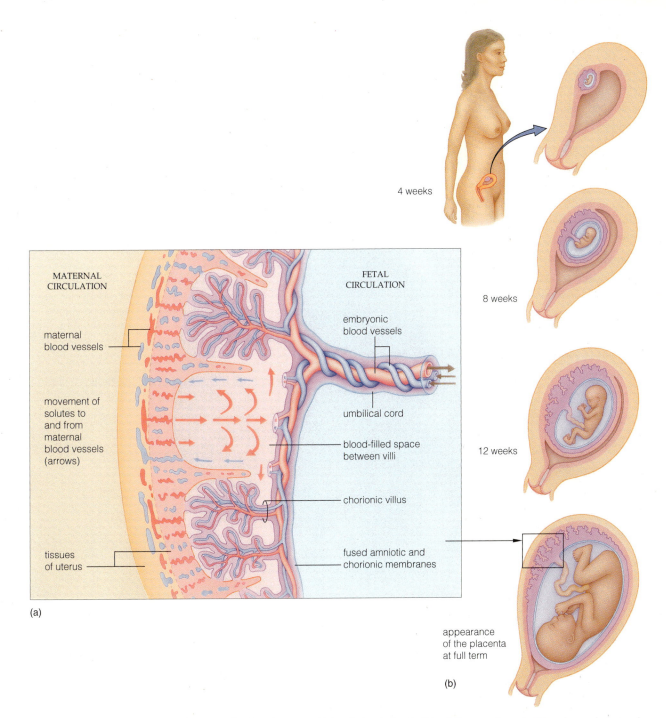

MATERNAL CIRCULATION

maternal blood vessels

movement of solutes to and from maternal blood vessels (arrows)

tissues of uterus

FETAL CIRCULATION

embryonic blood vessels

umbilical cord

blood-filled space between villi

chorionic villus

fused amniotic and chorionic membranes

(a)

4 weeks

8 weeks

12 weeks

appearance of the placenta at full term

(b)

FIGURE 16-18 (a) Relationship between fetal and maternal blood circulation in (b) a full-term placenta. Blood vessels extend from the fetus, through the umbilical cord, and into chorionic villi. Maternal blood spurts into spaces between the villi, but the two bloodstreams do not intermingle. Oxygen, carbon dioxide, and other small solutes diffuse across the placental membrane surface. *(After Starr, 2000.)*

MATERIALS

Per student:

- Compound light microscope
- Lens paper
- Bottle of lens cleaning solution (optional)
- Lint-free cloth (optional)

- Prepared slide of a whole mount of sea star development through gastrulation
- Prepared slides with whole mounts of chicken embryos at 18 and 24 hours of incubation

Per student pair:

- Two Syracuse dishes
- Two blue camel-hair brushes (optional)

- Two red camel-hair brushes
- Dissection microscope
- Fertile hen's eggs incubated for 18 and 24 hours (optional)

Per student group (bench or table):

- Models of human gastrulation

Per lab section:

- Frog embryos (early gastrulae) in pond water from eggs fertilized 21 hours before the lab (optional)
- Frog embryos (late gastrulae) in pond water from eggs fertilized 36 hours before the lab (optional)
- Pond water

Per lab room:

- Preserved specimens of early and late gastrulae of frog in easily accessible screw-top containers
- A source of dH_2O
- Pond water (optional)
- Models of early and late gastrulae of frog
- Boxes of different sizes of lab gloves
- Safety goggles

PROCEDURE

A. Gastrulation in Sea Stars

1. Reexamine with the compound microscope the prepared slide bearing whole mounts of the early stages of sea star development. Using the low- and medium-power objectives, refind a blastula (Figure 16-19a).

2. Locate an early gastrula (Figure 16-19b). As gastrulation starts, the vegetal pole flattens and folds in like a pocket to create a new cavity, the **archenteron** ("ancient gut").

3. Find a late gastrula (Figure 16-19c). The hole connecting the archenteron to the outside is called the **blastopore.** Gastrulation initially forms two layers of cells—the ectoderm covering the outside of the gastrula and the endoderm lining the archenteron. Then the mesoderm buds off from the inner tip of the archenteron, and its cells migrate to form a third layer of cells between the ectoderm and endoderm.

B. Gastrulation in Frogs

1. Gastrulation in frogs is affected by the large amount of yolk in the vegetal hemisphere. Because the

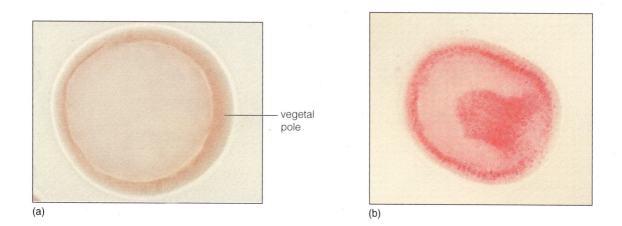

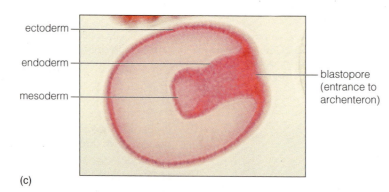

FIGURE 16-19 Gastrulation in the sea star (320X). (a) Late blastula, (b) early gastrula, and (c) late gastrula. *(Photos by D. Morton.)*

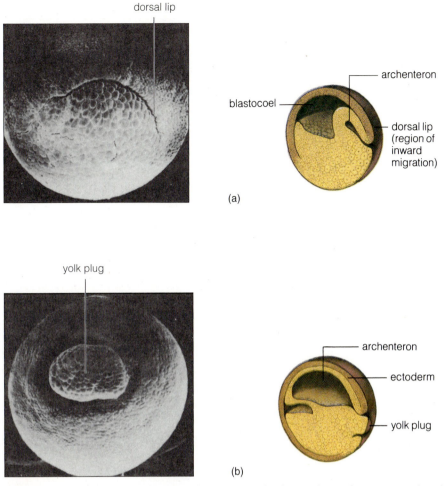

dorsal lip

archenteron

blastocoel

dorsal lip (region of inward migration)

(a)

yolk plug

archenteron

ectoderm

yolk plug

(b)

FIGURE 16-20 (a) Early and (b) late gastrulation in the frog. *(Photos from R. Kessel and C. Shih, Scanning Electron Microscopy in Biology, Springer-Verlag, 1974. Diagrams after Starr and Taggart, 2001.)*

pigmented cells of the animal pole divide faster, they partially overgrow the yolk-laden cells of the vegetal pole. Examine a living or preserved gastrula using the same protocol previously described for earlier developmental stages.

2. Examine models of an early and a late gastrula. Note that gastrulation does not occur simultaneously over the surface of the vegetal pole. Rather, it starts at a point just under what will be the anus of the adult frog and continues, forming a crescent (Figure 16-20a) that will close to form a circle around a plug of yolk-laden cells, the *yolk plug* (Figure 16-20b). The initial point of the folding-in is referred to as the *dorsal lip of the blastopore.*

C. Gastrulation in Chickens

1. Obtain a prepared slide bearing a whole mount of a chicken gastrula (18 hours of incubation) and one of an embryo incubated for 24 hours.

CAUTION Do not use the high-power objectives when examining whole mounts of thick material. Its use will break the coverslip.

2. Observe the gastrula using only the low- and medium-power objectives. Similar to cleavage, gastrulation in chickens is influenced by the large amount of yolk. The gastrula does not fold in like a pocket through a blastopore; rather, cells move or migrate into a groove on the blastoderm called the **primitive streak** (Figure 16-21a). At one end of the primitive streak, the migrating cells pile up to form *Hensen's node.* This is thought to be equivalent to the dorsal lip of the blastopore.

3. Examine the slide of the embryo at 24 hours of incubation. The three-layered embryo forms in the same axis as the primitive streak but in front of Hensen's node (Figure 16-21b). The three layers from the top of the embryo toward the yolk are the ectoderm,

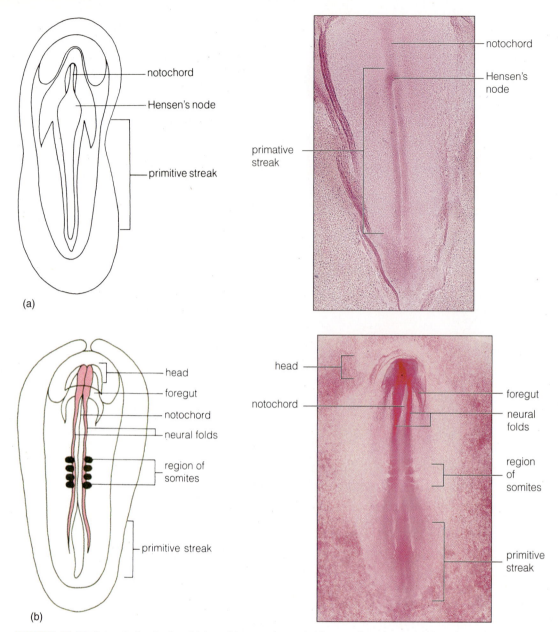

FIGURE 16-21 Gastrulation in the chicken. (a) gastrulation (18 hours of incubation) (28X); (b) early embryo (24 hours of incubation) (16X). The embryos have been removed from the surface of the yolk. *(Photos courtesy Biodisc, Inc.)*

mesoderm, and endoderm. One of the first recognizable structures in the embryo is the mesodermal **notochord.** The notochord induces the formation of the embryo's nervous system from the ectoderm. Find *neural folds* on either side of the notochord. With time, the neural folds will fuse like a zipper, anterior to posterior, and in so doing will form the **neural tube** (Figure 16-22). Likewise, the head of the embryo has lifted off the surface of the yolk and has thereby formed the anterior portion of the digestive tract, the *foregut.*

4. Note the two rows of **somites** on either side of the neural folds (Figure 16-21b). These are segmental condensations of mesoderm that will later develop into the skeleton, the skeletal muscles of the trunk, and the dermis of the skin. Count the number of pairs of somites present in this embryo and record this number:

_____

5. *Optional.* If your instructor has fertile hen's eggs incubated for 18 and 24 hours, examine them under the dissection microscope.

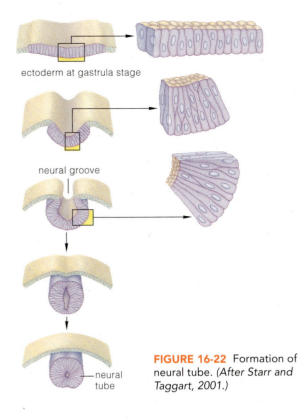

ectoderm at gastrula stage

neural groove

neural tube

FIGURE 16-22 Formation of neural tube. *(After Starr and Taggart, 2001.)*

D. Gastrulation in Humans

Examine the models of human gastrulation. Human gastrulation follows much the same scenario as that of chickens. However, before gastrulation occurs, a cavity forms between the inner cell mass and the trophoblast. This is the **amniotic cavity** of the *amnion* (Figure 16-23). Also, cells from the inner cell mass grow downward along the inner surface of the trophoblast, fuse, and form the **yolk sac cavity**. Between the amniotic and yolk sac cavities is the embryonic disk. The **embryonic disk** in mammals is the equivalent of the blastoderm in chickens. Thus, gastrulation commences in the embryonic disk with the formation of a primitive streak.

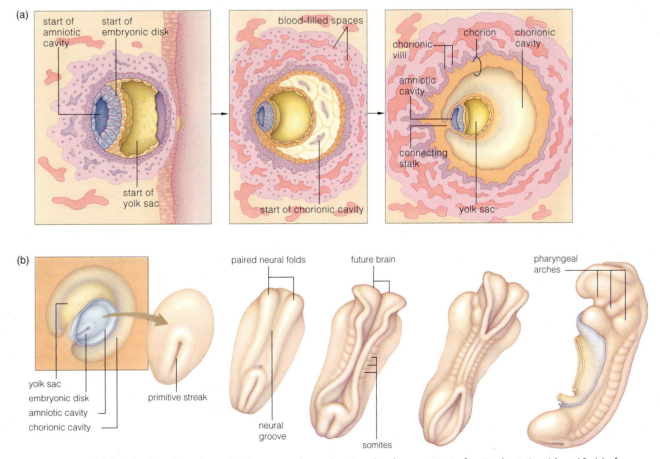

(a)

start of amniotic cavity

start of embryonic disk

start of yolk sac

blood-filled spaces

start of chorionic cavity

chorionic villi

chorion

chorionic cavity

amniotic cavity

connecting stalk

yolk sac

(b)

yolk sac
embryonic disk
amniotic cavity
chorionic cavity

primitive streak

paired neural folds

neural groove

future brain

somites

pharyngeal arches

FIGURE 16-23 (a) Series of sections through a human embryo showing development just after implantation (days 10–14 after fertilization). (b) Formation of the primitive streak (day 15 after fertilization). *(After Starr, 1999.)*

_____ **1.** Sperm
   (a) are male gametes
   (b) are female gametes
   (c) contain yolk
   (d) are produced in the ovaries

_____ **2.** Ova are
   (a) male gametes
   (b) female gametes
   (c) specialized for motility
   (d) both a and c

_____ **3.** The formation of gametes in the gonads can best be called
   (a) gametogenesis
   (b) spermatogenesis
   (c) oogenesis
   (d) none of the above

_____ **4.** Which structure is found in a section of the testis?
   (a) secondary spermatocytes
   (b) sperm
   (c) Sertoli cells
   (d) all of the above

_____ **5.** Oocytes are found in an ovary in
   (a) seminiferous tubules
   (b) follicles
   (c) corpora lutea
   (d) none of the above

_____ **6.** One primary oocyte will form
   (a) one ovum
   (b) four ova
   (c) up to three polar bodies
   (d) both a and c

_____ **7.** Which process results in a zygote?
   (a) meiosis
   (b) mitosis
   (c) fertilization
   (d) none of the above

_____ **8.** The entrance into the cavity formed during gastrulation in sea stars and frogs is called the
   (a) archenteron
   (b) blastocoel
   (c) blastocyst
   (d) blastopore

_____ **9.** The three primary germ layers are formed during
   (a) fertilization
   (b) cleavage
   (c) organ formation
   (d) gastrulation

_____ **10.** Muscle tissues, connective tissues, and the epithelia lining the urinary and reproductive systems arise from
   (a) mesoderm
   (b) endoderm
   (c) ectoderm
   (d) two of the above

EXERCISE 16 Reproduction and Development

POST-LAB QUESTIONS

16.1 Gametes

1. Define and characterize the following:

   a. gametes

   b. gonads

   c. gametogenesis

2. Describe the similarities and differences between sperm and ova.

3. Compare chicken, frog, and mammalian ova as to size and the amount of yolk present.

16.2 Gametogenesis

4. Why do you think four sperm cells are produced as a result of gametogenesis but only one ovum is produced?

5. Are oogonia present in adult women? _____ (yes or no) Explain your answer.

16.3 Sperm Penetration and Fertilization in the Frog

6. Read about the gray crescent in your textbook. How does its location relate to the site of sperm penetration?

16.4 Cleavage

7. Identify the following sea star developmental stages (185X). (*Photos by D. Morton.*)

a. _____   b. _____

c. _____   d. _____

8. How do the amount and distribution of yolk in animal zygotes affect cleavage?

16.5 Gastrulation

9. List the primary germ layers formed by gastrulation. What tissues will they form in the adult?

Food for Thought

10. As various newspaper articles, books, and movies suggest, the cloning of human beings—producing new individuals from activated somatic cells, perhaps followed by uterine implant—is now possible. Can you suggest any biological advantages or disadvantages to having the earth populated with clones of a few of the best examples of our species?

Cell Reproduction

OBJECTIVES

After completing this exercise, you will be able to

1. define *chromosome, diploid, homologous chromosome (homologue), sister chromatid, mitosis, cytokinesis, meiosis, haploid, gamete, ovum, sperm, fertilization, zygote, chromatin, centriole, spindle, centromere, gene, allele, gene pair, locus, synapsis, genotype, crossing over, segregation, independent assortment, linked genes, nondisjunction, spermatogenesis,* and *oogenesis.*

2. identify the stages of the cell cycle.

3. distinguish between mitosis and cytokinesis.

4. identify the structures involved in nuclear and cell division (those in **boldface**) and describe the role each plays.

5. describe the process of meiosis and recognize events that occur during each stage.

6. describe and discuss the consequences of crossing over, segregation, and independent assortment.

7. describe the process of nondisjunction and chromosome number abnormalities in resulting gametes and zygotes.

8. list the differences and similarities between meiosis and mitosis.

9. describe the processes of gametogenesis and identify the meiotic products in males and females.

Introduction

"All cells arise from preexisting cells." This is one part of the cell theory. It's easy to understand this concept if you think of a single-celled *Amoeba* or bacterium. Each cell divides to give rise to two entirely new individuals, and it is fascinating that each of us began life as *one* single cell and developed into this astonishingly complex animal, the human. Our first cell has *all* the hereditary information we'll ever get.

In eukaryotic organisms, the **chromosomes** consist of DNA and proteins complexed together within the nucleus. Human body cells (somatic cells) have 46 chromosomes, consisting of 23 pairs. These cells are said to be **diploid,** also written as 2*n,* which means that there are two of each of the 23 kinds of chromosome, or two complete sets. The "look-alike" chromosome pairs are called **homologous chromosomes.**

Somatic cells divide to produce more cells to support growth and for repair and renewal of tissues. Cell division

is preceded by duplication of the chromosomes; for much of the cell division process, the two identical copies are attached together and are called **sister chromatids.**

Cell division of somatic cells usually involves two processes: **mitosis** (nuclear division) and **cytokinesis** (cytoplasmic division). During mitosis, the duplicated sister chromatids are separated into two newly formed daughter nuclei; daughter nuclei are diploid, each containing two full sets of homologous chromosomes identical to each other *and* to the parent cell that gave rise to them. Cytokinesis then ensures that each new cell contains all the metabolic machinery necessary for life.

Mitosis alone is not adequate for every nuclear division job required to complete a human life cycle. Sexual reproduction requires **meiosis,** which, similar to mitosis, is a process of nuclear division usually followed by cytokinesis. In meiosis, however, the chromosome number is halved, resulting in daughter nuclei containing one of each kind of chromosome. Those nuclei are said to be **haploid,** or *n.* Moreover, whereas mitosis produces daughter nuclei with identical chromosomes, each nucleus produced by meiosis is genetically unique.

In animals, the cells containing the daughter nuclei produced by meiosis are called **gametes:** *ova* (singular is *ovum*) if the parent is female, **sperm** cells if male. As you probably know, gametes are produced in the gonads—the ovaries and testes, respectively. In fact, this is the *only* place where meiosis occurs in higher animals. Each human ovum contains 23 chromosomes, and each sperm 23 contains chromosomes.

Fertilization, the fusion of egg and sperm nuclei, produces a single-celled, genetically unique, diploid **zygote.** The zygote divides into two cells via mitosis and cytokinesis, those two into four, and so on to produce a multicellular organism. During cell division, each new cell receives a complete set of hereditary information and an assortment of cytoplasmic components.

In this exercise, you will explore the nuclear and cellular division processes of mitosis, cytokinesis, and meiosis.

17.1   The Cell Cycle

Dividing somatic cells pass through a regular sequence of events called the *cell cycle* (Figure 17-1). Notice that the majority of the time is spent in interphase and that actual nuclear division—mitosis—is but a brief portion of a cell's lifespan.

Interphase is composed of three parts (Figure 17-1): the G1 period, during which cytoplasmic growth takes place; the S period, when the DNA is duplicated; and the G2 period, when structures directly involved in mitosis are synthesized. Mitosis and cytokinesis follow, with each new daughter cell then entering interphase.

In animal cells, a zygote undergoes a special type of cell division, *cleavage*, in which no increase in cytoplasm occurs between divisions. A ball of cells called a *blastula* is produced by cleavage early in development. Within the blastula, repeated nuclear and cytoplasmic divisions take place. Because of this, the whitefish blastula is an excellent sample in which to observe the cell cycle of an animal.

MATERIALS

Per student:

- Prepared slide of whitefish blastula mitosis
- Compound microscope

PROCEDURE

A. Interphase

Obtain a slide labeled "whitefish blastula." Scan it with the low-power objective and then at medium power. This slide has numerous thin sections of a blastula, each with dozens of individual cells. Select one section (Figure 17-2) and then switch to the high-dry objective for detailed observation.

As you proceed with this activity, draw and label the cells in Figure 17-3 to show the correct sequence of events in the cell cycle of whitefish blastula.

1. *Interphase.* Locate a cell in interphase (see Figure 17-4). Note the presence of a typical nucleus with no distinctive structures within. During interphase, the DNA–protein complex of each chromosome is extended as an extremely thin strand, called **chromatin**. Before the onset of mitosis, in S-phase, the DNA duplicates.

2. Draw an interphase cell above the word "Interphase" in Figure 17-3 and label the cytoplasm, nucleus, and plasma membrane.

Refer to Figure 17-5 as you identify and study cells in various phases of mitosis and cytokinesis.

B. Mitosis

1. *Prophase.* During **prophase**, chromatin folds and condenses, and the **chromosomes** become visible as threadlike structures. Within the cytoplasm are two pairs of **centrioles**. Although the centrioles are too small to be resolved with your light microscope, you can see a starburst pattern of spindle fibers (microtubules) that appear to radiate from the centrioles. Other microtubules extend between the centrioles, forming the **spindle** (Figure 17-6).

Find a prophase cell, identifying the spindle and starburst cluster of fibers around the centriole. Draw the prophase cell in the proper location on Figure 17-3. Label the spindle, chromosomes, cytoplasm, and position of the plasma membrane.

The transition from prophase to the next stage (metaphase) is marked by the breakdown and disappearance of the nuclear envelope.

2. *Metaphase.* The key thing to look for in a metaphase cell is the position of the chromosomes. During metaphase, the spindle fiber microtubules attach to the centromere region of each chromosome (the indented area where sister chromatids are attached to each other). The fibers move the duplicated chromosomes (each consisting of two **sister chromatids**) so they line up on the **spindle equator**, the imaginary midline of the cell (Figure 17-6).

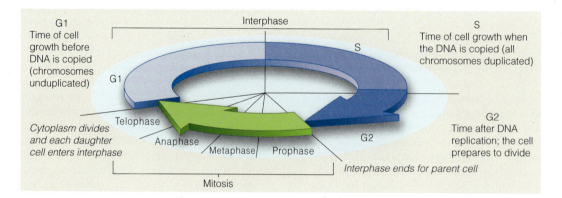

FIGURE 17-1 The cell cycle.

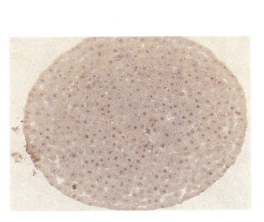

FIGURE 17-2 Section of a whitefish blastula (75X). (*Photo by J. W. Perry.*)

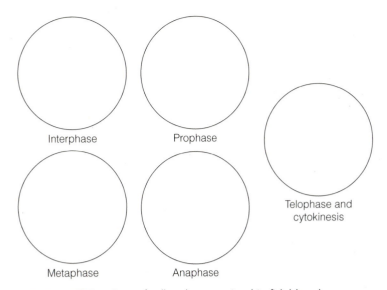

Interphase   Prophase

Metaphase   Anaphase

Telophase and cytokinesis

FIGURE 17-3 Drawings of cell cycle stages in whitefish blastula. Labels: cytoplasm, nucleus, plasma membrane, spindle, chromosomes, spindle equator, sister chromatids, daughter nuclei, chromatin, furrow. (Note: some terms are used more than once.)

Locate a metaphase cell. Draw it in the proper location on Figure 17-3 and label the chromosomes, spindle equator, spindle, and plasma membrane.

3. *Anaphase.* Anaphase begins with the separation of sister chromatids into individual (daughter) chromosomes. Spindle fibers attached at the centromeres shorten, pulling daughter chromosomes toward opposite poles. Note that the daughter chromosomes consist of a single DNA molecule, a single chromatid.

Observe a blastula cell in anaphase and draw it in the proper location on Figure 17-3. Label the separating daughter chromosomes, spindle, cytoplasm, and plasma membrane.

4. *Telophase.* Telophase is characterized by the arrival of the individual daughter chromosomes at the poles. A nuclear envelope forms around each daughter nucleus. Find a telophase cell.

FIGURE 17-4 Micrograph of whitefish blastula cell in interphase of cell cycle.

Is the spindle still visible? _____

Can you see evidence of a nuclear envelope forming around the chromosomes?_____

Draw the telophase cell in Figure 17-3. Label the daughter nuclei, chromatin, cytoplasm, and plasma membrane.

C. Cytokinesis

Cytokinesis, division of the cytoplasm into two separate daughter cells, takes place in animal cells by *furrowing.* To visualize how furrowing takes place, imagine wrapping a string around a balloon and slowly tightening the string until the balloon has been pinched in two. In life, the human cell is pinched in two, forming two discrete cytoplasmic entities, each with a single nucleus. Figure 17-7 illustrates cytokinesis and the cleavage furrow in an animal cell.

Find a cell in the blastula undergoing cytokinesis. The telophase cell that you drew in Figure 17-3 may also show an early stage of cytokinesis. Label the cleavage furrow if it does.

17.2   Simulating Mitosis

Understanding chromosome movements is crucial to understanding mitosis. This is a simple activity but a valuable one. It will be especially helpful when comparing the events of mitosis with those of meiosis later.

MATERIALS

Per student pair:

- 30 pop beads each of two colors
- Eight magnetic centromeres

(a) Early Prophase

Mitosis begins in the nucleus. The chromatin begins to appear grainly as it organizes and condenses. The centromere is duplicated.

(b) Prophase

The chromosomes become visible as discrete structures as they condense further. Microtubules assemble and move one of the two centrosomes to the opposite side of the nucleus, and the nucleus envelope breaks up.

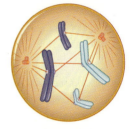

(c) Transition to Metaphase

The nuclear envelope is gone, and the chromosomes are at their most condensed. Microtubules of this bipolar spindle assemble and attach sister chromatids to opposite spindle poles.

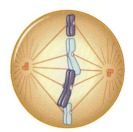
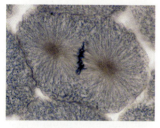

(d) Metaphase

All of the chromosomes are aligned midway between the spindle poles. Microtubules attach each chromatid to one of the spindle poles, and its sister to the opposite pole.

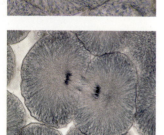

(e) Anaphase

Motor proteins moving along the spindle microtubules drag the chromatids toward the spindle poles, and the sister chromatids separate. Each sister chromatid is now a separate chromosome.

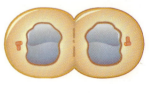

(f) Telophase

The chromosomes reach the spindle poles and decondense. A nuclear envelope begins to form around each cluster: new plasma membrane may assemble between them. Mitosis is over.

FIGURE 17-5 Drawings and micrographs of whitefish blastula cells in phases of mitosis. *(Photo credit: (a-f) left, Michael Clayton/University of Wisconsin, Deparment of Botany; (a-d) right, Ed Reschke.)*

PROCEDURE

1. Build the components for two pairs of chromosomes by assembling strings of pop beads as follows:

   (a) Assemble two strands of pop beads with four pop beads of one color on one arm and five beads of the same color on the other arm, with a magnetic centromere connecting the two arms.

   (b) Repeat step a but use pop beads of the second color.

   (c) Assemble two more strands of pop beads but with three pop beads of one color on each arm.

   (d) Repeat step c using pop beads of the second color.

   You should have four long strings, two of each color, and four short strings, with two of each color. Thus, you will have created two homologous pairs of chromosomes. Each pop-bead string should have a magnetic centromere at its midpoint. Note that pop-bead strings are able to attach to each other at the magnetic centromere. Each pop-bead string represents a single molecule of DNA plus proteins.

2. In the center of your workspace, place **one** of each kind of strand, which represents the nucleus. You have created a diploid interphase nucleus with four "chromosomes," two long and two short. Each chromosome consists of a single DNA molecule at this stage.

3. Manipulate these model chromosomes through the phases of the cell cycle, *beginning in G1 of interphase* and proceeding through the rest of interphase, mitosis, and cytokinesis. Check with your instructor to ensure that you understand the mechanics of chromosomal duplication, mitosis, and cytokinesis.

4. Describe the number and kinds of chromosomes found in each daughter nucleus after mitosis and cytokinesis, as well as whether they are diploid or haploid. How do the chromosomes of the daughter nuclei compare with those of the starting parental nucleus?

   _____

   _____

   _____

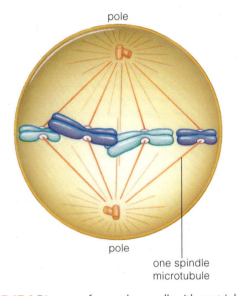

pole

pole

one spindle
microtubule

FIGURE 17-6 Diagram of metaphase cell, with centrioles and spindle apparatus.

17.3 Meiosis

Similar to mitosis, meiosis is a process of nuclear division. During mitosis, the number of chromosomes in the daughter nuclei remains the same as that in the parental nucleus. In meiosis, however, the genetic complement is halved, resulting in daughter nuclei containing only half the number of chromosomes as the parental nucleus. Thus, although mitosis is sometimes referred to as an *equational division*, meiosis is often called *reduction division*.

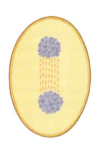
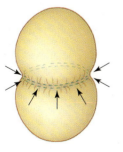

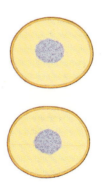

1 Mitosis is over, and the spindle is disassembling.

2 At the former spindle equator, a ring of microfilaments attached to the plasma membrane contracts.

3 As the microfilament ring shrinks in diameter, it pulls the cell surface inward.

4 Contractions continue; the cell is pinched in two.

FIGURE 17-7 Cytokinesis of an animal cell.

Moreover, whereas mitosis is completed after a single nuclear division, two divisions, called meiosis I and meiosis II, occur during meiosis. A diploid (2n) parental nucleus begins the process of meiosis; four daughter nuclei contain the haploid (n) number of chromosomes at the completion of meiosis. Note this important concept: *Meiosis always halves the chromosome number. The diploid chromosome number is eventually restored when two haploid gamete nuclei fuse during fertilization.*

Examine Figure 17-8, which shows the life cycle of a higher animal. As you learned in Exercise 16, meiosis occurs in human gonads, where ova are produced in ovaries and sperm are produced in testes.

Within the nucleus of a cell, each chromosome bears **genes,** which are sections of DNA that are units of inheritance. Genes may exist in two or more alternative forms called **alleles.** Each homologue bears genes for the same traits; these are the **gene pairs.** However, the homologues may or may not have the same alleles. An example will help here.

Suppose the trait in question is earlobe shape, with only two possible options, free or attached (Figure 17-9a and 17-9b). The gene codes (provides the information) for earlobe shape. There are two homologues in the same nucleus, so each bears the gene for earlobe shape. But on one homologue, the fallele might code for free earlobes, but the allele on the other homologue might code for attached earlobes, *f* (Figure 17-9c). There are two other possibilities. The alleles on *both* homologues might code for free earlobes (Figure 17-9d), or they *both* might code for attached earlobes (Figure 17-9e). These three possibilities are mutually exclusive.

Understanding meiosis is absolutely necessary for understanding the patterns of inheritance in Mendelian genetics. Gregor Mendel, an Austrian monk, spent years deciphering the complexity of simple genetics. Although he knew nothing of genes and chromosomes, he noted certain patterns of inheritance and formulated three principles, now known as Mendel's principles of recombination, segregation, and independent assortment. The following activities demonstrate the events of meiosis and the genetic basis for Mendel's principles.

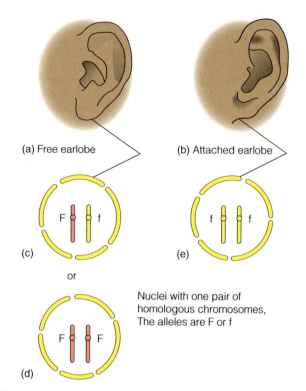

FIGURE 17-9 Chromosomal control of earlobe shape. (c – e show a nucleus with one pair of homologous chromosomes) The alleles for earlobe shape are F or f.

A. Construction of Simulated Homologous Chromosomes

MATERIALS

Per student pair:

- 30 pop beads each of two colors (e.g., red and yellow)
- Eight magnetic centromeres
- Marking pens
- Eight pieces of string, each 40 cm long
- Meiotic diagram cards similar to those used here
- Colored pencils

Per student group (table):

- Bottle of 95% ethanol to remove marking ink
- Tissues

PROCEDURE

Work in pairs.

1. Build the components for two pairs of homologous chromosomes by assembling strings of pop beads as you did in Section 17.2:

   (a) Assemble two strands of pop beads with four pop beads of one color on one arm, five beads on the second arm, and a magnetic centromere connecting the two arms.

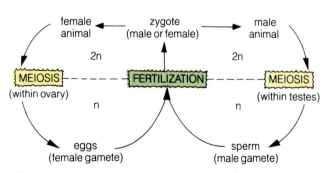

FIGURE 17-8 Life cycle of higher animals.

(b) Repeat step (a) but use pop beads of the second color.

(c) Assemble two more strands of pop beads but with three pop beads of one color on each arm.

(d) Repeat step (c) using pop beads of the second color.

You should have four long strings, two of each color, and two short strings, also two of each color. Each pop-bead string should have a magnetic centromere at its midpoint. Each pop-bead string represents a single molecule of DNA plus proteins, with each bead representing a gene.

2. In the center of your workspace, place **one** of each kind of strand, which represents the interphase nucleus of a cell that will undergo meiosis. You have created a nucleus with four "chromosomes," two long and two short. The long strands represent one homologous pair, and the short strands represent a second homologous pair of chromosomes.

Assume that these chromosomes represent the diploid condition. The two colors represent the origin of the chromosomes: One homologue (color _____) came from the male parent, and the other homologue (color _____) came from the female parent.

3. The four single-stranded chromosomes represent four unduplicated chromosomes. Now simulate DNA duplication during the S-phase of interphase (Figure 17-1), whereby each DNA molecule and its associated proteins are copied exactly. The two copies, called sister chromatids, remain attached to each other at their centromeres (Figure 17-10). During chromosome replication, the genes also duplicate. Thus, alleles on sister chromatids are identical.

How many sister chromatids are there in a duplicated chromosome? _____

How many chromosomes are represented by four sister chromatids? _____ By eight? _____

What is the diploid number of this starting (parental) nucleus? (Hint: Count the number of homologues to obtain the diploid number.) _____

4. As mentioned previously, genes may exist in two or more alternative forms, called alleles. The location of an allele on a chromosome is its **locus** (plural: *loci*). Using the marking pen, mark two loci on each long chromatid with letters to indicate alleles for a common trait. Suppose the long pair of homologous chromosomes codes for two traits, skin pigmentation and the presence of attached earlobes in humans. We'll let the capital letter *A* represent the allele for normal pigmentation and a lowercase *a* the allele for albinism (the absence of skin pigmentation); *F* will represent free earlobes and *f* attached earlobes as before. A suggested marking sequence is illustrated in Figure 17-10.

5. Let's assign a gene to our second homologous pair of chromosomes, the short pair. We'll suppose this gene codes for the production of an enzyme necessary for metabolism. On one homologue (consisting of two chromatids), mark the letter *E*, representing the allele causing enzyme production. On the other homologue, *e* represents the allele that prevents normal enzyme production.

6. Obtain a meiotic diagram card similar to the one in Figure 17-11. Manipulate your model chromosomes through the stages of meiosis described below, locating the chromosomes in the correct diagram circles (representing nuclei at various stages) as you go along. Reference to Figure 17-11 will be made at the proper steps. *Do not* draw on the meiotic diagram cards.

B. Simulation of Meiosis without Crossing Over

Although crossing over is a nearly universal event during meiosis, we will first work with a simplified model to illustrate chromosomal movements and separations during meiosis. Refer to Figure 17-12 as you manipulate your model.

1. *Late interphase.* During interphase, the nuclear envelope is intact, and the chromosomes are randomly distributed throughout the nucleus. All four duplicated chromosomes (with eight chromatids total) should be in the parental nucleus, indicating that DNA duplication took place. The sister chromatids of each homologue should be attached by their magnetic centromeres, but the four homologues should be separate. Your model nucleus contains a diploid number 2n = 4.

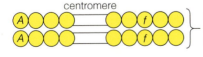
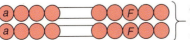

FIGURE 17-10 One homologous pair of pop bead chromosomes.

The pop-bead chromosomes should appear during interphase in the parental nucleus as shown in Figure 17-11. Be sure to mark the locations of the alleles on that Figure if yours differ from those shown. You will draw chromosomes as you work through the process of meiosis. Use different pencil or pen colors to differentiate the homologues on your drawings.

2. *Meiosis I.* During meiosis I, homologues are separated from each other into different nuclei. Daughter nuclei created are thus haploid.

   (a)*Prophase I.* During the first prophase, the parental nucleus contains four duplicated homologous chromosomes, each composed of two sister chromatids joined at their centromeres. The chromatin condenses to form discrete, visible chromosomes. The homologues pair with each other. This pairing is called **synapsis.** Slide the two homologues together.

   Twist the chromatids about one another to simulate synapsis.

   The nuclear envelope disorganizes at the end of prophase I.

   (b)*Metaphase I.* Homologous chromosomes now move toward the spindle equator, the centromeres of each homologue coming to lie *on either side of the equator.* Spindle fibers, consisting of aggregations of microtubules, attach to the centromeres. One homologue attaches to microtubules extending from one pole, and the other homologue attaches to microtubules extending from the opposite spindle pole.

   To simulate the spindle fibers, attach one piece of string to each centromere. Then lay the free ends of strings from two homologues toward one spindle pole and the ends of the other homologues toward the opposite pole.

   (c)*Anaphase I.* During anaphase I, the homologous chromosomes separate, one homologue moving toward one pole and the other toward the opposite pole. The movement of the chromosomes is apparently the result of shortening of some spindle fibers and lengthening of others. Each homologue is still in the duplicated form, consisting of two sister chromatids.

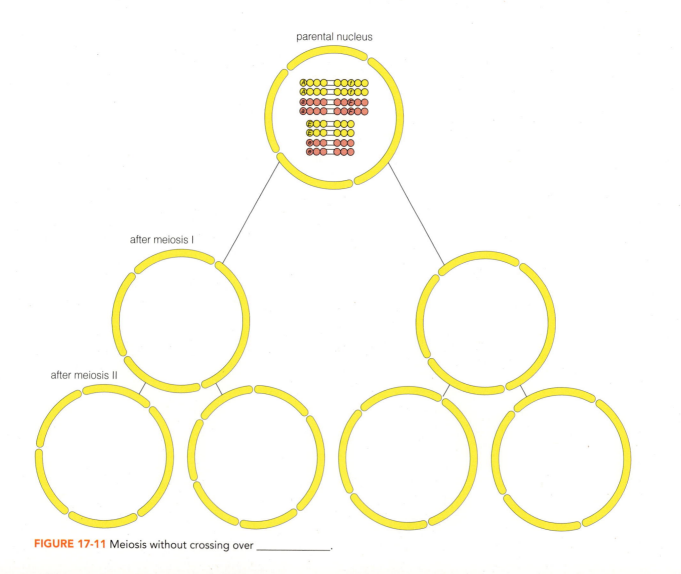

parental nucleus

after meiosis I

after meiosis II

FIGURE 17-11 Meiosis without crossing over _____.

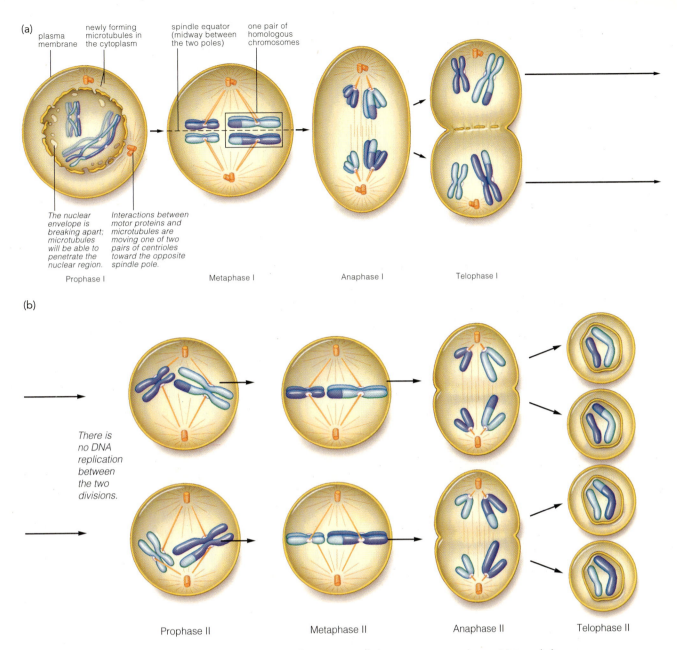

(a)

plasma membrane

newly forming microtubules in the cytoplasm

spindle equator (midway between the two poles)

one pair of homologous chromosomes

The nuclear envelope is breaking apart; microtubules will be able to penetrate the nuclear region.

Interactions between motor proteins and microtubules are moving one of two pairs of centrioles toward the opposite spindle pole.

Prophase I   Metaphase I   Anaphase I   Telophase I

(b)

There is no DNA replication between the two divisions.

Prophase II   Metaphase II   Anaphase II   Telophase II

FIGURE 17-12 Meiosis in a generalized animal germ cell. Two pairs of chromosomes are shown. Maternal chromosomes are shaded purple. Paternal chromosomes are shaded light blue. *(After Starr, 2000.)*

Pull the two strings of one homologous pair toward its spindle pole and the other toward the opposite spindle pole, separating the homologues from one another. Repeat with the second pair of homologues.

(d) *Telophase I.* Continue pulling the string spindle fibers until each homologue is at its respective pole. The first meiotic division is now complete. You should have two nuclei, each containing two chromosomes (one long and one short) and with each consisting of two sister chromatids.

Draw your pop-bead chromosomes as they appear after meiosis I on the two nuclei labeled "after meiosis I" on Figure 17-11. Depending on

the organism involved, an interphase (interkinesis) and cytokinesis may precede the second meiotic division *or* each nucleus may enter directly into meiosis II. The chromosomes decondense into chromatin form.

Important: Note here that DNA synthesis *does not* occur between meiosis I and meiosis II.

Before meiosis II, the spindle is rearranged into two spindles, one for each nucleus.

3. *Meiosis II.* During meiosis II, sister chromatids are separated into different daughter nuclei. The result is four haploid nuclei.

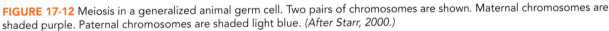

(a) *Prophase II.* At the beginning of the second meiotic division, the sister chromatids are still attached by their centromeres. During prophase II, the nuclear envelope disorganizes, and the chromatin recondenses.

(b) *Metaphase II.* Within each nucleus, the duplicated chromosomes align with the equator, the centromeres lying *on the equator.* Spindle fiber microtubules attach the centromeres of each chromatid to opposite spindle poles.

Your string spindle fibers should be positioned so that the two spindle fiber strings from sister chromatids lie toward opposite poles. Note that each nucleus contains only *two* duplicated chromosomes (one long and one short), each consisting of *two* sister chromatids.

(c) *Anaphase II.* The centromeres divide and sister chromatids separate, moving to opposite poles. Pull on the string until the two sister chromatids of each duplicated chromosome separate. After the sister chromatids separate, each is an individual (not duplicated) daughter chromosome.

(d) *Telophase II.* Continue pulling on the string spindle fibers until the daughter chromosomes are at opposite poles. The nuclear envelope reforms around each chromosome set, and the chromosomes decondense back into chromatin form. Four daughter nuclei now exist. Note that each nucleus contains two individual unduplicated chromosomes (each formerly a chromatid) originally present within the parental nucleus. These nuclei and the cells they're in generally undergo a differentiation and maturation process to become gametes in humans.

Draw your pop-bead chromosomes as they appear after meiosis II in the "gamete nuclei" of Figure 17-11. Your diagram should indicate the genetic (chromatid) complement *before* meiosis (in the parental nucleus) and *after* each meiotic division, not during all the stages of each division.

Remember that meiosis takes place in both male and female organisms (see Figure 17-8).

If the parental nucleus was from a male, what is the gamete called? _____

If female? _____

Is the parental nucleus diploid or haploid? _____

Are the nuclei produced after the *first* meiotic division diploid or haploid? _____

Are the nuclei of the gametes diploid or haploid?

_____

What is the **genotype** of each gamete nucleus after meiosis II? (The genotype is the genetic composition of an organism, or the alleles present. Another way to ask this question is: What alleles are present in each gamete nucleus? Write these in the format: *AFE, afe,* and so on.)

_____

If you answered the preceding questions correctly, you might logically ask, "If the chromosome number of the gametes is the same as that produced after the first meiotic division, why bother to have two separate divisions? After all, the genes present are the same in both gametes and first-division nuclei."

There are two answers to this apparent paradox. The first, and perhaps the most obvious, is that the second meiotic division ensures that a *single* chromatid (nonduplicated chromosome) is contained within each gamete. After gametes fuse, producing a zygote, the genetic material duplicates before the zygote undergoes mitosis. If gametes contained two chromatids per chromosome, the zygote would have four copies of each chromosome, and duplication before zygote mitotic division would produce eight, twice as many as the organism should have. If DNA duplication within the zygote were not necessary for the onset of mitosis, this problem would not exist. Alas, DNA synthesis apparently is a necessity to initiate mitosis.

You can discover the second answer for yourself by continuing with the exercise because although you have simulated meiosis, you have done so without showing what happens in *real* life. That's the next step.

C. Simulation of Meiosis with Crossing Over

A very important event that results in a reshuffling of alleles on the chromatids occurs during prophase I. Recall that synapsis results in pairing of the homologues. During synapsis, the chromatids break, and portions of chromatids bearing genes for the same characteristic (but perhaps *different* alleles) are exchanged between *nonsister* chromatids. This event is called **crossing over,** and it results in **recombination** (shuffling) of alleles.

1. Look again at Figure 17-10. Distinguish between sister and nonsister chromatids. Now look at Figure 17-13, which demonstrates crossing over in one pair of homologues.

2. Return your chromosome models to the parental nucleus condition with two pairs of homologues entering prophase I. (Recall that DNA duplication will have occurred during interphase.)

3. To simulate crossing over, break the four-bead arms from the centromeres of two *nonsister* chromatids in the long homologue pair, exchanging bead colors between the two arms. During actual crossing over, the chromosomes may break anywhere within the arms.

   Crossing over is virtually a universal event in meiosis. Each pair of homologues may cross over in several places simultaneously during prophase I.

4. Manipulate your model chromosomes through meiosis I and II again and watch what happens to the distribution of the alleles as a consequence of the

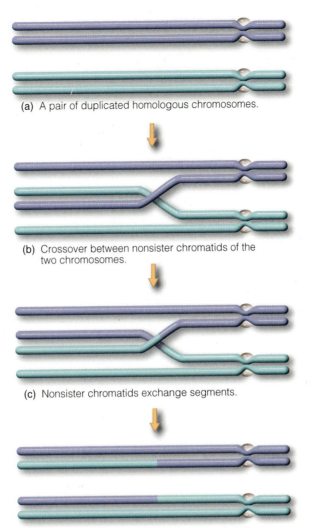

(a) A pair of duplicated homologous chromosomes.

(b) Crossover between nonsister chromatids of the two chromosomes.

(c) Nonsister chromatids exchange segments.

(d) Homologues have new combinations of alleles.

FIGURE 17-13 Crossing over in one pair of homologues. Maternal chromosomes are shaded purple. Paternal chromosomes are shaded light blue. (*After Starr, 2000.*)

crossing over. Fill in Figure 17-14 as you did before but this time show the effects of crossing over. Again, use different colors in your sketches.

What are the genotypes of the gamete nuclei?

_____

Is the distribution of alleles present in the gamete nuclei after crossing over the same as that which was present without crossing over?

_____

Is the distribution of alleles present in the gamete nuclei after crossing over the same as that in the nuclei after the first meiotic division?

_____

Crossing over provides for genetic recombination, resulting in increased variety. How many different genetic *types* of daughter chromosomes are present in the gamete nuclei without crossing over (Figure 17-11)?

_____

How many different types are present with crossing over (Figure 17-14)?

_____

We think you would agree that a greater number of *types* of daughter chromosomes indicates greater *variety*.

Recall that the parental nucleus contained a pair of homologues, each homologue consisting of two sister chromatids. Because sister chromatids

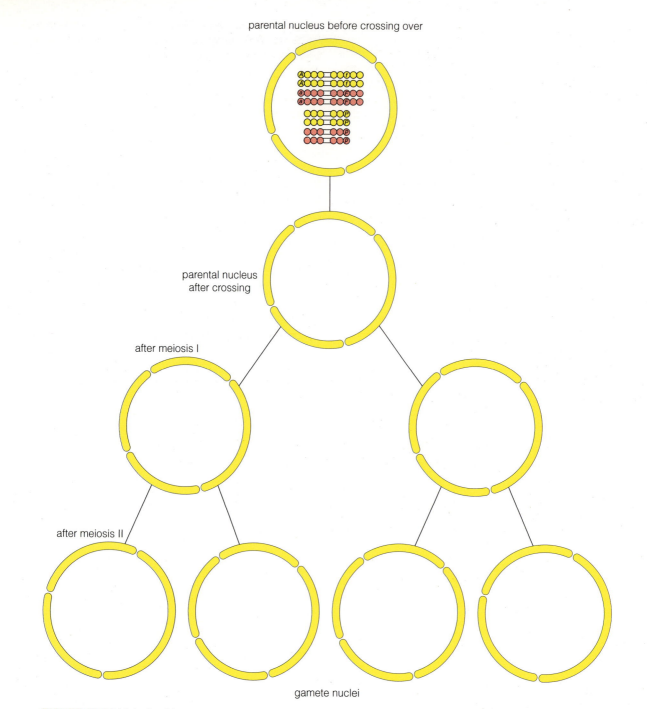

parental nucleus before crossing over

parental nucleus
after crossing

after meiosis I

after meiosis II

gamete nuclei

FIGURE 17-14 Meiosis with crossing over.

are identical in all respects, they have the same alleles of a gene (see Figure 17-10). As your models showed, the alleles on nonsister chromatids may not (or may) be identical; they bear the same genes but may have different alleles, different forms of some genes.

What is the difference between a gene and an allele?

_____

_____

Let's look at a single set of alleles on your model chromosomes that are, say, the alleles for pigmentation, *A* and *a*. Both alleles were present in the parental nucleus. How many are present in the gametes?

_____

This illustrates Mendel's first principle, **segregation.** Segregation means that during gamete formation, pairs of alleles are separated (segregated) from each other and end up in different gametes.

D. Demonstrating Independent Assortment

Manipulate your model chromosomes again through meiosis with crossing over (Figure 17-14), searching for different possibilities in chromosome distribution that would make the gametes genetically different.

Does the distribution of the alleles for enzyme production to different gametes on the second set of homologues have any bearing on the distribution of the alleles on the first set (alleles for skin pigmentation and earlobe condition)? _____

This distribution demonstrates the **principle of independent assortment**, which states that segregation of alleles into gametes is independent of the segregation of alleles for other traits, *as long as the genes are on different sets of homologous chromosomes.* Genes that are on different (nonhomologous) chromosomes are said to be **nonlinked.** By contrast, genes for different traits that are on the same chromosome are **linked.**

Because the genes for enzyme production and those for skin pigmentation and earlobe attachment are on different homologous chromosomes, these genes are _____, but the genes for skin pigmentation and earlobe attachment are _____ because they are on the same chromosome.

In reality, most organisms have many more than two sets of chromosomes. Humans have 23 pairs (2n = 46), but some plants literally have hundreds!

A thorough understanding of meiosis is necessary to understand genetics. With this foundation, you'll find that problems involving Mendelian genetics are easy and fun to do. Without an understanding of meiosis, Mendelian genetics will be hopelessly confusing.

E. Nondisjunction and the Production of Gametes with Abnormal Chromosome Number

Errors in the process of meiosis can occur in many ways. Perhaps the best understood error process is that of **nondisjunction,** when one or more pairs of homologous chromosomes fails to separate in anaphase I. The result is gamete nuclei with too few or too many chromosomes.

1. Begin to manipulate your model chromosomes to show meiosis without crossing over (as in Figure 17-11). In modeling events at metaphase I, however, arrange the spindle fiber threads for the long pair of homologues so that they all extend to the same pole.

2. Model anaphase I, pulling the chromosomes toward their respective poles. Nondisjunction occurs in the long pair of homologues, with both duplicated chromosomes being pulled to the same pole (see Figure 17-15).

3. Continue to manipulate the model chromosomes through the remainder of the meiotic process.

How many chromosomes are found in gamete nuclei?

_____

How does this compare with the chromosome number in normal gametes? _____

chromosome number in gametes:

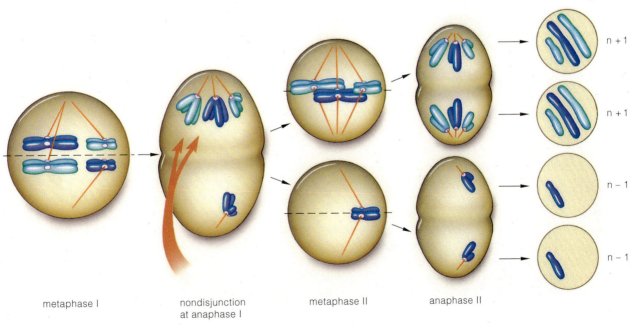

metaphase I   nondisjunction at anaphase I   metaphase II   anaphase II

n + 1
n + 1
n − 1
n − 1

FIGURE 17-15 Nondisjunction. Of the two pairs of homologous chromosomes shown, one pair fails to separate at anaphaseI of meiosis. The chromosome number changes in the gametes. *(After Starr, 2000.)*

Recall that each chromosome bears a unique set of genes and speculate about the effect of nondisjunction on the resulting zygotes formed from fertilization with such a gamete.

_____

_____

_____

One of the most common human genetic disorders arises from nondisjunction during gamete (usually ovum) formation. Down syndrome results from nondisjunction in one of the 23 pairs of human chromosomes, chromosome 21. An individual with Down syndrome has three copies of chromosome 21 instead of the normal two copies. Although symptoms of this genetic disorder vary greatly, most individuals show moderate to severe mental impairment and a host of associated physical defects. Relatively few other human genetic disorders arise from nondisjunction, probably because the consequences of abnormal chromosome number are often lethal.

Note: **Remove marking ink from the pop beads with 95% ethanol and tissues.**

17.4 Comparison of Mitosis and Meiosis

Review what you've learned about the two methods of cellular reproduction, mitosis and meiosis. Fill in table 17-1 below, indicating the differences between these processes.

17.5 Meiosis in Human Cells

Now that you have a conceptual understanding of meiosis, let's consider where and when the actual divisions occur in human cells.

A. Spermatogenesis in Males

In animals, as mentioned previously, meiosis results in the production of gametes—ova in females and sperm in males.

Examine Figure 17-16, which depicts sperm formation in a seminiferous tubule within human testes. A diploid reproductive cell, the _spermatogonium_, first enlarges into a _primary spermatocyte_. The primary spermatocyte undergoes meiosis I to form two haploid _secondary spermatocytes_. After meiosis II, four haploid _spermatids_ are produced, which develop flagella during differentiation into four mature _sperm cells_. This process is called **spermatogenesis**. Spermatogenesis begins at puberty and continues throughout the rest of a man's life, producing millions of sperm per day.

B. Oogenesis in Females

In the ovaries of female animals, _ova_ (eggs) are produced by meiosis during the process called **oogenesis**. Unlike spermatogenesis, only one of the meiotic products becomes a gamete. Oogenesis begins in a female embryo (see Figure 17-17).

Within a follicle of a developing ovary, the diploid reproductive cell, called an _oogonium_, grows into a _primary oocyte_. The primary oocyte begins meiosis I but arrests development in prophase I. At puberty, meiosis resumes, in usually one oocyte per month until menopause.

After meiosis I, one product is the _secondary oocyte_; the other is a _polar body_. Notice the difference in size of the secondary oocyte and the polar body. This is because the secondary oocyte ends up with nearly all of the cytoplasm after meiosis I.

Secondary oocytes are released from the ovary. If the secondary oocyte is fertilized, meiosis II continues, producing one large cell that will become the mature ovum and three very small polar bodies that have no role in reproduction.

TABLE 17-1	Comparison of Mitosis and Meiosis		
Characteristic		**Mitosis**	**Meiosis**
Purpose of process			
Type(s) of cell in which process occurs			
Amount of genetic material in product cells compared with that in parent cell			
Similarity of genetic material of product cells compared with that of parent cell			
Number of cells usually produced for each parent cell			

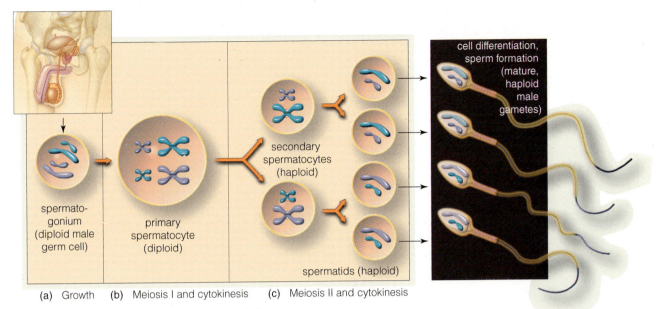

(a) Growth   (b) Meiosis I and cytokinesis   (c) Meiosis II and cytokinesis

FIGURE 17-16 Spermatogenesis within testes in humans.

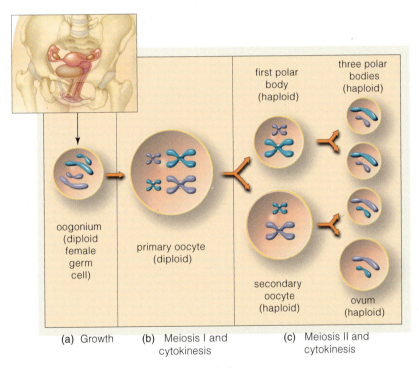

(a) Growth   (b) Meiosis I and cytokinesis   (c) Meiosis II and cytokinesis

FIGURE 17-17 Oogenesis within ovaries in humans.

_____ 1. The process of cytoplasmic division is known as
   (a) meiosis
   (b) cytokinesis
   (c) mitosis
   (d) fission

_____ 2. The products of chromosome duplication are
   (a) two chromatids
   (b) two nuclei
   (c) two daughter cells
   (d) two spindles

_____ 3. The correct sequence of stages in _mitosis_ is
   (a) interphase, prophase, metaphase, anaphase, telophase
   (b) prophase, metaphase, anaphase, telophase
   (c) metaphase, anaphase, prophase, telophase
   (d) prophase, telophase, anaphase, interphase

_____ 4. In the process of mitosis, chromatids separate during
   (a) prophase
   (b) telophase
   (c) cytokinesis
   (d) anaphase

_____ 5. In meiosis, the number of chromosomes _____, but in mitosis, it _____.
   (a) is halved/is doubled
   (b) is halved/remains the same
   (c) is doubled/is halved
   (d) remains the same/is halved

_____ 6. In humans, meiosis results in the production of
   (a) egg cells (ova)
   (b) gametes
   (c) sperm cells
   (d) all of the above

_____ 7. Recombination of alleles on nonsister chromatids occurs during
   (a) anaphase I
   (b) meiosis II
   (c) telophase II
   (d) crossing over

_____ 8. Alternative forms of genes are called
   (a) homologues
   (b) locus
   (c) loci
   (d) alleles

_____ 9. If both homologous chromosomes of each pair exist in the same nucleus, that nucleus is
   (a) diploid
   (b) unable to undergo meiosis
   (c) haploid
   (d) none of the above

_____ 10. Nondisjunction
   (a) results in gametes with abnormal chromosome numbers
   (b) occurs at anaphase I of meiosis
   (c) results when homologues fail to separate properly in meiosis
   (d) is all of the above

EXERCISE 17 CELL REPRODUCTION

POST-LAB QUESTIONS

Introduction

1. Define and distinguish among interphase, mitosis, cytokinesis, and meiosis.

2. a. If the chromosome number of a human somatic cell is 46 before mitosis, what is the chromosome number of each newly formed nucleus after mitosis has taken place?

   b. If the chromosome number of a human cell destined to give rise to gametes is 46 before meiosis, what is the chromosome number of each newly formed nucleus after meiosis is completed?

17.1 The Cell Cycle

3. Describe differences between the structure of a duplicated chromosome before mitosis and the chromosome produced by separation of two chromatids during mitosis.

4. The cells in the following photomicrographs have been stained to show microtubules comprising the spindle apparatus. Identify the stage of mitosis in each:

(b) _____

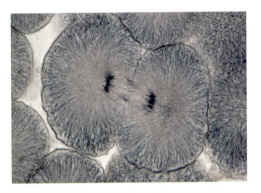

(c) _____

17.3 Meiosis

5. Suppose one sister chromatid of a chromosome has the allele *H*. What allele will the other sister chromatid have? (Assume crossing over has not taken place.) _____

6. Suppose that two alleles on one homologous chromosome are *A* and *B*, and the other homologous chromosome's alleles are *a* and *b*.

   a. How many different genetic types of gametes would be produced *without* crossing over?

   _____

   b. What are the genotypes of the gametes?

   _____

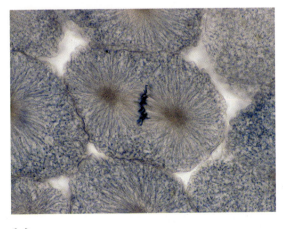

(a) _____

c. If crossing over were to occur, how many different genetic types of gametes could occur?

_____

d. List them. _____

_____

7. Examine the meiotic diagram at the right. Describe in detail what's wrong with it.

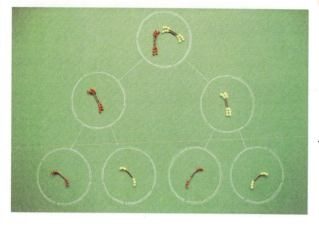

17.4 Comparison of Mitosis and Meiosis

8. Describe two features that distinguish mitosis from meiosis.

Food for Thought

9. Why must the DNA be duplicated during the S-phase of the cell cycle before mitosis? Why must DNA be duplicated prior to meiosis?

10. What would happen if a cell underwent mitosis but not cytokinesis?

Human Heredity

OBJECTIVES

After completing this exercise, you will be able to

1. define *true-breeding, hybrid, monohybrid cross, law of segregation, genotype, phenotype, dominant, recessive, complete dominance, homozygous, heterozygous, probability, incomplete dominance, codominance, sex-linked, X-linked, Y-linked, dihybrid cross. multiple alleles, cystic fibrosis,* and *hemophilia.*

2. draw Punnett squares to determine possible offspring genotypes.

3. determine the probability of specific offspring genotypes.

4. solve monohybrid and dihybrid cross problems, including those with incomplete dominance, codominance, and X-linkage, as well as multiple alleles.

5. determine your phenotype and probable genotype for some common traits.

6. understand the genetic basis of some human diseases and disorders.

Introduction

Our understanding of inheritance began in 1866 when an Austrian monk, Gregor Mendel, presented the results of painstaking experiments on the inheritance patterns of the garden pea. The principles generated by Mendel's pioneering experiments are the foundation for the genetic counseling that is so important today to families with genetically based health disorders. These principles are also the framework for the modern research that is making inroads into treating diseases previously believed to be incurable. These activities and genetics problems should give you a basic understanding of "Mendelian" inheritance.

18.1 Monohybrid Crosses

Garden peas have both male and female parts in the same flower and are able to self-fertilize. For his experiments, Mendel chose parental plants that were **true breeding,** meaning that all self-fertilized offspring displayed the same form of a trait as their parent. For example, if a true-breeding purple-flowered plant self-fertilizes, all of its offspring will have purple flowers.

When parents that are true breeding for *different* forms of a trait are crossed—for example, purple flowers and white flowers—the offspring are called **hybrids.** When only one trait is being studied, the cross is a **monohybrid cross.** We'll begin this exercise considering monohybrid problems and crosses, first when the genes are located on autosomes and then when they are on the sex chromosomes.

 NOTE As you complete this exercise, always distinguish clearly between upper- and lowercase letters.

A. Monohybrid Problems with Complete Dominance

MATERIALS

Optional:

- Simulated chromosomes (consisting of pop beads with magnetic centromeres) and meiotic diagram cards (as for Exercise 17)

- Bottle of 70% ethanol

PROCEDURE

1. As you learned in Exercise 17, most higher organisms are diploid—that is, they contain homologous chromosomes with genes for the same traits. Chromosomes have numerous genes, many with different alternatives forms (*alleles*) as shown in Figure 18-1.

   Let's consider one gene pair at the *F* locus. There are three possibilities for the alleles at the *F* locus. Both alleles are *FF*:

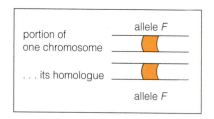

Both alleles are *ff*:

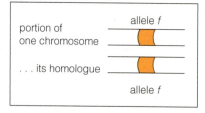

TABLE 18-1 Possible Gamete Genotypes	
Diploid Genotype	**Potential Gamete Genotype(s)**
FF	_____
Ff	_____
Ff	_____ , _____

One allele is *F*, and the other is *f*:

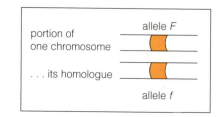

Gametes, on the other hand, are haploid; they contain only one of the two homologues and thus only one of the two alleles for a specific trait. According to Mendel's *law of segregation*, each diploid organism contains two alleles for each trait, and the alleles segregate (separate) during the formation of gametes during meiosis. Each gamete then contains only one allele of the pair.

The **genotype** of an organism represents its genetic constitution—that is, the alleles present, either for each locus, or taken cumulatively as the genotype of the entire organism.

For each of the diploid genotypes indicated in Table 18-1, list all possible genotypes of the gametes that can be produced by the organism.

If you don't understand the process that gives rise to the gamete genotypes, manipulate the pop-bead models that you used in Exercise 17. Using a marking pen, label one bead of each chromosome and go through the meiotic division process that gives rise to the gametes. *It is imperative that you understand meiosis before you attempt to do genetics problems.*

2. During fertilization, two haploid gamete nuclei fuse, and the diploid condition is restored. In Table 18-2, write the diploid zygote genotypes produced by fusion of the following gamete genotypes.

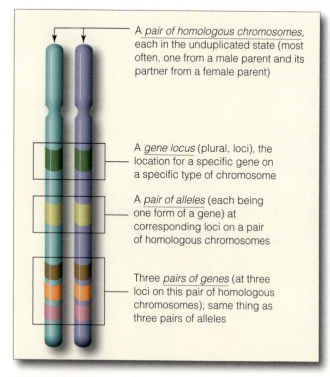

FIGURE 18-1 Review of genetic terms.

Chromosome labels:
- A *pair of homologous chromosomes,* each in the unduplicated state (most often, one from a male parent and its partner from a female parent)
- A *gene locus* (plural, loci), the location for a specific gene on a specific type of chromosome
- A *pair of alleles* (each being one form of a gene) at corresponding loci on a pair of homologous chromosomes
- Three *pairs of genes* (at three loci on this pair of homologous chromosomes); same thing as three pairs of alleles

TABLE 18-2	Diploid Zygote Genotypes Resulting from Fertilization			
Gamete Genotype	**X**	**Gamete Genotype**	**Diploid Genotype**
F	X	F X F	_____
F	X	f	_____
F	X	f	_____

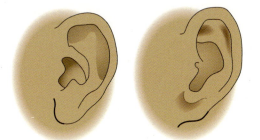

FIGURE 18-2 Free (left) and attached (right) earlobes.

3. The genotype is the actual genetic makeup of the organism. The **phenotype** is the outward expression of the genotype—that is, what the organism looks like because of its genotype, as well as its physiologic traits and behavior. (Although phenotype depends on genotype, in many instances, interactions and environmental factors can modify phenotype. We will not deal with that complexity in this exercise.)

Human earlobes are either attached or free (Figure 18-2). This trait is determined by a single gene consisting of two alleles, *F* and *f*.

An individual whose genotype is *FF* or *Ff* has the free earlobe phenotype. This is the **dominant** condition. Note that the presence of one *or* two *F* alleles results in the dominant phenotype, free earlobes. The allele *F* is said to be dominant over its allelic partner, *f*. The **recessive** phenotype—attached earlobes—occurs only when the genotype is *ff*. In the case of **complete dominance,** the dominant allele completely masks the expression or effect of the recessive allele.

When both alleles in a nucleus are identical, the nucleus is **homozygous.** Those with both dominant alleles are *homozygous dominant.* "True-breeding" parents can be assumed to be homozygous for the trait in question.

When both recessives are present in the same nucleus, the individual is said to be *homozygous recessive* for the trait.

When both the dominant and recessive alleles are present in a single nucleus, the individual is **heterozygous** for that trait.

(a) A man has the genotype *FF*. What is the genotype of his gamete (sperm) nuclei? _____

(b) A woman has attached earlobes. What is her genotype? _____

(c) What allele(s) do(es) her gametes (ova) carry?

_____

(d) These two individuals produce a child. Determine the genotype of the child by doing the cross:

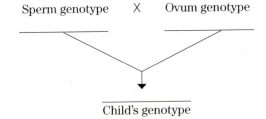

(e) What is the phenotype of the child (i.e., does this child have attached or free earlobes)?

_____

4. In garden peas, purple flowers are dominant over white flowers. Let *A* represent the allele for purple flowers and *a* the allele for white flowers.

(a) Fill in Table 18-3 to list the phenotype (color) of the flowers with the following genotypes:

A white-flowered garden pea is crossed with a homozygous dominant purple-flowered plant.

(b) Name the genotype(s) of the gametes of the white-flowered plant. _____

(c) Name the genotype(s) of the gametes of the purple-flowered plant. _____

(d) Name the genotype(s) of the plants produced by the cross. _____

(e) Name the phenotype(s) of the plants produced by the cross. _____

(f) The Punnett square is a convenient way to perform the mechanics of a cross. The circles along the top and side of the Punnett square represent the possible gamete nuclei genotypes. Insert the proper letters indicating the genotypes of the possible gamete nuclei for the white- and purple-flowered cross in the circles. Fill in the Punnett square for all of the possible outcomes.

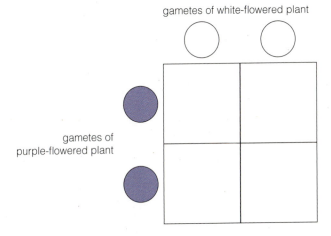

(g) A heterozygous plant is crossed with a white-flowered plant. Fill in the Punnett square on next page and then list the genotypes and phenotypes of all the possible genetic outcomes.

TABLE 18-3 Flower Phenotypes

Genotype	Phenotype
AA	
Aa	
Aa	

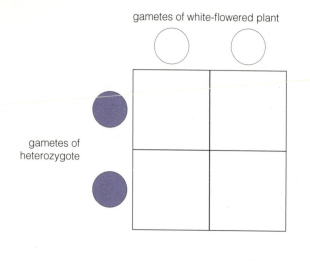

gametes of white-flowered plant

gametes of
heterozygote

Possible genotypes: _____

Possible phenotypes: _____

Draw a line from each possible genotype to its
respective phenotype.

(h) It's unlikely that every cross between two pea
plants will produce four seeds that will in turn
grow into four offspring every time. Rather, these
problems allow you to *predict the chances* of a
particular genetic outcome for each offspring. Genetics is really a matter of **probability**, the likeli-
hood of the occurrence of a particular outcome.

To take a simple example, consider that the
probability of coming up with heads in a single
toss of a coin is one chance in two, or 1/2. Now
let's apply this idea to the probability that offspring will have a certain genotype. Look at your
Punnett square above in part **(g).** The probability
of having a genotype is the sum of all occurrences
of that genotype. For example, the genotype Aa
occurs in two of the four boxes. The probability
that the genotype Aa will be produced from that
particular cross is thus 2/4, or 50%.

(i) What is the probability of an individual from part
(g) having the genotype aa? _____

For the remaining problems, you may wish to draw your
own Punnett squares on a separate sheet of paper.

5. In mice, black fur (B) is dominant over brown fur
(b). Breeding a brown mouse to a homozygous black
mouse produces all black offspring.

(a) What is the genotype of *the gametes* produced by
the brown-furred parent? _____

(b) What is the genotype of the brown-furred parent?

_____

(c) What is the genotype of the black-furred parent?

_____

(d) What is the genotype of the black-furred offspring? _____

By convention, P stands for the parental generation. The offspring are called the *first filial generation*,

abbreviated F_1. If these F_1 offspring are crossed, their
offspring are called the *second filial generation*, designated F_2.

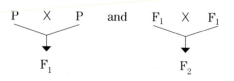

P   X   P   and F_1 X F_1

F_1 F_2

(e) If two of the F_1 mice are bred with one another,
what will the phenotype(s) of the F_2 be, and in
what proportion?

Phenotype(s): _____

Proportion: _____

6. The presence of horns on Hereford cattle is controlled by a single gene. The hornless (H) condition
is dominant over the horned (h) condition. A hornless cow was crossed repeatedly over the years with
the same horned bull. The following results were obtained in the F_1 offspring:

Eight hornless cattle _____

Seven horned cattle _____

What are the genotypes of the parents?

Cow _____

Bull _____

B. AN OBSERVABLE MONOHYBRID CROSS WITH COMPLETE DOMINANCE

MATERIALS

Per student group (table):

- Genetic corn ears illustrating monohybrid cross
- Hand lens (optional)

PROCEDURE

Examine the monohybrid genetic corn demonstration
ears. These ears resulted from a monohybrid cross
between plants producing purple kernels and ones producing yellow kernels.

1. Note that whereas all the first-generation kernels (F_1)
in the genetic corn demonstration are purple, the
second-generation ear (F_2) has both purple and yellow kernels. Count the purple kernels and then the
yellow ones in the F_2 ear: _____ purple; _____
yellow. When reduced to the lowest common de-
nominator, is this ratio closest to 1:1, 2:1, 3:1, or 4:1?
_____. This ratio is the *phenotypic ratio*.

2. A corncob with kernels represents the products of
multiple instances of sexual reproduction. Each kernel represents a single instance; fertilization of one
egg by one sperm produced *each* kernel. Thus, each
kernel represents a different cross.

Let the letter P represent the gene for kernel color.

(a) What genotypes produce a purple phenotype? _____

(b) Which allele is dominant? _____

(c) What is the genotype of the yellow kernels on the F_2 ear? _____

(d) You are given an ear with purple kernels. How could you determine its genotype with a single cross? (The cross would be performed by growing plants from those purple kernels and crossing them with plants grown from kernels of another ear.)

_____

_____

C. MONOHYBRID PROBLEMS WITH INCOMPLETE DOMINANCE

1. Petunia flower color is governed by two alleles, but neither allele is truly dominant over the other. Petunias with the genotype R^1R^1 are red flowered, those that are heterozygous (R^1R^2) are pink, and those with the R^2R^2 genotype are white. This is an example of **incomplete dominance.** (Note that superscripts are used rather than uppercase and lowercase letters to describe the alleles.)

(a) If a white-flowered plant is crossed with a red-flowered petunia, what is the genotypic ratio of the F_1? _____

(b) What is the phenotypic ratio of the F_1? _____

(c) If two of the F_1 offspring were crossed, what phenotypes would appear in the F_2? _____

(d) What would be the genotypic ratio in the F_2 generation? _____

D. MONOHYBRID PROBLEMS ILLUSTRATING CODOMINANCE

1. Another type of monohybrid inheritance involves the expression of *both* phenotypes in the heterozygous situation. This is called **codominance.**

One of the well-known examples of codominance occurs in the coat color of Shorthorn cattle. Those with reddish-gray (roan) coats are heterozygous (RR'), and result from a mating between a red (RR) Shorthorn and one that's white ($R'R'$). Roan cattle do not have roan-colored hairs, as would be expected with incomplete dominance, but rather appear roan as a result of both red and white hairs occurring on the same animal. Thus, the roan coloration is not a consequence of blending of pigments within each hair. Because the R and R' alleles are *both* fully expressed in the heterozygote, they are codominant.

(a) If a roan Shorthorn cow is mated with a white bull, what will be the genotypic and phenotypic ratios in the F_1 generation?

Genotypic ratio _____

Phenotypic ratio _____

(b) List the parental genotypes of crosses that could produce at least some:

White offspring _____

Roan offspring _____

E. MONOHYBRID, SEX-LINKED PROBLEMS

1. In humans, as well as in other primates, sex is determined by special sex chromosomes. Whereas an individual containing two X chromosomes is a female, an individual possessing an X and a Y chromosome is a male.

I am genetically a male/female (circle one).

(a) What sex chromosomes do you have? _____

(b) What sex chromosome would be present in the gametes (ova) a female produces? _____

(c) What sex chromosomes would be present in a male's sperm cells? _____

(d) Draw a Punnett square to show possible combinations of sex chromosomes in offspring.

(e) The gametes of which parent determine the sex of the offspring? _____

2. The sex chromosomes bear alleles for traits, just like the other chromosomes in our bodies. Genes that occur on the sex chromosomes are said to be **sex-linked.** More specifically, the genes present on the X chromosome are said to be **X-linked,** and those on the Y chromosome are **Y-linked.**

The Y chromosome is smaller than its homologue, the X chromosome. Consequently, some of the loci present on the X chromosome are absent on the Y chromosome and vice versa. Although the X and Y chromosomes are not true homologues, they are moved as homologues during the process of meiosis.

In humans, color vision is X-linked; the gene for red-green color vision is located on the X chromosome but is absent from the Y chromosome.

Normal color vision (X^N) is dominant over color blindness (X^n). Suppose a colorblind man fathers children of a woman with the genotype X^NX^N.

(a) What is the genotype of the father? _____

(b) What proportion of daughters would be colorblind? _____

(c) What proportion of sons would be colorblind? _____

3. One of the daughters from problem 10 marries a colorblind man.

(a) What proportion of their sons will be colorblind? (Another way to think of this is to ask what the *chances* are that their sons will be colorblind.) _____

(b) Explain how these parents might give birth to a colorblind daughter.

_____

_____

18.2 Dihybrid Inheritance

All of the problems so far have involved the inheritance of only one trait; that is, they are monohybrid problems. Now we'll examine cases in which two traits are involved, **dihybrid crosses.**

> **NOTE** We will assume that the genes for these traits are carried on different (nonhomologous) chromosomes.

A. DIHYBRID PROBLEMS

1. Let's consider these two traits: In humans, a pigment in the front part of the eye masks a blue layer at the back of the iris. The dominant allele P causes production of this pigment. Those who are homozygous recessive (pp) lack the pigment, and the back of the iris shows through, resulting in blue eyes. (Other genes determine the color of the pigment, but in this problem, we'll consider only the presence or absence of *any* pigment at the front of the eye.)

Dimpled chins (D = allele for dimpling) are dominant over undimpled chins (d = allele for lack of dimple).

(a) List all possible genotypes for an individual with pigmented irises *and* a dimpled chin.

_____

(b) List the possible genotypes for an individual with pigmented irises but lacking a dimpled chin.

_____

(c) List the possible genotypes of a blue-eyed, dimple-chinned individual. _____

(d) List the possible genotypes of a blue-eyed individual lacking a dimpled chin. _____

2. An individual is heterozygous for both traits (eye pigmentation and chin form).

(a) What is the genotype of such an individual?

_____

(b) What are the possible genotypes of that individual's gametes? (Remember that gametes are haploid.) _____

If determining the answer for question 2 was difficult, recall from Exercise 17 that the principle of independent assortment states that genes on different (nonhomologous) chromosomes are separated out into gamete nuclei independently of one another during meiosis. That is, the occurrence of an allele for eye pigmentation in a gamete has no bearing on which allele for chin form will occur in that same gamete.

There is a useful method for determining possible gamete genotypes produced during meiosis from a given parental genotype. Using the genotype $PpDd$ as an example, follow the four arrows to determine the four gamete genotypes:

(c) Two individuals heterozygous for both eye pigmentation and chin form have children. What are the possible genotypes of those F_1 offspring?

_____

You can set up a Punnett square to do dihybrid problems just as you did with monohybrid problems. However, depending on the parental genotypes, the square may have as many as 16 boxes rather than just four. Insert the possible genotypes of the gametes from one parent in the top circles and the possible gamete genotypes of the other parent in the circles to the left of the box. Remember that every gamete will have a single copy of each gene.

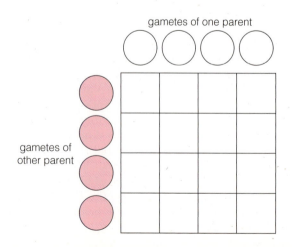

(d) Possible genotypes of children produced by two parents heterozygous for both eye pigmentation and chin form: _____

What is the ratio of the genotypes? _____

What is the phenotypic ratio? _____

(e) Recalling the discussion of probability in Section 18.1.A, state the probability of a child from part **(d)** having the following genotypes:

ppDD _____

PpDd _____

PPDd _____

To extend the probability discussion, let's reconsider flipping a coin by asking the question, what is the probability of flipping heads twice in a row? The chance of flipping heads the first time is 1/2. The same is true for the second flip. The chance (probability) that we'll flip heads twice in a row is $1/2 \times 1/2 = 1/4$. The probability that we could flip heads three times in a row is $1/2 \times 1/2 \times 1/2 = 1/8$.

(f) State the probability that three children born to the parents in part **(d)** will have the genotype *ppdd*. _____

What is the probability that three children born to these parents will have dimpled chins and pigmented eyes? _____

(g) What is the genotype of the F1 generation when the father is homozygous for both pigmented eyes and a dimpled chin but the mother has blue eyes and no dimple? _____

What is the phenotype of this individual?

_____

B. AN OBSERVABLE DIHYBRID CROSS

MATERIALS

Per student group (table):

- Genetic corn ears illustrating a dihybrid cross

PROCEDURE

1. Examine the demonstration of dihybrid inheritance in corn. Notice that not only are the kernels two different colors (one trait), but they are also differently shaped (second trait). Whereas kernels with starchy endosperm (the carbohydrate-storing tissue) are smooth, those with sweet endosperm are shriveled. Notice that all *four* possible phenotypic combinations of color and shape are present in the F$_2$ generation.

The *P* gene is involved in pigment production, with two alleles *P* and *p*. The *S* gene determines carbohydrate (starch or sugar) storage, with two alleles *S* and *s*.

Which genotypes of the parents produced the F$_2$ generation kernels? _____

TABLE 18-4 Phenotypes in Dihybrid Corn Cross

Ear	Number of Kernels with Phenotypes			
	Yellow Smooth	Yellow Shriveled	Purple Smooth	Purple Shriveled
1				
2				
3				
Totals				

2. Set up a Punnett square of this dihybrid cross.

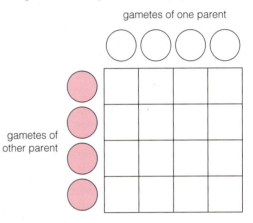

gametes of one parent

gametes of other parent

What is the predicted phenotypic ratio? _____

3. Count the number of kernels of each possible phenotype and record in Table 18-4. To increase your sample size and improve the reliability of your conclusions, count three ears.

Which traits seem dominant? _____

Which traits seem recessive? _____

4. Calculate the actual phenotypic ratio you observed:

_____

Do your observed results differ from the expected results? _____

If you answered "yes" to the previous question, why might that have been the case?

_____

_____

18.3   Multiple Alleles

The major blood groups in humans are determined by **multiple alleles**, that is, there are *more than* two possible alleles, any one of which can occupy a locus. In this ABO blood group system, a single gene can exist in any of three allelic forms: I^A, I^B, or i. The alleles *A* and *B* code

TABLE 18-5 The ABO Blood Groups

Blood Type	Antigens Present	Antibodies Present	Genotype
O	Neither A nor B	Anti-A and Anti-B	ii
A	A	Anti-B	$I^A I^A$ or $I^A i$
B	B	Anti-A	$I^B I^B$ or $I^B i$
AB	AB	none	$I^A I^B$

for production of antigen A and antigen B (two proteins) on the surface of red blood cells. The i allele produces no antigen. Alleles A and B are codominant, and allele i is recessive. Four blood groups (phenotypes) are possible from combinations of these alleles (Table 18-5).

Blood cell antigens are markers that the immune system recognizes as "self." The blood contains antibodies, receptors that recognize "non-self" antigens and trigger an immune response against the non-self material. Knowledge of ABO blood groups is necessary to ensure that blood transfusions are compatible.

1. Is it possible for a child with blood type O to be produced by two AB parents? Explain.

_____

_____

2. In a case of disputed paternity, the child is type O, and the mother type A.
   (a) What is the child's genotype?

   _____

   (b) What are the mother's possible genotypes?

   _____

(c) Could an individual of the following blood types be the father? Explain.

O _____

A _____

B _____

AB _____

18.4 Some Readily Observable Human Traits

In the preceding sections, we examined several human traits that are fairly simple and that follow the Mendelian pattern of inheritance. Most of our traits are much more complex, involving many genes or interactions among genes. For example, hair color is determined by at least four genes, each one coding for the production of melanin, a brown pigment. Because the effect of these genes is cumulative, hair color can range from blond (little melanin) to very dark brown (much melanin).

Clearly, human traits are of great interest to us. Table 18-6 lists a number of traits that exhibit Mendelian inheritance. For each trait, consult Figure 18-3 and work with a lab partner to determine your phenotype. Fill in Table 18-6 by listing your possible genotype(s) for each trait. When possible, examine your parents' phenotypes and attempt to determine your actual genotype.

1. *Mid-digital hair* (Figure 18-3a). Examine the joint of your fingers for the presence of hair, the dominant condition (MM, Mm). Complete absence of hair is caused by the homozygous-recessive condition (mm). You may need a hand lens to determine your phenotype. Even the slightest amount of hair indicates the dominant phenotype.

2. *Tongue rolling* (Figure 18-3b). The ability to roll one's tongue is caused by a dominant allele, T.

TABLE 18-6 Summary of My Mendelian Traits

Trait	My Phenotype	Mom's Phenotype	Mom's Possible Genotype	Dad's Phenotype	Dad's Possible Genotype	My Possible or Probable Genotypes
Mid-digital hair						
Tongue rolling						
Widow's peak						
Earlobe attachment						
Hitchhiker's thumb						
Relative finger length						

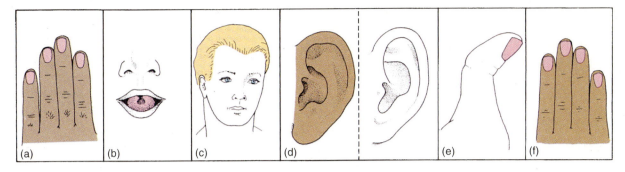

FIGURE 18-3 Some readily observable human Mendelian traits.

FIGURE 18-4 A student at San Diego State University exhibiting the tongue-rolling trait for the benefit of a tongue-roll-challenged student. *(Photo courtesy Evan Cerasoli.)*

The homozygous-recessive condition, *tt*, results in inability to roll the tongue (Figure 18-4).

3. *Widow's peak* (Figure 18-3c). A widow's peak is a distinct downward point in the frontal hairline and is caused by the dominant allele, *W*. The recessive allele, *w*, results in a continuous hairline. (Omit study of this trait if baldness is affecting your hairline.)

4. *Earlobe attachment* (Figure 18-3d). Most individuals have free (unattached) earlobes (*FF*, *Ff*). Homozygous recessives (*ff*) have earlobes attached directly to the head.

5. *Hitchhiker's thumb* (Figure 18-3e). Although considerable variation exists in this trait, we'll consider individuals who *cannot* extend their thumbs backward to approximately 45 degrees to be carrying the dominant allele, *H*. Homozygous-recessive persons (*hh*) can bend their thumbs at least 45 degrees , if not farther.

6. *Relative finger length* (Figure 18-3f). An interesting sex-influenced (*not* sex-linked) trait relates to the relative lengths of the index and ring finger. In males, the allele for a shorter index finger (*S*) is dominant. In females, it's recessive. In rare cases, each hand is different. If one or both index fingers are greater than or equal to the length of the ring finger, the recessive genotype is present in males, and the dominant one is present in females.

18.5 Hereditary Diseases

Many diseases or harmful conditions are hereditary, caused by specific allele combinations or abnormalities of the chromosomes inherited from one's parents. We'll briefly consider two such disease which have different inheritance patterns.

A. CYSTIC FIBROSIS

Cystic fibrosis is inherited as an autosomal recessive trait on chromosome 7. This is the most common lethal genetic disorder in the United States. The disease allele causes production of a defective cell membrane protein unable to effectively transport chloride ions. A person with two copies of the recessive allele has abnormal glandular secretions and thick, sticky mucus production in the lungs and digestive organs. The mucus interferes with breathing along with digestive system functions and causes a susceptibility to frequent bacterial infections. There is no cure for this disease, but special diets, antibiotics, and physical therapies to clear the mucus from the lungs can extend life to an average age of 30 years. Untreated children often die before the age of 6 years.

Cystic fibrosis is most common in people of Northern European descent, with an estimated one in 29 Caucasians heterozygous for the trait. Such a person who is heterozygous for a recessive disorder is called a **carrier** of the trait.

1. Designate the normal allele *C*, and the disease allele *c*. In the space below, draw a Punnett square showing inheritance probabilities when two cystic fibrosis carriers have a child.

2. What is the probability that the child will have cystic fibrosis? _____

3. What is the probability that the child will be a cystic fibrosis carrier? _____

4. Draw a Punnett square showing possible outcomes if a healthy man with no history of cystic fibrosis in his family (and who can be assumed to be homozygous dominant) has a child with a woman who is a carrier.

5. What is the probability that the child will have cystic fibrosis? _____

6. If this couple had three children, what is the probability that they will *all* be cystic fibrosis carriers? _____

B. HEMOPHILIA

Hemophilia is a group of disorders that results from lack of blood clotting factors. In serious cases, uncontrolled bleeding poses a threat to life; milder cases may cause severe bruising, bleeding into joints and muscles, and prolonged bleeding. The disease results from mutated clotting proteins produced by genes on the X chromosome; hemophilia is an X-linked disorder. The mutated allele is recessive to the normal allele *when two alleles are present;* when only one allele is present (only one X chromosome present), then that allele determines the phenotype.

Standard treatment is to replace the missing clotting factor via infusion into blood vessels. The clotting factors are derived from donated blood and plasma.

X^H is the normal allele, and X^h identifies the hemophilia allele.

1. What is the genotype of a man with hemophilia? _____

2. What is the genotype of a woman who is a hemophilia carrier? _____

3. Can a father without hemophilia have a daughter with hemophilia? Explain.

_____

_____

4. A father without hemophilia has a son with hemophilia. Draw one or more Punnett squares in the space below to identify the possible genotypes of the mother.

5. What is the genotype of a woman with normal blood clotting but whose father had hemophilia? _____

6. The woman in (e) has a child fathered by a hemophiliac man. Draw a Punnett square in the space below to identify the probabilities that the child will have hemophilia.

_____ **1.** In a monohybrid cross
   (a) only one trait is being considered
   (b) the parents are always dominant
   (c) the parents are always heterozygous
   (d) no hybrid is produced

_____ **2.** The genetic makeup of an organism is its
   (a) phenotype
   (b) genotype
   (c) locus
   (d) gamete

_____ **3.** An allele whose expression is completely masked by the expression or effect of its allelic partner is
   (a) homologous
   (b) homozygous
   (c) dominant
   (d) recessive

_____ **4.** The physical appearance and physiology of an organism, resulting from interactions of its genetic makeup and its environment, is its
   (a) phenotype
   (b) hybrid vigor
   (c) dominance
   (d) genotype

_____ **5.** When both dominant and recessive alleles are present within a single nucleus, the organism is ____ for the trait.
   (a) diploid
   (b) haploid
   (c) homozygous
   (d) heterozygous

_____ **6.** A Punnett square is used to determine
   (a) probable gamete genotypes
   (b) possible parental phenotypes
   (c) possible parental genotypes
   (d) possible genetic outcomes of a cross

_____ **7.** A gene located only on the X chromosome and have no allelic partner on the Y chromosome would be:
   (a) incompletely dominant
   (b) codominant
   (c) sex-linked
   (d) homozygous

_____ **8.** The sex chromosome determining maleness is:
   (a) the Y chromosome
   (b) the X chromosome
   (c) sex-linked
   (d) heterozygous

_____ **9.** Possible gamete genotypes produced by an individual of genotype $PpDd$ are
   (a) Pp and Dd
   (b) all $PpDd$
   (c) PD and pd
   (d) PD, Pd, pD, and pd

_____ **10.** If you can roll your tongue
   (a) you have at least one copy of the dominant allele T
   (b) you have two copies of the recessive allele t
   (c) you must be male
   (d) you are haploid

EXERCISE 18 Human Heredity

POST-LAB QUESTIONS

18.1 Monohybrid Crosses

1. Explain how Mendel's law of segregation applies to the distribution of alleles in gametes.

2. Production of freckles on skin is controlled by a single gene on an autosome, with two alleles *F* (dominant) and f (recessive). Freckled skin is dominant to nonfreckled skin.

   a. List the genotype(s) of a nonfreckled person.

   b. List the genotype(s) of a person with freckles.

   c. What are the possible genotypes of gametes produced by the nonfreckled person?

   d. What are the possible genotypes of gametes produced by the freckled person?

   e. Suppose a freckled man and a nonfreckled woman have a child. Draw a Punnett square for that cross in the space below and determine the probability that their child will have freckles.

      Probability of child with freckles:

3. What is the probability that parents will have three sons and no daughters?

4. Duchenne muscular dystrophy is an X-linked, recessive disorder in which muscles waste away early in life, resulting in death in the teens or twenties. A man and woman in their late thirties have five children—three boys (ages 1, 3, and 10 years) and two girls (ages 5 and 7 years). The oldest boy shows symptoms of the disease. What are the probabilities that their other children will develop the disease? Explain.

18.2 Dihybrid Inheritance

5. Suppose you have two traits controlled by genes on separate chromosomes. If sexual reproduction occurs between two heterozygous parents, what is the genotypic ratio of all possible gametes?

6. In the situation in question 5, what is the phenotypic ratio of all possible offspring?

18.2 Multiple Alleles

7. A woman with type A blood bears a child with type O blood. The man she thinks is the father has type B blood. On the basis of ABO blood types, is it possible to prove that he could not be the father?

Food for Thought

8. Suppose students in previous semesters had removed some of the corn kernels from the genetic corn ears before you counted them. What effect would this have on your results?

9. Assume that one allele is completely dominant over the other for the following questions.

   a. Two individuals heterozygous for a *single* trait have children. What is the expected phenotypic ratio of the possible offspring?

   b. Two individuals heterozygous for *two* traits have children. What would be the expected phenotypic ratio of the possible offspring?

   c. Crossing two individuals heterozygous for two traits results in the same phenotypic ratio as for a single trait. Are the genes for these two traits on separate chromosomes or on the same chromosome? Explain your answer. (Remember that the gene for each trait is located at a locus, a physical region on the chromosome.)

10. Search the Internet for sites about phenylketonuria (PKU) *or* Huntington's disease. List two sites below and briefly summarize their contents. What is the inheritance pattern for the disease? How does the defective gene cause the disease? What, if any, treatments are possible?

    http://

    http://

DNA, Genes, Cancer, and Biotechnology

OBJECTIVES

After completing this exercise, you will be able to

1. define *DNA, RNA, deoxyribose, principle of base pairing, ribose, replication, DNA polymerase, transcription, messenger RNA, RNA polymerase, translation, transfer RNA, codon, anticodon, peptide bond, gene, tumor, biotechnology, genome, polymerase chain reaction (PCR), restriction digestion of DNA,* and *agarose gel electrophoresis.*

2. identify and distinguish between the components of deoxyribonucleotides and ribonucleotides.

3. distinguish between DNA and RNA based on their structure and function.

4. describe the processes of DNA replication, transcription, and translation.

5. give the base sequence of DNA or RNA when presented with the complementary strand.

6. identify a codon and anticodon on RNA models and describe the location and function of each.

7. give the base sequence of an anticodon when presented with that of a codon and vice versa.

8. distinguish between benign and cancerous tumors and recognize chromosome number abnormalities of HeLa cells.

9. describe biotechnology techniques useful in forensic science.

Introduction

By 1900, Gregor Mendel had demonstrated patterns of inheritance based solely on careful experimentation and observation. Mendel had no clear idea how the traits he observed were passed from generation to generation. Subsequently, **deoxyribonucleic acid (DNA)** was identified as the genetic material. This breakthrough and determination of its molecular structure were among the most significant discoveries of the twentieth century.

This exercise will familiarize you with the basic structure of nucleic acids and their roles in the cell. Understanding the function of nucleic acids—both DNA and **ribonucleic acid (RNA)**—is central to understanding life itself.

Cancer develops after genetic changes accumulate. Cancerous cells differ from normal cells in several ways. In this exercise, you will see abnormalities in human cancer cells.

Finally, knowledge of molecular biology has given rise to the rapidly growing field of biotechnology. You will experience a common technique used for many applications of biotechnology, including in the field of forensic science.

We hope you will gain an understanding of nucleic acids and processes that will allow you to form educated opinions concerning how we should use our knowledge and this technology.

19.1 Isolation and Identification of Nucleic Acids

In this section, you'll isolate and identify a nucleic acid component of strawberries (*Fragaria ananassa*). Like all organisms, strawberries are composed of cells containing genetic material. Commercial garden strawberries are especially interesting because they are 8N, that is, each nucleus has eight copies of each of its seven different chromosomes.

MATERIALS

Per student:

- Frozen strawberry, thawed
- Ziploc bag
- Funnel with cheesecloth square
- Wooden skewer or wire loop
- 9 mL of ice cold 95% ethanol in test tube
- Two test tubes
- Test tube rack
- Agar gel plate with methylene blue stain
- Dropping pipettes

Per student group (4):

- Detergent (10% Woolite) and salt (1% NaCl) solution in flask fitted with a 5-mL pipet and Pi-pump
- 10-mL graduated cylinder
- TBE (Tris/borate/EDTA buffer solution
- Fine-pointed marker
- DNA standard solution in dropper bottle
- 1% albumin solution in dropper bottle
- paper towels

Per lab room:

- Source of dH$_2$O
- White light transilluminator or other bottom source of white light

PROCEDURE

A. Isolation of Nucleic Acids

1. Thoroughly mash up a thawed, frozen strawberry by kneading the berry in a Ziploc bag with 5 mL of the detergent and salt solution. Continue kneading for at least 2 minutes. This action exposes the strawberry fruit cells to the detergent and salt solution.

   Review from Exercise 5: What are the major molecular components of cell membranes?

   _____

Detergents break down lipids, and salt causes proteins that would otherwise obscure the nucleic acids to precipitate out of solution.

2. Filter the liquefied berry through cheesecloth into a test tube.

3. Slowly pour the ice cold ethanol down the side of the test tube so the ethanol forms a clear layer on top of the watery strawberry filtrate. B*e careful not to mix the water and alcohol layers.*

4. Allow the test tube to sit for several minutes. You should observe a frothy material begin to form at the interface between the two layers.

   Nucleic acids precipitate at the boundary between alcohol and water. The nucleic acids you have extracted are not pure; they contain cellular debris as well as adhering proteins. You will test your extracted precipitate to identify its major component.

B. Identification—Test for DNA

1. Use a 10-mL graduated cylinder to measure 3 mL of TBE buffer solution into a clean test tube.

2. Slowly twirl the skewer or loop to wrap up the precipitated material and withdraw the skewer from the test tube.

3. Place the slimy material from the skewer into the TBE buffer. Swirl the skewer in the buffer solution to thoroughly dissolve the viscous material in the buffer.

4. Obtain an agar plate with methylene blue stain. (Agar is an inert gel-like substance; methylene blue stain binds specifically with DNA, causing a visible purplish color.) Turn the unopened plate over and use the marking pen to draw four small, widely separated circles on the *underside* of the plate. Label the circles U, D, A, and C. These will be visible when looking at the plate from above to mark the locations where different test substances will be applied.

5. Open the plate, and with a dropping pipette, apply a drop of the precipitated strawberry material dissolved in TBE buffer solution onto the area marked U, for "unknown."

6. Similarly, apply a drop of DNA standard solution onto the area marked D.

7. Apply a drop of 1% albumin protein solution onto the area marked A.

8. Apply a control drop of TBE buffer onto the area marked C.

9. Set the plate aside for 20 to 30 minutes to allow time for color development.

FIGURE 19-1 Agarose-methylene blue plate showing positive (left) and negative (right) tests for DNA. *(Photo by J. W. Perry.)*

TABLE 19-1	Identification of Contents Extracted from Strawberry	
Droplet Code	**Droplet Contents**	**Color**
U	Unknown strawberry precipitate in TBE buffer	
D	DNA standard	
A	1% albumin protein	
C	TBE buffer	

10. View the petri plate on the white light transilluminator or other source of bottom light. A purple color indicates the presence of DNA (Figure 19-1). Record your results in Table 19-1.

Name the substance(s) present in the material you isolated from the strawberry.

_____

What is the purpose of the DNA standard droplet?

_____

What is the purpose of the TBE buffer droplet?

_____

19.2 Modeling the Structure and Function of Nucleic Acids and Their Products

MATERIALS

Per student group:
- DNA puzzle kit

Per lab room:
- DNA model

PROCEDURE

Work in groups.

> **NOTE** Clear your work surface of everything except your lab manual and the DNA puzzle kit.

In this section, we are concerned with three processes: *replication, transcription,* and *translation.* But before we study these, let's visualize the structure of DNA itself.

A. Nucleic Acid Structure

1. Obtain a DNA puzzle kit. It should contain the following parts:

- Four adenine bases
- Four thymine bases
- Six cytosine bases
- Six guanine bases
- 18 deoxyribose sugars
- 18 phosphate groups
- Nine ribose sugars
- Two uracil bases
- Three transfer RNA (RNA)
- Three amino acids
- Ribosome template sheet

- 18 deoxyribose sugars
- Four thymine bases
- Nine ribose sugars
- Two uracil bases
- 18 phosphate groups
- Three transfer RNA (tRNA)
- Four adenine bases
- Three amino acids
- Six guanine bases
- Ribosome template sheet
- Six cytosine bases

2. Group the components into separate stacks. Select a single deoxyribose sugar, an adenine base (labeled A), and a phosphate, fitting them together as shown in Figure 19-2. This is a single nucleotide (specifically, a deoxyribonucleotide), a unit consisting of a sugar (deoxyribose), a phosphate group, and a nitrogen-containing base (adenine).

Let's examine each component of the nucleotide.

Deoxyribose (Figure 19-3) is a sugar compound containing five carbon atoms. Four of the five are joined by covalent bonds into a ring. Each carbon is given a number, indicating its position in the ring. (These numbers are read "1 prime," "2 prime," and so on. "Prime" is used to distinguish the carbon atoms from the position of atoms that are sometimes numbered in the nitrogen-containing bases.) This structure is usually drawn in a simplified manner without actually showing the carbon atoms within the ring (Figure 19-4).

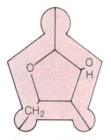

Ribose unit (pink)

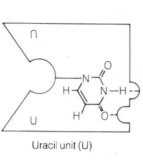

Uracil unit (U)

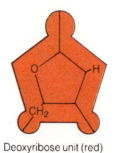

Deoxyribose unit (red)

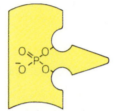

Phosphate unit

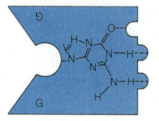

Adenine (A) unit

Cytosine (C) unit

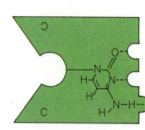

Guanine (G) unit

Thymine (T) unit

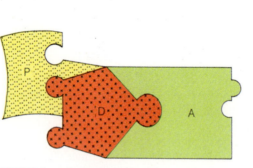

FIGURE 19-2 One deoxyribonucleotide.

There are four kinds of nitrogen-containing bases in DNA. Two are double-ring structures, *adenine* and *guanine* (abbreviated A and G, respectively; Figure 19-5).

The other two nitrogen-containing bases are single-ring compounds, *cytosine* and *thymine* (abbreviated C and T, respectively, Figure 19-6).

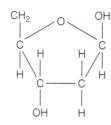

FIGURE 19-3 Deoxyribose.

FIGURE 19-4 Simplified representation of deoxyribose.

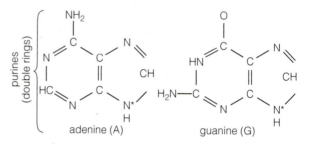

FIGURE 19-5 Double-ring bases found in DNA.

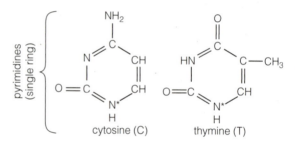

FIGURE 19-6 Single-ring bases found in DNA.

The symbol * indicates where a bond forms between each nitrogen-containing base and a carbon atom of the sugar ring structure.

The final component of a nucleotide is a phosphate group (Figure 19-7).

There are four kinds of deoxyribonucleotides, each differing only in the type of base it possesses. A simple way to depict the deoxyribonucleotide containing adenine is shown in Figure 19-8a. Construct the four kinds of deoxyribonucleotides with your model pieces and then draw them in Figures 14-9b to d. Use D for deoxyribose; P for a phosphate group; and A, C, G, and T for the different bases as shown in Figure 19-8a.

FIGURE 19-7 Phosphate group found in nucleic acids.

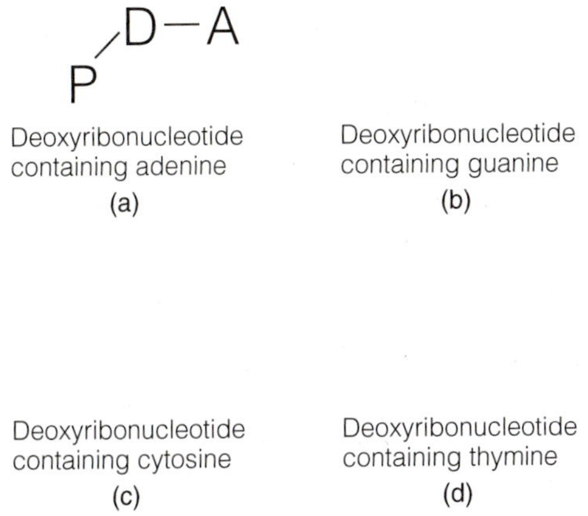

P
D—A

Deoxyribonucleotide
containing adenine
(a)

Deoxyribonucleotide
containing guanine
(b)

Deoxyribonucleotide
containing cytosine
(c)

Deoxyribonucleotide
containing thymine
(d)

FIGURE 19-8 Drawings of deoxyribonucleotides containing guanine, cytosine, and thymine.

Note the small notches and projections in the nitrogen-containing base models. Will the notches of adenine and thymine fit together? _____

Will guanine and cytosine? _____

Will adenine and cytosine? _____

Will thymine and guanine? _____

The notches and projections represent bonding sites. Make a prediction about which bases will bond with one another. _____

Will a double-ring base bond with another double-ring base? _____

Will a double-ring base bond with both types of single-ring base? _____

3. Linking the four nucleotide models to form a nucleotide strand of DNA as shown in Figure 19-9. The base sequence will be A-T-G-C, from top to bottom. Note that the sugar backbone is bonded together by phosphate groups.

4. Now assemble a second four-nucleotide strand, similar to that of Figure 19-9. However, this time make the base sequence T-A-C-G, from bottom to top. DNA molecules consist of *two* strands of nucleotides, each strand the *complement* of the other.

5. Assemble the two strands by attaching (bonding) the nitrogen bases of complementary strands. Note that the adenine of one nucleotide always pairs with the thymine of its complement; similarly, guanine always pairs with cytosine. This phenomenon is called the **principle of base pairing.** On Figure 19-10, attach letters to the model pieces indicating the composition of your double-stranded DNA model.

What do you notice about the *direction* in which each strand is running (i.e., are both phosphate groups at the same end of the strands)?

_____

_____

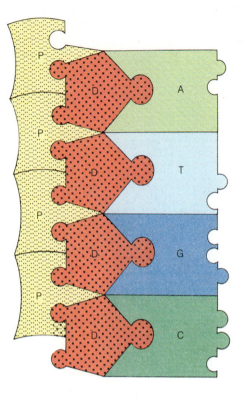

FIGURE 19-9 Four-nucleotide model strand of DNA.

In life, the complementary bases are joined together by hydrogen bonds. Note again that the sugar backbone is linked by phosphate groups. Your model illustrates only a very small portion of a DNA molecule. The entire molecule may be tens of thousands of nucleotides in length!

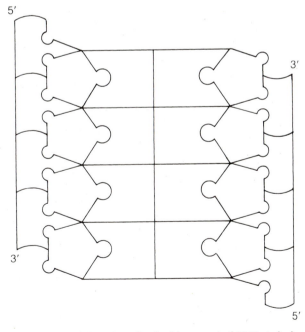

FIGURE 19-10 Drawing of a double-strand of DNA. **Labels:** A, T, G, C, D, P (all used more than once).

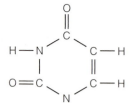

FIGURE 19-13 The single-ring base, uracil, found in RNA.

Ribonucleotide containing adenine
(a)

Ribonucleotide containing guanine
(b)

Ribonucleotide containing cytosine
(c)

Ribonucleotide containing uracil
(d)

FIGURE 19-14 Drawings of four possible ribonucleotides.

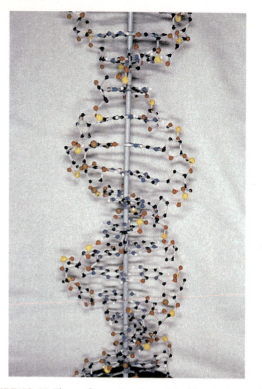

FIGURE 19-11 Three-dimensional model of DNA. *(Photo by J. W. Perry.)*

6. Slide your double-stranded DNA segment aside for the moment.

7. Examine the three-dimensional model of DNA on display in the laboratory (Figure 19-11). Notice that the two strands of DNA are twisted into a spiral staircase–like pattern. This is why DNA is known as a *double helix.* Identify the deoxyribose sugar, nitrogen-containing bases, hydrogen bonds linking the bases, and phosphate groups.

The second type of nucleic acid is RNA. There are three important differences between DNA and RNA:

1. RNA is a *single strand* of nucleotides.

2. The sugar of RNA is **ribose** (Figure 19-12), which contains one more oxygen atom than deoxyribose (Figure 19-3).

3. RNA lacks the nucleotide that contains thymine. Instead, it has one containing the single-ringed *uracil* (U) (see Figure 19-12).

8. From the remaining pieces of your model kit, select four ribose sugars—one each of adenine, uracil, guanine,

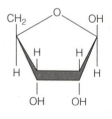

FIGURE 19-12 Ribose.

and cytosine, and four phosphate groups. Assemble the four *ribonucleotides* and draw each in Figure 19-14. (Use the convention illustrated in Figure 19-8 rather than drawing the actual shapes.)

Disassemble the RNA models after completing your drawing.

B. Modeling DNA Replication

DNA **replication** takes place during the S-stage of interphase of the cell cycle (see Exercise 17). Recall that each chromosome is made of a single DNA molecule plus proteins. Before mitosis, the chromosomes duplicate themselves so that the daughter nuclei formed by mitosis will have the same number of chromosomes (and hence the same amount of DNA) as did the parent cell.

Replication begins when hydrogen bonds between nitrogen bases break and the two DNA strands "unzip." Unattached nucleotides within the nucleus bond to the exposed bases, thus creating *two* new strands of DNA (as described below). The process of replication is controlled by enzymes called **DNA polymerases.**

1. Construct eight more deoxyribonucleotides (two of each kind) but don't link them into strands.

2. Now return to the double-stranded DNA segment you constructed earlier. Separate the two strands, imagining the zipperlike fashion in which this occurs within the nucleus.

3. Link the free deoxyribonucleotides to each of the "old" strands. When you are finished, you should have two double-stranded segments.

FIGURE 19-15 Drawing of two replicated DNA segments, illustrating their semiconservative nature.

Note that one strand of each is the parental ("old") strand, and the other is newly synthesized from free nucleotides. This illustrates the *semiconservative* nature of DNA replication. Each of the parent strands remains intact—it is *conserved*—and a new complementary strand is formed on it. Two "half-old, half-new" DNA molecules result.

4. Draw the two replicated DNA molecules in Figure 19-15, labeling the old and new strands. (Again, use the convention shown in Figure 19-8.)

C. Transcription: DNA to RNA

DNA is an "information molecule" residing *within* the nucleus. The information it provides is for assembling proteins *outside* the nucleus within the cytoplasm. The information does not go directly from the DNA to the cytoplasm. Instead, RNA serves as an intermediary, carrying the information from DNA to the cytoplasm.

Synthesis of RNA takes place within the nucleus by **transcription.** During transcription, the DNA double helix unwinds and unzips, and a single strand of RNA, designated **messenger RNA (mRNA)**, is assembled using the nucleotide sequence of *one* of the DNA strands as a pattern (template).

1. Disassemble the replicated DNA strands into their component deoxyribonucleotides.

2. Construct a new DNA strand consisting of nine deoxyribonucleotides. With the bases pointing away from you, lay the strand out horizontally in the following base sequence, from left to right: T-G-C-A-C-C-T-G-C.

3. Now assemble RNA ribonucleotides complementary to the exposed nitrogen bases of the DNA strand. Do not forget to use ribose as the sugar and to substitute uracil instead of thymine. **RNA polymerase** enzymatically links the adjacent ribonucleotides together to form a single strand.

What is the sequence from left to right of nitrogen bases on the mRNA strand?

_____

4. As the mRNA is synthesized within the nucleus, the hydrogen bonds between the nitrogen bases of the deoxyribonucleotides and ribonucleotides break to release the mRNA transcript. Separate your mRNA strand from the DNA strand. (You can disassemble the deoxyribonucleotides now.) Further modification of the mRNA transcript occurs as enzymes snip out non-coding sections, called *introns*, and stitch together the protein-coding sections, called *exons*. Then the mRNA moves out of the nucleus and into the cytoplasm.

Set the mRNA strand aside; you will use it in the next section.

How can the mRNA leave the nucleus? (*Hint:* Re-examine the structure of the nuclear membrane, as described in Exercise 4.)

_____

To *transcribe* means to "make a copy of." Is transcription of RNA from DNA the formation of an *exact* copy? _____
Explain.

_____

_____

D. Translation—RNA to Polypeptides

In the cytoplasm, mRNA strands attach to *ribosomes*, on which translation occurs. To *translate* means to change from one language to another. In the biologic sense, **translation** is the conversion of the linear message encoded on mRNA to a linear strand of amino acids to form a polypeptide. (A *peptide* is two or more amino acids linked by a peptide bond; review Exercise 3.)

Translation is accomplished by the interaction of mRNA, ribosomes, and **transfer RNA (tRNA)**, another type of RNA. The tRNA molecule is formed into a four-cornered loop that acts as a baggage-carrying molecule. Within the cytoplasm, each tRNA attaches to specific amino acids. The amino acid–carrying tRNAs position themselves one at a time on ribosomes, where the amino acids become linked together to form polypeptides.

1. Obtain three tRNA pieces, three amino acid units, and three activating units.

2. Join the amino acids first to the activating units and then to the tRNA. Will a particular tRNA bond with *any* amino acid, or is each tRNA specific?

_____

3. Now let's do some translating. In the space below, list the sequence of bases on the *messenger* RNA strand from part C, starting at the left.

(Left end) _____ (Right end)

Translation occurs when a *three-base sequence* on mRNA is "read" by tRNA. This three-base sequence on mRNA is called a **codon.** Think of a codon as a three-letter word read *right to left*. What is the order of the rightmost (first) mRNA codon? (Remember to list the letters in the *reverse* order of that in the mRNA sequence.)

The first codon on the mRNA model is (right end)_____(left end)

4. Slide the mRNA strand onto the ribosome template sheet with the first codon at the right end.

5. Find the tRNA–amino acid complex that complements (will fit with) the first codon. The complementary three-base sequence on the tRNA is the **anticodon.**

6. Move the correct tRNA–amino acid complex onto the P site on the small ribosome subunit template sheet and fit the codon and anticodon together. In the boxes below, indicate the codon, anticodon, and specific amino acid attached to the tRNA.

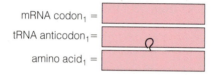

mRNA codon$_1$ =

tRNA anticodon$_1$ =

amino acid$_1$ =

7. Now identify the second mRNA codon and fill in the boxes below.

mRNA codon$_2$ =

tRNA anticodon$_2$ =

amino acid$_2$ =

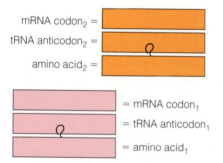

= mRNA codon$_1$

= tRNA anticodon$_1$

= amino acid$_1$

8. Position the second tRNA–amino acid complex onto the A site of the small ribosome subunit.

9. An enzyme catalyzes a condensation reaction between the two adjacent amino acids, forming a **peptide bond** and linking the two amino acids into a dipeptide. (Water, HOH, is released by this condensation reaction.) Separate amino acid$_1$ from its tRNA and link it to amino acid$_2$.

One tRNA–amino acid complex remains. It must occupy the A site of the ribosome in order to bind with its codon. Consequently, the dipeptide must move to the right.

10. Slide the mRNA to the right (so that tRNA2 is on the P site) and fit the third mRNA codon and tRNA anticodon into the A site. Another peptide bond forms, creating a tripeptide model. At about the same time that the second peptide bond is forming, the first

tRNA is released from both the mRNA and the first amino acid. Eventually, that tRNA will pick up another specific amino acid to ferry into position again.

What amino acid will tRNA$_1$ pick up? _____.

Record the amino acid sequence of the tripeptide that you have just modeled. _____

You have created a short polypeptide. Polypeptides may be thousands of amino acids in length. As you see, the amino acid sequence is ultimately determined by DNA, because it was the original source of information.

Finally, let's turn our attention to the concept of a gene. A **gene** is a unit of inheritance. Our current understanding of a gene is that a gene codes for a polypeptide. Given this concept, do you think a gene consists of one, several, or many deoxyribonucleotides?

A gene probably consists of _____ deoxyribonucleotides.

NOTE Please disassemble your models and return them to the proper location.

19.3 Cancer

Certain genes control cell division and lifespan, including when and how often cells divide and when they die. Cancers are fundamentally disorders caused by faulty control mechanisms over the genes that control these activities.

A **tumor,** a tissue mass, results from excess cell divisions. A *benign tumor* is often contained within connective tissue and expands slowly without spreading. It may or may not cause health impact. A *cancerous tumor,* in contrast, grows rapidly, invades surrounding tissue, and spreads (metastasizes) to other body regions. Cancerous tumors result from out-of-control cell division and delayed cell death.

Cancer cells display a number of abnormal properties, including lack of specialization and large nuclei. Additionally, many cancerous cells have abnormal numbers of chromosomes.

In 1951, Henrietta Lacks died from cervical cancer. Before her death, some of the cancerous cells were removed from her body and grown in culture. The cells were allowed to divide repeatedly in this artificial culture environment and resulted in the formation of what is known as a *cell line.* These HeLa cells (the abbreviation coming from *He*nrietta *La*cks) live on today and are used for research on cell and tumor growth. (It's worth noting that neither Ms. Lacks nor her family were consulted or gave informed consent for this action.)

Recall that normal human body cells contain 23 pairs of chromosomes (46 total chromosomes). Human cells of tumor origin (e.g., the HeLa cells) often produce greater chromosome numbers when grown in culture.

HeLa cells containing up to 200 chromosomes have been observed, although 50 to 70 per cell are most frequently found. In this activity, you will make a stained preparation to observe and count the chromosomes of the descendant tumor cells of Henrietta Lacks.

NOTE There is no danger from this activity. The tumor-producing properties are nontransmissible, and the cells have been killed and preserved.

MATERIALS

Per student:

- Clean microscope slide prechilled in cold 40% methanol
- Coverslip
- Disposable dropping pipet
- Paper towels
- Compound microscope (with oil-immersion objective preferable)

Per student pair:

- Metric ruler
- Tube of HeLa cells
- Dropper bottle of dH$_2$O
- Clear nail polish (optional)

Per lab room:

- Coplin staining jars containing stains 1 and 2
- 1-L beaker containing dH$_2$O

PROCEDURE

Refer to Figure 19-16.

1. Place a paper towel on your work surface.

2. Remove a clean glass microscope slide from the cold methanol and lean it against a surface at a 45-degree angle with one short edge resting on the paper towel.

3. Most of the cells are at the bottom of the culture tube. Gently resuspend the cells by inserting your pipet and squeezing the bulb. This expels air from the pipet, which disperses the cells. Remove a small cell sample with the pipet. Holding the pipet about 18 to 36 cm above the slide, allow 8 to 10 drops of the cell suspension to "splat" onto the upper edge of the wet slide and to roll down the slide.

NOTE The slide must be wet when the "splatting" takes place.

4. Allow the cells to *air-dry completely.*

   You will now stain the slide three times with both stains, for *1 second each time.* The time in the stain is critical. Count "one thousand one"; this will be 1 second.

5. Go to the location of the staining solutions and dip the slide in stain 1 for 1 second. Withdraw the slide and then dip it twice more. Drain the slide of the excess stain, blotting the bottom edge of the slide on the paper towel before proceeding.

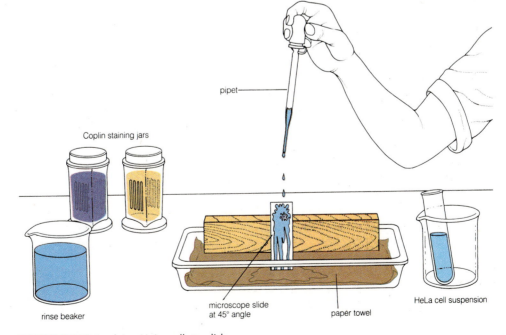

FIGURE 19-16 Applying HeLa cells to slide.

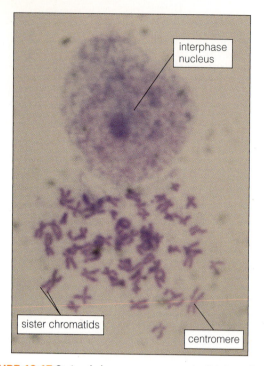

FIGURE 19-17 Stained chromosome spread of HeLa cells. Sister chromatids connected at their centromeres are clearly visible (w.m., 1000×). (Photo by J. W. Perry.)

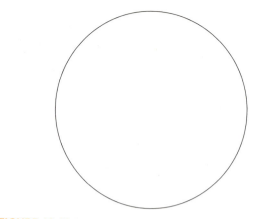

FIGURE 19-18 Drawing of human chromosomes of HeLa cells (___ ×) **Labels:** chromosome, sister chromatid, centromere.

NOTE If your microscope has an oil-immersion objective, proceed to step 11. If not, skip to step 12.

6. Immediately dip the slide in stain 2 for 1 second. Repeat both stains twice more. Drain and blot the excess stain as before.

7. Rinse the slide in dH2O by swishing it gently back and forth in the beaker.

8. Allow the slides to *air dry completely.*

9. Make a wet-mount preparation on the slide in the region of the stained cells. Using the technique illustrated in Figure 2-9, place a coverslip on the slide.

    If desired, you may preserve your slide indefinitely by blotting any excess water from the edges of the coverslip. Then apply a continuous layer of nail polish around the coverslip. Allow it to dry thoroughly.

10. Place the slide on the microscope stage and observe your chromosome spread, focusing first with the medium-power objective and then with the high-dry objective. Locate cells that appear to have burst and have the chromosomes spread out (see Figure 19-17). The number of good spreads will be low, so careful observation of many cells is necessary. Be careful to use only the fine-focus when working at high magnifications.

11. After a good spread has been located, rotate the high-dry objective out of the light path and place a drop of immersion oil on that spot. Rotate the oil-immersion objective into the light path.

    The cells with obvious chromosomes were "captured" while undergoing division (see Exercise 17). Recall that before mitosis and cytokinesis, DNA replication occurs. Each replicated chromosome consists of two identical sister chromatids held together at the centromere.

12. Observe the structure of the chromosomes, identifying sister chromatids and centromeres. Draw several different chromosomes in Figure 19-18, labeling these parts.

13. Select 10 cells in which the chromosomes are visible and count the number of chromosomes you observe in each cell. Record your results in Table 19-2. When you have counted the chromosomes in 10 individual cells, calculate the average number of chromosomes per cell.

19.4 Biotechnology in Forensic Science

We have gained much in understanding the processes of DNA replication, transcription, and translation, as well as the mechanisms by which gene activity is controlled. This knowledge has been translated into a host of new

TABLE 19-2	Chromosome Numbers in HeLa Cells										
Cell Number	1	2	3	4	5	6	7	8	9	10	Average
Number of chromosomes											

technologies intended to provide benefits to humans and that are based on control and manipulation of these crucial cellular processes. **Biotechnology** uses living organisms to perform useful tasks, often involving manipulation of DNA.

Biotechnology methods are used to create *transgenic organisms*, those with useful genes from a different species inserted into their chromosomes; explore possibilities for *gene therapy*, replacing mutated, disease-causing alleles with normal ones; diagnose mutations and diseases; determine paternity; and inform criminal investigations as part of forensic science.

As you have learned, DNA is a very long, linear molecule with information encoded in the sequence of As, Ts, Cs, and Gs. This structure is analogous to that of a videotape, a long, linear series of still images. Just as a videotape can be cut, rearranged, and spliced, even from one videotape to another in a film studio, so can DNA in a lab.

The tools of biotechnology include A, T, C, and G nucleotides and a host of useful enzymes such as DNA polymerase and *restriction enzymes*, enzymes that recognize a specific base sequence in a DNA strand and cut the strand at each of those sites. Specialized equipment and techniques allow biotechnologists to distinguish even very small amounts of DNA from different people.

The human **genome**, all of the DNA in a haploid set of chromosomes, consists of about 3.2 billion nucleotides arranged into about 21,500 genes. More than 99.5% of the human base sequence is the same for every person. However, a small fraction of our DNA is made of short regions of repeated DNA bases; each individual has a unique combination of repeated sections. These unique regions are of great use in matching DNA from crime scene evidence with DNA from suspects.

In this section, you will use a biotechnology kit to perform DNA analyses on simulated criminal investigation samples. Although the samples are simulated, the methods are the same as those used in crime labs. Depending on your kit and equipment, you will perform one or more of the following three tests:

1. **Polymerase chain reaction (PCR)** produces millions of copies of a targeted fragment of DNA present in only trace amounts of a sample, such as from a drop of blood or single hair follicle. PCR provides a relatively simple, quick, and inexpensive initial step in the process of determining the precise genetic identity of an individual from a tiny amount of biological material. As such, it has become a mainstay of forensic analysis.

2. **Restriction digestion of DNA** is accomplished using restriction enzymes that cut DNA at particular 4-6 base pair "recognition sequences." This technique is used to develop a DNA "fingerprint" based on the sizes of the DNA fragments produced by the restriction enzymes and unique to each individual.

3. DNA pieces produced by each of the above techniques are separated based on size by **agarose gel electrophoresis**. The DNA is loaded into small wells in a slab of gel soaked in a conducting fluid. Electrical current passes through the gel, and the DNA pieces are drawn to the opposite side of the gel. The smallest fragments move farther than the larger fragments. The fragments form bands in the gel which can be seen after staining. The banding patterns produced from crime scene evidence are then compared with those of victim and suspects (Figure 19-19).

Follow the procedures provided by your instructor for this activity.

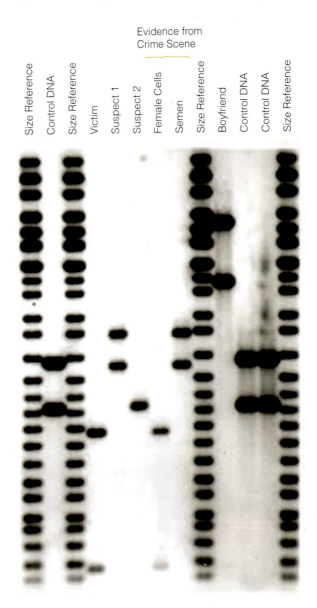

FIGURE 19-19 DNA fingerprints comparing DNA gathered during investigation of a sexual assault. Assay compared DNA from the victim and from the semen of two suspects and the victim's boyfriend. Three control samples were included to confirm that the test procedure was working correctly.

_____ **1.** Nucleic acid extraction begins by breaking down cell membranes with
  (a) cold ethanol
  (b) restriction enzymes
  (c) detergent and salt
  (d) methylene blue

_____ **2.** Deoxyribose is
  (a) a five-carbon sugar
  (b) present in RNA
  (c) a nitrogen-containing base
  (d) one type of nucleotide

_____ **3.** A nucleotide may consist of
  (a) deoxyribose or ribose
  (b) nitrogen-containing base
  (c) phosphate group
  (d) all of the above

_____ **4.** Which of the following is consistent with the principle of base pairing?
  (a) cytosine–cytosine
  (b) cytosine–thymine
  (c) adenine–thymine
  (d) guanine–thymine

_____ **5.** Nitrogen-containing bases between two complementary DNA strands are joined by
  (a) restriction enzymes
  (b) hydrogen bonds
  (c) phosphate groups
  (d) deoxyribose sugars

_____ **6.** The difference between deoxyribose and ribose is that ribose
  (a) is a six-carbon sugar
  (b) bonds only to thymine, not uracil
  (c) has one more oxygen atom than deoxy-ribose has
  (d) is all of the above

_____ **7.** Replication of DNA
  (a) takes place during interphase
  (b) results in two double helices from one
  (c) is semiconservative
  (d) is all of the above

_____ **8.** Transcription of DNA
  (a) results in formation of a strand of RNA
  (b) produces two new strands of DNA
  (c) produces proteins
  (d) is semiconservative

_____ **9.** An anticodon
  (a) is a three-base sequence of nucleotides on tRNA
  (b) is produced by translation of RNA
  (c) has the same base sequence as does the codon
  (d) is the same as a gene

_____ **10.** Polymerase chain reaction (PCR)
  (a) separates pieces of DNA based on their size
  (b) produces DNA fragments from intact strands
  (c) creates millions of copies of a target section of DNA
  (d) creates proteins from the base sequence on a target strand of DNA

EXERCISE 19 DNA, Genes, Cancer, and Biotechnology

POST-LAB QUESTIONS

19.1 Isolation and Identification of Nucleic Acids

1. The photo below shows the result of mixing part of a banana with detergent and salt and then layering ice-cold ethanol on top of the mixture. What is the composition of the material containing all the bubbles?

19.2 Modeling the Structure and Function of Nucleic Acids and Their Products

2. The following diagram represents some of the puzzle pieces used in this section.

   a. Assembled in this form, do they represent a(an) amino acid, base, portion of messenger RNA, *or* deoxyribonucleotide?

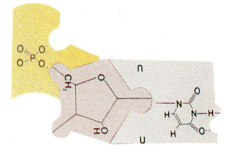

   b. Justify your answer.

3. State the following ratios.

   a. Guanine to cytosine in a double-stranded DNA molecule: _____

   b. Adenine to thymine: _____

4. Define the following terms.

   a. Replication

   b. Transcription

   c. Translation

   d. Codon

   e. Anticodon

5. What does it mean to say that DNA replication is *semiconservative*?

6. a. If the base sequence on one DNA strand is ATGGCCTAG, what is the sequence on the other strand of the helix?

   b. If the original strand serves as the template for transcription, what is the sequence on the newly formed RNA strand?

7. a. What amino acid would be produced if *transcription* takes place from a nucleotide with the three-base sequence ATA? _____

b. A genetic mistake takes place during *replication*, and the new DNA strand has the sequence ATG. What is the three-base sequence on an RNA strand transcribed from this series of nucleotides? _____

c. Which amino acid results from this codon?

19.3 Cancer

8. How does a cancerous tumor differ from a benign tumor?

19.4 Biotechnology in Forensic Science

9. How is a DNA fingerprint used to provide evidence in a murder trial?

Food for Thought

10. Why are DNA fingerprints constructed from only small portions of a genome rather than the entire 3.2 billion nucleotide base pairs?

Evidences of Evolution

OBJECTIVES

After completing this exercise, you will be able to

1. define *evolution, natural selection, population, species, fitness, fossil, hominid, Australopithecus, Homo,* and *Homo sapiens.*

2. explain how natural selection operates to alter the genetic makeup of a population over time.

3. describe the general sequence of evolution of life forms over geologic time.

4. recognize primitive and advanced characteristics of skull structure of human ancestors and relatives.

Introduction

The process by which the incredible diversity of mammals living and extinct species came to exist is called *evolution,* the focus of this exercise. **Evolution,** the process that results in changes in the genetic makeup of populations of organisms through time, is the unifying framework for the whole of biology. Less than 200 years ago, it seemed obvious to most people that living organisms had not changed over time—that oak trees looked like oak trees and humans looked like humans, year after year, generation after generation, without change. As scientists studied the natural world more closely, however, evidence of change and the relatedness of all living organisms emerged from geology and the fossil record, as well as from comparative morphology, developmental patterns, and biochemistry.

Charles Darwin (and, independently, Alfred Russel Wallace) postulated the major mechanism of evolution to be **natural selection,** the difference in survival and reproduction that occurs among individuals of a population that differ in one or more alleles. (Review Exercise 12 for the definition of *allele.*) A **population** is a group of individuals of the same species occupying a given area. A **species** is one or more populations that closely resemble each other, interbreed under natural conditions, and produce fertile offspring.

As we understand natural selection, genetic modifications that place an organism at a disadvantage in its environment will be "selected against," that is, those members of a population will die off, reproduce less, or both. Other individuals may have allele combinations that are advantageous; these are "selected for" and tend to live longer, have more offspring, or both. They have greater **fitness.** Their offspring will receive those advantageous alleles from their parents. As physical, chemical, or biological aspects of the populations' environment change, those modifications that are best adapted to that environment tend to be found more often within the population. The population evolves as some traits become more common and others decrease or disappear over time.

Many people think of evolution in historic terms—as something that produced the dinosaurs but that no longer operates in today's world. Remember, though, that *the process of genetic change in populations over time continues today* and is a dominant force shaping the living organisms of our planet.

The scientific evidence for evolution is overwhelming, although scientists continue to debate the exact mechanisms by which natural selection and other evolutionary agents change allele frequencies. In this exercise, you will track the effects of natural selection in a simulated population, consider the time over which evolution has occurred, and examine the fossil record for evidence of large-scale trends and change among our human ancestors and relatives.

20.1 How Natural Selection Works

In this experiment, you will simulate interacting populations of predators and prey. You will study how predators exert selective pressure on prey populations and vice versa and consider how the environment affects the interaction. The simulated prey are populations of different dry beans, and the simulated predator populations are familiar utensils.

This activity tests two hypotheses: (1) that *prey populations of different forms vary in fitness as a consequence of predation* and (2) *predator populations of different forms vary in fitness as a consequence of prey characteristics.* Each student group will act as one predator species and forage for prey over a short time period. Analysis of changes in prey population numbers will allow us to draw conclusions about the validity of these hypotheses.

MATERIALS

Per student group (4):

- Predator utensils of one form (knife, fork, spoon, forceps, *or* chopsticks)
- Collection of dry beans (50 each of five types)
- Four cups to hold prey
- Calculator

Per lab room:

- Patterned fabric or carpet pieces
- Reservoirs of additional beans of each type
- Stopwatch or timer

PROCEDURE

1. Obtain a fabric or carpet piece to serve as the habitat. Scatter the beans randomly over the habitat.

   What prey species (bean type) do you expect to be most able to survive predation? Why?

   _____

   _____

   Which prey species do you expect to be least successful? Why?

   _____

   _____

   What predator species (utensil type) do you expect to be most successful in capturing prey? Why?

   _____

   _____

   Which predator do you expect to be least successful? Why?

   _____

   _____

2. At the signal from your instructor, students (predators) will collect prey for 30 seconds. Prey may be collected only with your predator utensil and only one at a time. Place each prey bean into your cup and continue foraging until the signal to stop.

3. Record the number of each kind of prey in your cup after this first round foraging session:

   Black bean prey _____

   Garbanzo prey _____

   Lentil prey _____

   Lima bean prey _____

   Navy bean prey _____

   In Table 20-1, in the "first round" column, record the total number of each prey species taken by your whole group. Also record your group's results on the class master data sheets.

4. Calculate the number of each prey species remaining. These surviving prey reproduce so that the total number of each species in the habitat doubles. For example, if 10 garbanzo prey were captured from the habitat, the remaining 40 will reproduce and add 40 new garbanzo prey for a total of 80 in the habitat. Repeat this process for each species.

5. Repeat foraging and reproduction of the survivors for two more generations, entering group data in Table 20-1 and on class master data sheets. Double the number of surviving prey of each species in the habitat after each round of predation.

   My round two prey capture:

   Black bean prey _____

   Garbanzo prey _____

   Lentil prey _____

   Lima bean prey _____

   Navy bean prey _____

TABLE 20-1	Group Totals: Prey Population Changes Caused by Predation by _____ (Predator Species)									
Prey Species	**Black**		**Garbanzo**		**Lentil**		**Lima**		**Navy**		**Total Prey Captured**
	# Captured	# Remaining	# Captured	# Remaining	# Captured	# Remaining	# Captured	# Remaining	# Captured	# Remaining	
1st round											
2nd round											
3rd round											
Total											

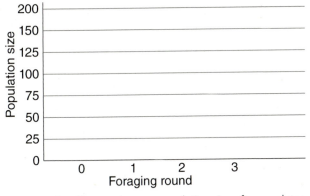

FIGURE 20-1 Changes in prey population size after predator foraging.

My round three prey capture:

Black bean prey _____

Garbanzo prey _____

Lentil prey _____

Lima bean prey _____

Navy bean prey _____

6. Graph the total population size over time for each prey species in Figure 20-1. Be sure to distinguish among the five species by using different symbols or colors for the trend lines for each.

   Do your group's results support the hypothesis that *prey populations of different forms vary in fitness as a consequence of predation*? Write a conclusion accepting or rejecting the hypothesis and explaining your reasoning.

   _____

   _____

Which prey species was most vulnerable to your team's predator species? _____ Which prey species was least vulnerable to your predator? _____ How well do these results agree with your expectations?

   _____

Study the class total results. Did any prey species become extinct? _____ If so, speculate why.

   _____

   _____

Which prey species appears to be most fit in all environments with all predator species? What trait(s) may contribute to that fitness?

   _____

   _____

Which predator species appears to be most fit? Which predator species appears to be least fit? Explain.

   _____

   _____

What effect did the habitat characteristics have on your group's results? If the habitat had been uniformly white, speculate on the effect of this environmental change on the outcome for predator and prey species. _____

   _____

   _____

If you failed to detect any differences in fitness among prey species, speculate on the reason(s) for this failure. Does this failure necessarily mean that natural selection would not occur in these populations?

   _____

   _____

Explain the results of this activity in terms of the effects of natural selection on the genetic makeup of the five prey populations.

   _____

   _____

   _____

20.2 Geologic Time

Earth is an ancient planet that formed from a cloud of dust and gas approximately 4.6 billion years ago. Life formed relatively quickly on the young planet, and the first cells emerged in the seas by 3.8 billion years ago. Initially, the only living organisms on Earth were prokaryotic bacteria and bacteria-like cells. Eventually, though, more complex, eukaryotic organisms evolved. The seas were colonized by single-celled organisms first followed by multicelled and colonial plants and animals (Figure 20-2). Only much later did life move onto the land.

Geologic time, "deep time," is difficult for humans to grasp given our recent evolution and 75-year life spans. In this section, you will construct a geologic timeline to help you gain perspective on the almost incomprehensible sweep of time over which evolution has operated.

MATERIALS

Per student pair:

- One 4.6-m rope or string
- Meter stick or metric ruler (or both)
- Masking tape
- Calculator

PROCEDURE

1. Obtain a 4.6-m length of rope or string. The string represents the entire length of time (4.6 billion years) since the Earth was formed. On this timeline, you will attach masking tape labels to mark the points when the events below occurred. Measure from the starting point with the meter stick or metric ruler.

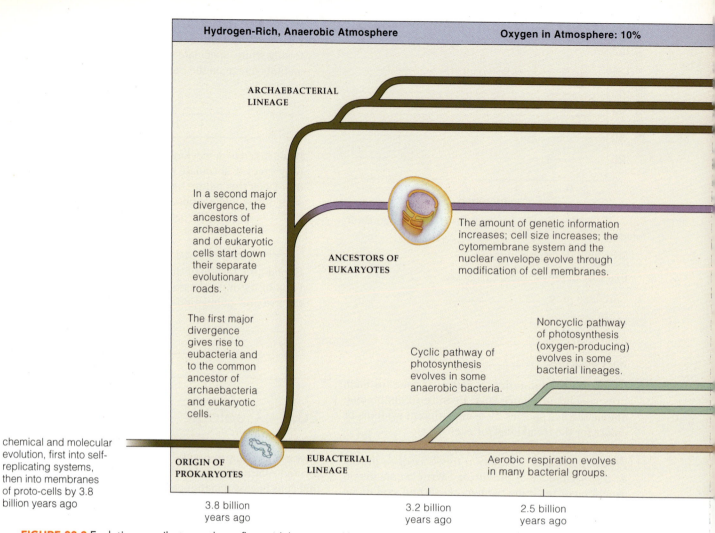

ARCHAEBACTERIAL
LINEAGE

In a second major divergence, the ancestors of archaebacteria and of eukaryotic cells start down their separate evolutionary roads.

ANCESTORS OF
EUKARYOTES

The amount of genetic information increases; cell size increases; the cytomembrane system and the nuclear envelope evolve through modification of cell membranes.

The first major divergence gives rise to eubacteria and to the common ancestor of archaebacteria and eukaryotic cells.

Cyclic pathway of photosynthesis evolves in some anaerobic bacteria.

Noncyclic pathway of photosynthesis (oxygen-producing) evolves in some bacterial lineages.

chemical and molecular evolution, first into self-replicating systems, then into membranes of proto-cells by 3.8 billion years ago

ORIGIN OF
PROKARYOTES

EUBACTERIAL
LINEAGE

Aerobic respiration evolves in many bacterial groups.

3.8 billion
years ago

3.2 billion
years ago

2.5 billion
years ago

FIGURE 20-2 Evolutionary milestones that reflect widely accepted hypotheses about connections among major groups of organisms. (Time line not to scale.)

2. Locate and mark with tape on the timeline the events listed in Table 20-2 (bya = billion years ago, mya = million years ago, ya = years ago):

Over what proportion of the Earth's history were there only single-celled living organisms? _____

Over what proportion of the Earth's history have multicelled organisms existed? _____

Over what proportion of the Earth's history have mammals been a dominant part of the fauna? _____

Over what proportion of the Earth's history have modern humans existed? _____

20.3 The Fossil Record and Human Evolution

The fossil record provides us with compelling evidence of evolution. Most **fossils** are parts of organisms, such as shells, teeth, and bones of animals and stems and seeds of plants. These parts are replaced by minerals to form stone or are surrounded by hardened material that preserves the external form of the organism.

Anthropologists and *archeologists*, scientists who study human origin and cultures, piece together the story of human evolution through the study of fossils, as well as through comparative biochemistry and anatomy. The fossil record of our human lineage is fragmentary and generates much discussion and differing interpretations. However, scientists generally agree that **hominids,** all species on the evolutionary branch leading to modern humans, arose in Africa by about 7 million years ago from the same genetic line that also produced great apes (gorillas, orangutans, and chimpanzees).

Many anatomic changes occurred in the course of evolution from hominid ancestor to modern humans. Arms became shorter, feet flattened and then developed arches, and the big toe moved in line with the other toes. The legs moved more directly under the pelvis. These features allowed upright posture and bipedalism (walking on two legs), and they also allowed the use of the hands for tasks other than locomotion.

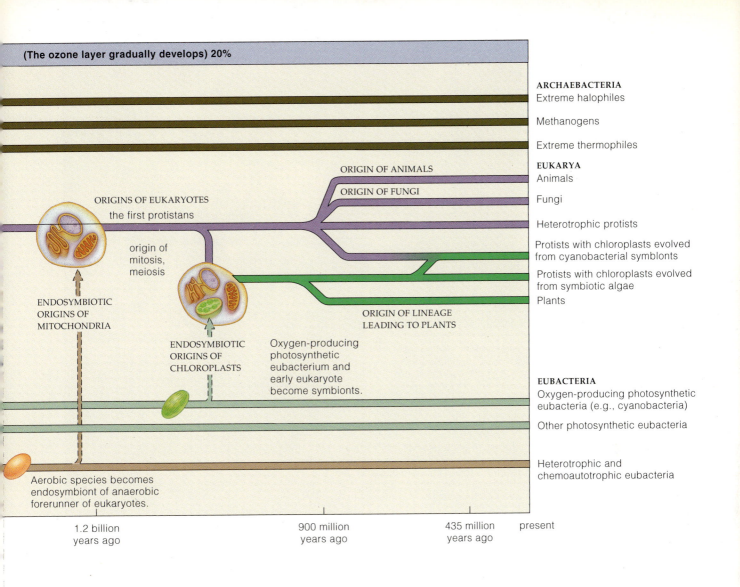

(The ozone layer gradually develops) 20%

ARCHAEBACTERIA
Extreme halophiles

Methanogens

Extreme thermophiles

EUKARYA
Animals

Fungi

ORIGIN OF ANIMALS

ORIGIN OF FUNGI

ORIGINS OF EUKARYOTES
the first protistans

Heterotrophic protists

origin of
mitosis,
meiosis

Protists with chloroplasts evolved
from cyanobacterial symbionts

Protists with chloroplasts evolved
from symbiotic algae

Plants

ENDOSYMBIOTIC
ORIGINS OF
MITOCHONDRIA

ORIGIN OF LINEAGE
LEADING TO PLANTS

ENDOSYMBIOTIC
ORIGINS OF
CHLOROPLASTS

Oxygen-producing
photosynthetic
eubacterium and
early eukaryote
become symbionts.

EUBACTERIA
Oxygen-producing photosynthetic
eubacteria (e.g., cyanobacteria)

Other photosynthetic eubacteria

Heterotrophic and
chemoautotrophic eubacteria

Aerobic species becomes
endosymbiont of anaerobic
forerunner of eukaryotes.

| 1.2 billion years ago | 900 million years ago | 435 million years ago | present |

Still other anatomic changes were associated with the head. In this section, you will observe some of the evolutionary trends associated with skull structure by studying images or reproductions of the skulls of several ancestors and relatives of modern humans.

MATERIALS

Per laboratory room:

- Images or sets of skull reproductions and replicas, including *Australopithecus afarensis*, *Homo habilis*, *Homo erectus*, *Homo sapiens*, and chimpanzee (*Pan troglodytes*)
- Collection of fossil plants and animals (optional)

PROCEDURE

1. If available, examine the fossil collection. Note which fossils lived in the seas and which lived on land.

   (a) Do any of the fossils resemble organisms now living? If so, describe the similarities between fossil and current forms.

   _____

   _____

   (b) Are there any fossils in the collection that are unlike any organisms currently living? If so, describe the features that appear to be most unlike today's forms.

   _____

   _____

2. Examine the hominid skulls in the laboratory or images available online as instructed. Be aware that physical skull specimens are not actual fossils but are plaster or plastic restoration designed partly from fossilized remains and partly from reconstruction. The *Homo sapiens* and chimpanzee skulls are replicas of a modern specimen.

   Identify the scientific name of the organism and approximate date of origin. As you can see, some of these hominids appear to have lived during the same approximate timespan.

TABLE 20-2 Events in Evolution of Life on Earth

Time Frame	Event
4.6 bya	Formation of Earth
3.8 bya	First living prokaryotic cells
2.5 bya	Oxygen-releasing photosynthetic pathway
1.2 bya	Origin of eukaryotic cells
580 mya	Origin of multicellular marine plants and animals
500 mya	First vertebrate animals (marine fishes)
435 mya	First vascular land plants and land animals
345 mya	Great coal forests; amphibians and insects undergo great diversification
240 mya	Mass extinction of nearly all living organisms
225 mya	Origin of mammals, dinosaurs; seed plants dominate
135 mya	Dinosaurs reach peak; flowering plants arise
65 mya	Mass extinction of dinosaurs and most marine organisms
63 mya	Flowering plants, mammals, birds, insects dominate
6 mya	First hominid (upright walking) human ancestors
2 mya	Humans (members of genus *Homo*) using stone tools, forming social life
200,000 ya	Evolution of *Homo sapiens* (modern humans)
6400 ya	Egyptian pyramids constructed
2000 ya	Peak of Roman Empire
A.D. 1776	Signing of the Declaration of Independence
A.D. 1969	Humans step on Earth's moon

Study the skulls and Figure 20-3. In Table 20-3, record your observations of the changes that occurred over time. Rate the following characteristics for each skull using this scale, (– – –) for the most primitive and (+ + +) for the most advanced.

Characteristic	Primitive	Advanced
Teeth	Many large, specialized	Fewer, smaller, less specialized
Jaw	Jaw and muzzle long	Jaw and muzzle short
Cranium (braincase)	Cranium small, forehead receding	Cranium large, prominent vertical forehead
Muscle attachments	Prominent eyebrow ridges and cranial keel (ridge)	Much reduced eyebrow ridges, no keel
Nose	Nose not protruding from face	Prominent nose with distinct bridge

3. The first true hominids found in the fossil record have been assigned to the genus **Australopithecus** (from the Greek *australis* for "southern" and *pithecus* for "ape" because the first specimens were found in southern Africa). Some of the oldest known fossils of *Australopithecus* are of the species *A. afarensis*. The most famous skeleton has been named Lucy. Lucy and others of her species lived approximately 3.6 to 2.9 million years ago.

4. Now turn your attention to the skulls of the *Homo* lineage. There is debate about the lineage of our own genus, *Homo* ("Man"). *Homo habilis*, the early toolmaker, lived in eastern and southern Africa from about 1.9 to 1.6 million years ago. *Homo erectus* ("upright man") also arose in Africa. However, it was *Homo erectus* whose populations left Africa in waves between 2 million and 500,000 years ago. Some still lived in southeast Asia until at least 53,000 years ago. Our own species, *Homo sapiens* ("wise man"), seems to have evolved from *Homo erectus* ancestors by 100,000 years ago.

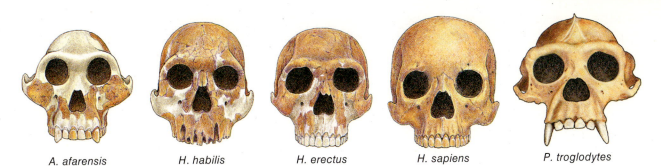

A. afarensis   H. habilis   H. erectus   H. sapiens   P. troglodytes

FIGURE 20-3 Comparison of skull shapes of hominid relatives and ancestors.

(a) What changes from primitive to advanced characteristics do you see in the evolution from *A. afarensis* to *Homo habilis*?

_____

_____

_____

_____

(b) What is the trend in cranial capacity between the ancestral *A. afarensis* and *Homo erectus*?

_____

_____

_____

_____

(c) Between *Homo erectus* and *Homo sapiens*?

_____

_____

_____

_____

6. Finally, study the skulls of the modern human form (*Homo sapiens*) and our nearest living relative, the

chimpanzee (*Pan troglodytes*). Fossil evidence, biochemistry, and genetic analyses indicate that chimpanzees and humans are the most closely related of all living primates. In fact, comparisons of amino acid sequences of proteins and DNA sequences show that chimpanzees and humans share about 99% of their genes! Other evidence shows that the separation from the hominid line of the lineage that led to chimpanzees and other great apes occurred about 6 million years ago.

(a) How does the chimpanzee skull compare with the skull of *Australopithecus afarensis*? What features are most alike? Which are most different?

_____

_____

_____

(b) How does the chimpanzee skull compare with the human skull with respect to primitive and advanced anatomical features?

_____

_____

_____

TABLE 20-3	Skull Characteristics of Human Ancestors and Relatives, from (– – –) = Most Primitive to (+ + +) = Most Advanced						
Genus and species	**Teeth**	**Jaw**	**Structure**	**Cranium**	**Muscle**	**Attachments**	**Nose**
Australopithecus afarensis							
Homo habilis							
Homo erectus							
Homo sapiens							
Pan troglodytes							

_____ 1. Evolution is
   (a) the survival of populations
   (b) the process that results in changes in the genetic makeup of a population over time
   (c) a group of individuals of the same species occupying a given area
   (d) change in an individual's genetic makeup over its lifetime

_____ 2. The major mechanism of evolution is
   (a) species development
   (b) predation
   (c) fitness
   (d) natural selection

_____ 3. An organism whose genetic makeup allows it to produce more offspring than another of its species is said to have greater
   (a) fitness
   (b) evolution
   (c) foraging
   (d) selection

_____ 4. The Earth formed approximately
   (a) 4600 years ago
   (b) 1 million years ago
   (c) 1 billion years ago
   (d) 4.6 billion years ago

_____ 5. The first living organisms appeared on Earth approximately
   (a) 4.6 billion years ago
   (b) 3.8 billion years ago
   (c) 1 billion years ago
   (d) 6400 years ago

_____ 6. Modern humans evolved about
   (a) 200,000 years ago
   (b) 2,000,000 years ago
   (c) 2,000,000,000 years ago
   (d) 4,600,000,000 years ago

_____ 7. Fossils
   (a) are remains of organisms
   (b) are formed when organic materials are replaced with minerals
   (c) provide evidence of evolution
   (d) are all of the above

_____ 8. The first true hominids in the fossil record are
   (a) *Australopithecus*
   (b) *Homo habilis*
   (c) the chimpanzee
   (d) *Homo sapiens*

_____ 9. Hominids with evolutionarily advanced skull characteristics would have
   (a) many large, specialized teeth
   (b) a large cranium with prominent vertical forehead
   (c) prominent eyebrow ridges
   (d) a long jaw and muzzle

_____ 10. The human ancestor whose populations dispersed from Africa to other parts of the world was
   (a) *Homo sapiens*
   (b) *Australopithecus afarensis*
   (c) *Pan troglodytes*
   (d) *Homo erectus*

EXERCISE 20 Evidences of Evolution

POST-LAB QUESTIONS

INTRODUCTION

1. Define the following terms:

a. Evolution

b. Natural selection

20.1 How Natural Selection Works

2. Describe how natural selection could operate to change the genetic makeup of a population of rabbits if a new, faster predator was introduced into their habitat.

20.2 Geologic Time

3. Which event occurred first in Earth's history?

a. Photosynthesis *or* eukaryotic cells?

b. Vertebrate animals *or* flowering plants?

c. Extinction of dinosaurs *or* origin of hominids?

20.3 The Fossil Record and Human Evolution

4. What primitive characteristics are visible in the skull pictured here?

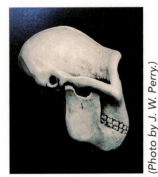

(Photo by J. W. Perry.)

5. Describe anatomical changes in the skull that occurred in human evolution between an *Australopithecus afarensis*–like ancestor and *Homo sapiens*.

6. Compare the slope of the forehead of chimpanzees with that of modern humans.

7. Compare the teeth of *Australopithecus afarensis* and *Homo sapiens*.

a. Describe similarities and differences between the two species.

b. Write a hypothesis about dietary differences between the two species based on their teeth.

Food for Thought

8. Why was the development of bipedalism a major advancement in human evolution?

9. Do you believe humans are still evolving? If not, why not? If yes, explain in what ways humans may be evolving.

10. Humans have selectively bred many radically different domestic animals (e.g., St. Bernard and Chihuahua dog breeds). Does this activity result in evolution? Why or why not?

Human Impact on the Environment

OBJECTIVES

After completing this exercise, you will be able to

1. define *runoff, evapotranspiration, fossil fuels, habitat loss, stream, pool, riffle, abiotic, biotic, watershed, point source pollution, nonpoint source pollution, pH, dissolved oxygen, nitrates, phosphates, chloride, fecal coliform bacteria, macroinvertebrate,* and *larvae.*

2. explain how human alterations of land affect water resources, temperature, atmospheric composition, pollution, and maintenance of species.

3. list and describe major types of water pollutants.

4. describe the effects of organic pollutants and nutrient pollution on stream organisms.

5. explain the importance of dissolved oxygen (DO) in stream water and pollutants that affect DO levels.

6. describe how taxa of aquatic macroinvertebrates present in a stream indicate water quality.

7. indicate the significance of finding fecal coliform bacteria in a water sample.

Introduction

Global warming, extinction of species, contamination with toxic chemicals, and water shortages. This short list of environmental effects linked to our expanding human population is diverse and alarming. Each of these issues is a complex problem with scientific, economic, philosophical, sociologic, and political aspects. No single exercise (or entire course!) can fully address any of these issues. However, this exercise, which focuses on some of the ways humans alter terrestrial (land) habitats and affect water quality, will give you an idea of how scientists build the database they use to assess environmental problems.

21.1 Impact of Land Use Changes

We depend completely on a healthy environment to provide clean air, water, and food; maintain the balance of the atmosphere and climate; decompose wastes; maintain wildlife and other organisms; and provide us with pharmaceuticals, building materials, and much more. Our economies and cultures rely on intact, healthy, functioning environments.

Although every organism impacts the environment, our human numbers and technologies make us the most significant force of landscape alteration as we convert natural landscapes to human uses. In this activity, you will analyze several locales typical in the United States to gauge some ways that human-caused changes affect their abilities to provide essential environmental services.

PROCEDURE

Study Figure 21-1, photos of a (a) suburban shopping area, (b) residential development, and (c) protected natural area, all in east-central Wisconsin. The mall and residential development occupy land that was similar to the natural area before their construction. (Alternatively, your instructor may provide information about three local areas.) As you read the questions and explanations below, fill in Table 21-1. Indicate briefly for each numbered statement how the characteristics of each site affect that feature.

(a)

(b)

(c)

FIGURE 21-1 East-central Wisconsin (a) strip mall, (b) residential development, (c) protected natural area. (*Photos by Joy B. Perry.*)

We rely on adequate fresh water from groundwater and surface sources (rivers and lakes) for irrigation water to grow crops, industrial processes, and personal and home usage.

1. Compacted, impenetrable, or steeply sloped ground surfaces cause most rainwater and snow melt to run off, flowing over the ground surface by gravity toward low points. Most **runoff** eventually reaches streams, carrying with it any pollutants picked up along the way. In contrast, level, porous ground surfaces with many plant roots allow rainfall to soak into the soil and eventually replenish groundwater supplies. To what extent does each site affect runoff, and how?

2. Wetlands and soils with good water infiltration are able to release water to support plant growth during dry periods. Plants are the base of food chains, with all animals (including humans) depending directly or indirectly on plants for their nutrition. To what extent does each site affect water availability during a drought, and how?

The ground surface and vegetation on a site can also modify daily and annual temperature changes.

3. Plants release water from pores in their leaves in a process called **evapotranspiration.** The evaporation creates a local cooling effect. To what extent does each site have a cooling effect caused by evapotranspiration?

4. Dark-colored solid surfaces such as building roofs, roads, and parking lots absorb solar radiation and release that energy as heat. These surfaces have a heating effect on their surroundings in contrast to vegetated or light-colored surfaces. To what extent does each site have a heating effect on the surrounding area?

Fossil fuels (petroleum, natural gas, and coal) provide most of the energy needed for transportation and electricity production in the United States. The energy is released when the fuels are burned; combustion products include carbon dioxide, carbon monoxide, soot, ozone, and other substances that damage the human respiratory system.

In addition, increasing carbon dioxide levels in the atmosphere accelerate global warming. Plants remove carbon dioxide from the atmosphere in the process of photosynthesis; as all organisms undergo cellular respiration and when microbes decompose dead organic matter, the carbon dioxide is returned to the atmosphere.

Additional human contributions to pollution include trash and waste production and disposal, pesticide use, and overfertilization of lawns and agricultural areas.

5. To what extent does each site degrade air quality by causing increased fossil fuel combustion?

6. Is each site a net producer *or* consumer of carbon dioxide?

7. What kinds of pollutants does each area release into the soil, water, or both?

Habitat loss is the single greatest cause for extinction today followed closely by the impact of invasive species. Every species has specific requirements for the range of physical and chemical properties of its habitat, as well as for interaction with many other species. Some species have very narrow requirements; even small environmental disturbances can overstress these organisms and cause their local or global extinction. Conversely, other species are adaptable to a broad range of conditions and have become widely distributed geographically. Some of these species, such as English sparrows, pigeons, Norway rats, German cockroaches, and fire ants, spread aggressively and become invasive, reducing native species.

8. Do you estimate the number of species living on each site to be greater than, similar to, or less than the number present 150 years ago?

9. What changes, if any, do you think there has there been in the types of species present over the past 150 years?

TABLE 21-1 Impact of Land Use Changes			
Feature	Site _____	Site _____	Site _____
Impact on Water Quantity			
1. Runoff during rainfall			
2. Water availability during drought			
Impact on Temperature Fluctuations			
3. Cooling by evapotranspiration			
4. More heat produced than absorbed from sun			
Impact on Pollution and Climate			
5. Increases air pollutants from fossil fuel combustion			
6. Net producer or consumer of carbon dioxide			
7. Pollutants added to soil, and/or water			
Impact on Species Diversity			
8. Number of species greater than, similar to, *or* less than 150 years ago			
9. Changes in predominance of native species from 150 years ago			

Which human activities or types of changes did you observe to have the greatest impact on the environment? Explain.

_____

_____

_____

Which environmental services seem to have been least impacted by commercial and housing development? Which have been most impacted? Explain.

_____

_____

_____

Which of the sites has the greatest ability to provide essential environmental services?

_____

Which of the sites has the least ability to provide essential environmental services?

_____

We will always need energy sources and homes, as well as other resources from the environment. Suggest three changes that would lessen the environmental impact of commercial and residential developments.

1. _____

2. _____

3. _____

21.2 Evaluation of Water Quality of a Stream

A **stream** is a body of running water flowing on the surface of the earth. Streams range from shallow, narrow brooks tumbling down mountainsides to broad, deep rivers such as the Mississippi. Regardless of its size, every stream has distinct major habitat types that influence the types of organisms living there. Whereas relatively slow-moving, deep waters are called **pools, riffles** are shallow areas with fast-moving current bubbling over rocks (Figure 21-2).

Riffles are places where stream water tumbling over rocks and gravel bottom recharges the water with oxygen. Because of the higher dissolved oxygen (DO) levels and the rocks and gravel that provide habitat, most stream inhabitants live in riffles. Pools, on the other hand, are still areas with lower oxygen content with fewer resident species.

Substances enter the stream and either dissolve in the water or are carried along as suspended particles.

(a)

(b)

FIGURE 21-2 Stream habitats. (a) Pool leading into a riffle, (b) Closer look at a riffle. *(Photos by E. F. Benfield.)*

The current thus transports these substances to where they may affect organisms downstream. The current speed, type of bottom structure, temperature, and dissolved substances are the **abiotic** (nonliving) components of the stream ecosystem. The abiotic components influence the life and habits of the **biotic,** or living organisms inhabiting the stream ecosystem.

Each stream and river has a **watershed,** all of the surrounding land area that serves as drainage for the stream, similar to a bowl with the stream flowing across the bottom. Your instructor will provide you with information about the watershed of the stream you will study in this activity.

The quality of a stream depends on its watershed and the human activities that occur there. In general, a stream is of the highest quality when the watershed and streamside vegetation remain in a natural state; most human influences tend to be detrimental to stream quality. These human-induced effects can be lumped under the general term *pollution*. The most common types of water pollutants are

- *Human pathogens*, disease-causing agents, such as bacteria and viruses.

- *Toxic chemicals*, which can harm humans as well as fish and other aquatic life.

- *Organic wastes*, such as sewage, food-processing and paper-making wastes, and animal wastes. Organic wastes can be decomposed by bacteria, but the process uses up oxygen dissolved in the water. The decomposition of large quantities of organic wastes can deplete the water of oxygen, causing fish, insects, and other aquatic oxygen-requiring organisms to die.

- *Plant nutrients*, such as nitrates and phosphates, that can cause too much growth of algae and aquatic plants. When the masses of algae and plants die, the decay process can again deplete the critical DO in the water.

- *Sediments*, particles of soil or other solids suspended in the water. Sediment originates where soil has been exposed to rainfall, most often from construction sites, forestry operations, and agriculture. This is the biggest pollutant by weight, and it can reduce photosynthesis by clouding the water and can carry pesticides and other harmful substances with the soil particles. When sediment settles, it can destroy fish feeding and spawning areas and clog and fill stream channels, lakes, and harbors.

- *Heat*, or thermal pollution. Heated water holds less DO than colder water; if DO levels become too low, stream animals are harmed or killed.

All of these substances enter water as either "point" or "nonpoint" pollution. **Point source pollution** occurs at a specific location where the pollutants are released from a pipe, ditch, or sewer. Examples of point sources include sewage treatment plants, storm drains, and factory discharges. Federal and state governments regulate the type and quantity of some types of pollutants for each point source discharge.

Nonpoint source pollution enters streams from a vast number of sources, with no specific point of entry, contributing a large amount of pollutants. Most nonpoint source pollution results when rainfall and runoff carry pollutants from large, poorly defined areas into a stream. Examples of nonpoint source pollution include the washing of fertilizers and pesticides from lawns and agricultural fields and soil erosion from construction sites.

What human activities can you name that might be affecting your study stream?

_____

_____

A. Streamside Evaluation and Sampling

In this section, you will take several measurements in the water and collect water samples for analysis. Your class will divide into groups, with each group performing different tasks. As a class you will measure the temperature and pH of the water, plus determine several nutrient/pollutant levels. Additionally, you will collect macroinvertebrates for classification and sample the water for the presence of potentially pathogenic (disease-causing) bacteria.

21.2.A.1. Physical and Chemical Sampling

MATERIALS

Per student:

- Disposable plastic gloves (optional)

Per student group (4):

- Nail clippers or scissors
- Plastic beaker
- Plastic sample bottles
- Watch or clock

Per lab room:

- Thermometer
- Portable pH meter or pH indicator paper
- Hach or other DO test kit
- Hach or other nitrate test kit
- Hach or other chloride test kit
- Hach or other phosphate test kit
- dH$_2$O
- Paper towels

PROCEDURE

Perform each following chemical analyses on three sites in the stream, average your readings, and record the average concentration in Table 21-2. Follow the instructions provided with each test kit.

1. **Temperature.** Stream organisms are typically more sensitive to higher temperatures. In addition to affecting the metabolic rate of the cold-blooded aquatic organisms, cooler water holds more DO than warmer water.

   Measure the water temperature by holding a thermometer about 10 cm below the water surface if possible. Keep the thermometer in the water for approximately 1 minute so the temperature can equilibrate. Record the temperature measurements from 3 locations, total them, and calculate the average temperature. Enter the average temperature in Table 21-2.

   Temperature 1: _____; 2: _____;

   3: _____; Average: _____

2. **pH. pH** is a measure of the concentration of hydrogen ions in the water. pH signifies the degree of acidity or alkalinity of water. The pH scale ranges from 0 (most acidic) to 14 (most basic). A solution with a pH of 7 is neutral. A change of only 1 on the scale means a 10-fold change in the hydrogen ion concentration of the water.

   Most aquatic organisms thrive in water of pH 6 to 8. pH that is too low can damage fish gills. Low water pH can also cause some minerals, such as aluminum, to become toxic. The higher the pH, the more carbonates, bicarbonates, and salts are dissolved in the water. Streams with moderate concentrations of these substances are nutritionally rich and support abundant life.

Use the pH meter or indicator paper to determine the pH of your stream water. Try to sample both riffle and pool areas. Determine and record the average pH in Table 46-1.

pH   1: _____ ;   2: _____ ;

3: _____ ;   Average: _____

3. **Dissolved oxygen.** Aquatic animals, similar to most living organisms, require oxygen for their metabolic processes. This oxygen is present in dissolved form in water. If much organic matter (e.g., sewage) is present in water, microorganisms use the DO in the process of decomposing the organic matter. In that case, the level of DO might fall too low to be available for aquatic animals. DO requirements vary by species and physical condition, but many studies suggest that 4 to 5 parts per million (ppm) is the minimum needed to support a diverse fish population. The DO of good fishing waters generally averages about 9 ppm. When DO levels drop below 3 ppm, even pollution-tolerant fish species may die.

DO   1: _____ ;   2: _____ ;

3: _____ ;   Average: _____

4. **Nitrate (NO_3^-).** Nitrogen (N) is a nutrient required for plant life. Nitrogen in the form of **nitrates (NO_3^-)** is a common component of the synthetic fertilizers used on lawns, golf courses, and farm fields to stimulate plant growth. Leaking sewage tanks, municipal wastewater treatment plants, manure from livestock and other animals, and discharges from car exhausts are all sources of stream nitrates. Excessive nitrates in streams can cause overgrowth of algae and other plant life. This plant growth may decrease the light available to other stream life but eventually dies and decays. The decay process can deplete water of the vital DO.

Nitrate   1: _____ ;   2: _____ ;

3: _____ ;   Average: _____

5. **Phosphate (PO_4^-).** Phosphorous (P), anotherz element crucial to plant life, is also a common component of commercial fertilizers in the form of **phosphates (PO_4^-)**. When it rains, varying amounts of phosphates wash from farm fields and other fertilized areas into nearby waterways. Similar to nitrates, phosphate enrichment also causes nutrient problems in streams. Excessive levels indicate pollution by nutrient sources or organic substances (e.g., fertilizers, sewage, detergents, animal wastes).

Phosphate   1: _____ ;   2: _____ ;

3: _____ ;   Average: _____

6. **Chloride (Cl^-). Chloride** is one of the major ions in sewage; it also enters waterways when washed from roadways that have been treated with salt (NaCl, sometimes $CaCl_2$) in the winter to melt ice and snow. Agricultural chemicals are other sources of chloride in water. Water with a high chloride content is toxic to plant life.

Chloride   1: _____ ;   2: _____ ;

3: _____ ;   Average: _____

TABLE 21-2	**Water Quality Values**			
Physical or Chemical Measurement	**Average Value**	**Water Quality Standards**		**Probable Cause of Degradation[a]**
		General[b]	**Trout Stream**	
Temperature		<32°C	<20–23°C	Lack of shade
pH		6.0–8.5	6.5–8.5	Acid drainage
Dissolved oxygen (DO)		>5 mg/L	>6 mg/L	Organic and nutrient sources
Nitrate (NO_3^-)		<2 mg/L	<10 mg/L3	Organic and nutrient sources; fertilizers
Phosphate (PO_4^-)		<0.1 mg/L		Organic and nutrient sources; fertilizers
Chloride (Cl^-)		<30 mg/L		Road salt, sewage, agricultural chemicals
Total coliform bacteria		0 colonies[c]		Sewage, animal wastes

[a] Probable cause(s) of values failing to meet standards.
[b] Standards for swimming and to protect general aquatic life. (*Source:* Wisconsin Department of Natural Resources.)
[c] Standards for drinking water. (*Source:* Wisconsin Department of Natural Resources.)

2. Biological Sampling

MATERIALS

Per student:

- Hip boots or other footwear that can get wet
- Forceps or tweezers

Per student group (4):

- Sterile water sample collection bottles or containers
- Kickseine or D-net sampler
- White enamel pan
- Bottle of alcohol
- Sample collecting jars
- Waterproof markers

Per lab room:

- One or more Coliscan dish plus Easygel bottle
- 1-mL sterile dropping pipet
- Nail clippers

PROCEDURE

1. **Fecal coliform bacteria.** Fecal coliform bacteria are tiny, rod-shaped bacteria that live in the intestines of humans and all warm-blooded animals and pass out of the body in feces. They can cause human disease if they enter a drinking water supply, and they are also associated with other pathogenic bacteria. You will sample the water of your stream to determine whether any fecal coliform bacteria are present.

    a. Use a marker to label the edge of a Coliscan plate with the date and other identifying information.

    b. Collect a water sample from beneath the surface of the stream.

    c. Use a sterile pipet and transfer 1 mL of the water into the Easygel bottle. Be careful not to touch the inside of the bottle or cap with anything to maintain sterility. Swirl the bottle to mix and then pour into the labeled petri dish. Place a lid on the dish and gently swirl the dish until the entire dish is covered with liquid. Be careful not to splash over the side or on the lid.

    d. Place dishes right side up on a level spot, preferably a warm spot, until solid.

    e. Incubate the sample at 35° to 40°C for 24 hours *or* at room temperature for 48 hours.

2. **Aquatic macroinvertebrates. Macroinvertebrates,** animals that are large enough to be seen with the naked eye and that lack a backbone, are an important part of the stream ecosystem. Nearly all of the macroinvertebrates present in the water are immature insects, called **larvae,** which mature and then spend their adult lives on land or in the air. Most aquatic larvae feed on decaying organic matter, but some graze on algae or pursue other animals. Other aquatic macroinvertebrates include mollusks (shelled organisms such as snails and clams), arthropods (relatives of shrimp and lobsters such as crayfish and scuds), and various types of worms.

Most aquatic macroinvertebrates, especially those that live on or in the bottom of riffle areas, are good indicators of water quality. Aquatic macroinvertebrates can be separated into three general groups on the basis of their tolerance to pollution. Some bottom dwellers such as mayfly larvae need a high-quality, unpolluted environment and are placed in group 1 (Figure 21-3). Dominance of group 1 organisms in a stream signifies good water quality. Group 2 organisms can exist in a wide range of water quality conditions and include crayfish and dragonfly larvae. Group 3 organisms are tolerant of pollution and include midge and blackfly larvae and leeches.

By looking at the diversity of macroinvertebrates present, you will be able to get a general idea of the health of a stream. Typically, waters of higher quality support a greater diversity (more kinds) of pollution-intolerant macroinvertebrates. As water becomes more polluted, fewer kinds of organisms thrive. You will collect samples of macroinvertebrates and separate them into general types to determine a stream quality index (SQI) value.

Try to use the kickseine or D-frame net to sample from riffle areas only because these areas usually have a higher quality environment for macroinvertebrates. Sample from pools only if most of the stream is pool area.

Work in groups.

1. Select a riffle area with a depth of 7 to 30 cm and stones that are 5 to 25 cm in diameter or larger if possible. Move into the sample area from the downstream side to minimize disturbance.

2. One student should hold the kickseine or D-frame net at the downstream edge of the riffle. Be sure that the bottom edge of the screen is held tightly against the stream bottom. Do not allow any water to flow over the top of the seine or you might lose specimens.

3. Other group members should walk in front of the net and begin the collection procedure. Pick up all rocks fist sized or larger and rub their surfaces thoroughly in front of the net to dislodge all organisms into the seine. Then disturb the streambed for a distance of 1 m upstream of the net. Stir up the bed with hands and feet until the entire 1-m2 area has been worked over. Kick the streambed with a sideways motion to allow bottom-dwelling organisms to be carried into the net.

4. Remove the net from the water with a forward-scooping motion so that no specimens are washed from the surface. Carry the seine to a flat area of the stream bank for examination.

5. Rinse the organisms and debris collected to the bottom of the screen. Many aquatic macroinvertebrates are very small, so be certain to remove all of them from the screen. Use forceps to remove any that cling to the screen. Place the collection in a collecting jar; drain off as much water as possible and cover the material with alcohol to kill and preserve the organisms. Label each jar to identify the area sampled.

Group 1 Taxa: Pollution-sensitive organisms found in good-quality water.

Stonefly Larvae: Order Plecoptera. 1.25–3.5 cm, 6 legs with hooked tips, antennae, 2 hairlike tails. Smooth (no gills) on lower half of body. (See arrow.)

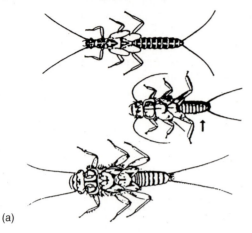

(a)

Water Penny Larva: Order Coleoptera. 0.5 cm, flat saucer-shaped body with a raised bump on one side and 6 tiny legs on the other side. Immature beetle.

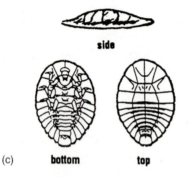

(c)

Mayfly Larvae: Order Ephemeroptera. 0.5–2.5 cm, brown, moving, platelike or feathery gills on sides of lower body (see arrow), 6 large hooked legs, antennae, 2 or 3 long, hairlike tails. Tails may be webbed together.

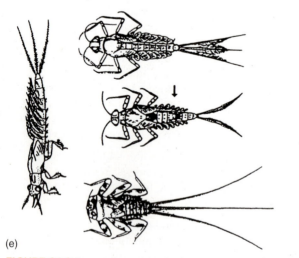

(e)

Caddisfly Larvae: Order Plecoptera. Up to 2.5 cm, 6 hooked legs on upper third of body, 2 hooks at back end. May be in a stick, rock or leaf case with its head sticking out. May have fluffy gill tufts on lower half.

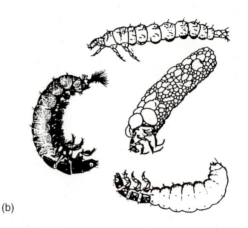

(b)

Dobsonfly (Hellgrammite) Larva: Family Corydalidae. 2–10 cm, dark-colored, 6 legs, large pinching jaws, eight pairs feelers on lower half of body with paired cottonlike gill tufts along underside, short antennae, 2 tails and 2 pairs of hooks at back end.

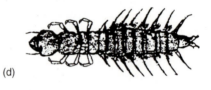

(d)

Gilled Snail: Class Gastropoda. Shell opening covered by thin plate called operculum. Shell usually opens on right.

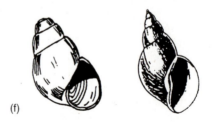

(f)

Riffle Beetle: Order Coleoptera. 0.5 cm, oval body covered with tiny hairs, 6 legs, antennae. Walks slowly underwater. Does not swim on surface.

(g)

FIGURE 21-3 Stream insects and other macroinvertebrates. *(Continued)*

Group 2 Taxa: Somewhat pollution-tolerant organisms can be in good- or fair-quality water.

Crayfish: Order Decapoda. Up to 15 cm, 2 large claws, 8 legs, resembles small lobster.

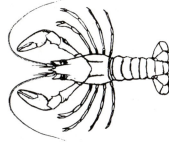

(h)

Scud: Order Amphipoda. 0.5 cm, white to gray, body higher than it is wide, swims sideways, more than 6 legs, resembles small shrimp.

(j)

Fishfly Larva: Family Corydalidae. Up to 3.5 cm long. Looks like small hellgrammite but often a lighter reddishtan color, or with yellowish streaks. No gill tufs underneath.

(k)

Damselfly Larvae: Suborder Zygoptera. 1.25–2.5 cm, large eyes, 6 thin hooked legs, 3 broad oar-shaped tails, positioned like a tripod. Smooth (no gills) on sides of lower half of body. (See arrow.)

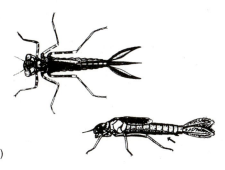

(n)

Beetle Larva: Order Coleoptera. 0.5–2.5 cm, light colored, 6 legs on upper half of body, feelers, antennae.

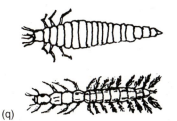

(q)

Sowbug: Order Isopoda. 0.5–2 cm gray oblong body wider than it is high, more than 6 legs, long antennae.

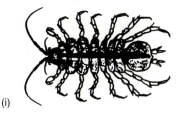

(i)

Alderfly Larva: Family Sialidae. 2.5 cm long. Looks like small hellgrammite but has one long, thin, branched tail at back end (no hooks). No gill tufts underneath.

(l)

Clam. Class Bivalvia.

(m)

Watersnipe Fly Larva: Family Athericidae (Atherix). 0.5–2.5 cm, pale to green, tapered body, many caterpillarlike legs, conical head, feathery "horns" at back end.

(o)

Crane Fly Larva: Suborder Nematocera. 1.25–5 cm, milky, green, or light brown, plump caterpillarlike segmented body, 4 fingerlike lobes at back end.

(p)

Dragon Fly Larvae: Suborder Anisoptera. 1.25–5 cm, large eyes, 6 hooked legs. Wide oval to round abdomen.

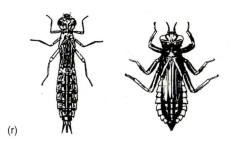

(r)

FIGURE 21-3 *Continued.*

Group 3 Taxa: Pollution-tollerant organisms can be in any quality of water.

Aquatic Worm: Class Oligochaeta. 0.5–5 cm, can be very tiny; thin wormlike body.

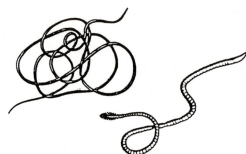

(s)

Leech: Order Hirudinea. 0.5–5 cm, brown slimy body, ends with suction pads.

(v)

Pouch Snail and Pond Snails: Class Gastropoda. No operculum. Breath air. Shell usually opens on left.

(w)

Midge Fly Larva: Suborder Nematocera. Up to 1 cm, dark head, wormlike segmented body, 2 tiny legs on each side.

(t)

Blackfly Larva: Family Simulidae. Up to 0.75 cm, one end of body wider. Black head, suction pad on end.

(u)

Other Snails: Class Gastropoda. No operculum. Breathe air. Snail shell coils in one plane.

(x)

FIGURE 21-3 *Continued.*

21.2.B. Laboratory Analysis— Evaluating Biological Tests and Samples

1. **Fecal Coliform Bacteria.** Inspect the dishes and count the *purple* colonies (*not* light blue, blue-green, white, or pink). Multiply that number by 100 to determine the number of fecal coliform bacterial colonies per 100 mL of water. Record your results in Table 21-2.

   Fecal coliform bacteria in excess of 200 colonies per 100 mL (2 colonies per plate) indicate contamination of water with animal or human wastes and are unsafe for human contact.

2. **Macroinvertebrates.** Identification of aquatic organisms can be difficult. However, biological sampling results can be rapidly translated into a water quality assessment without actually identifying each insect specimen but by determining the number of kinds of certain collected *macroinvertebrates.*

MATERIALS

Per student group:

- Petri dishes
- Forceps or tweezers
- White paper *or* white enamel pan
- Dissection microscope *or* magnifying glass

PROCEDURE

1. Pour a portion of a macroinvertebrate sample into a petri dish. Place a piece of white paper onto the stage of a dissecting microscope to provide an opaque white background and then observe the specimens in the petri dish. Alternatively, you might pour a sample into a white enamel pan and observe the specimens with a magnifying glass.

2. Use forceps or tweezers to sort all the organisms into "look-alike" groups, or taxa (Figure 21-3). Look primarily at body shape and number of legs and "tails" to aid in identification. Note the size range of the

pictured taxa. Record each taxonomic group present in your sample under the appropriate group number in Table 21-3 (group 1 = pollution intolerant; group 2 = somewhat pollution tolerant; group 3 = pollution tolerant). *For purposes of this lab, disregard organisms that are not pictured in Figure 21-3.*

3. Determine the SQI value, a measure of stream ecosystem health, as follows:

   a. The organisms in each group are assigned a group value as follows:

   Group 1 = group value 3

   Group 2 = group value 2

   Group 3 = group value 1

   b. Count the number of taxa (*types of organisms*) in each group and multiply that number by the group value. Record the resulting group index value in Table 21-3.

   Example:

Group 1 Taxa	Group 2 Taxa	Group 3 Taxa
Caddisfly	Damselfly	Blackfly
Mayfly	Crane Fly	Midge
Stonefly	Clam	
3 taxa × group value 3 = 9	3 taxa × group value 2 = 6	2 taxa × group value 1 = 2

   c. Add the respective group index values of the three groups to find the cumulative SQI value; record your results in Table 21-3. (In the preceding example, 9 + 6 + 2 = 17.)

Table 21-4 provides guidance in interpreting the SQI value.

TABLE 21-4

Stream Quality Index Value	Water Quality
≥ 23	Excellent
17–22	Good
11–16	Fair
≤10	Poor

In addition to indicating general water quality, the aquatic macroinvertebrate survey can provide you with information regarding the types of pollutants that may be affecting the stream. If you found few taxa of pollution-tolerant invertebrates but with great numbers of individuals of these taxa, the stream water may be overly enriched with organic matter or have severe organic pollution. Conversely, if you found relatively few taxa and only a few individuals of each kind (or if you found *no* macroinvertebrates but the stream appears clean), you might suspect the presence of toxic chemicals. A diversity of taxa, with an abundance of individuals for many of the taxa, indicates a healthy stream with good to excellent water quality.

21.3   Analysis of Results

Table 23-2 lists acceptable levels of a few chemicals in water to protect aquatic life, as well as the probable causes of test values that fail to meet those standards. Compare your test results with those values. Also consider your findings from biological observations.

Are your observations, chemical measurements, and biological observations consistent? (Refer to Tables 21-2 and 21-3.) That is, do they *all* indicate that your stream is healthy, with good water quality? Or conversely, do they *all* point to a stream in trouble, with poor water quality? Or do your results conflict, with some indicators of good quality and some indicators of poor water quality?

_____

_____

If your results aren't clear-cut, speculate on why this might be so.

_____

_____

What might you do to increase the consistency and reliability of your results?

_____

_____

TABLE 21-3	Taxa Present in Stream Samples	
Group 1 (Pollution Intolerant)	Group 2 (Somewhat Pollution Tolerant)	Group 3 (Pollution Tolerant)
No. taxa _____ × Group value 3 = Group index value	No. taxa _____ × Group value 2 = Group index value	No. taxa _____ × Group value 1 = Group index value
_____	_____	_____
Stream quality index = _____ + _____ + _____ = _____		

Which of your observed or measured factors indicate(s) higher quality water?

_____

_____

Which of your observed or measured factors indicate(s) lower water quality?

_____

_____

Why might you find a low SQI value even if the chemical and physical parameters you measured are within limits for "good" water quality?

_____

_____

If you found indications of organic or nutrient pollution, what might be some of the sources for this pollution in your stream's watershed?

_____

_____

Do you think most of the pollutants in your stream result from point source or nonpoint source pollutants?

_____

_____

Briefly describe three ways that the negative impacts on your stream could be reduced or eliminated.

_____

_____

_____

_____

_____

_____

_____ 1. Runoff of rainwater is minimized on
   (a) hard surfaces such as parking lots
   (b) compacted ground
   (c) level, spongy ground
   (d) steep slopes

_____ 2. Dark-colored solid surfaces
   (a) have a heating effect on their surroundings
   (b) have a cooling effect on their surroundings
   (c) release carbon dioxide
   (d) increase evapotranspiration

_____ 3. The greatest cause of extinction of species is
   (a) runoff
   (b) habitat loss
   (c) invasive species
   (d) evapotranspiration

_____ 4. Which of these is *not* a type of water pollutant?
   (a) dissolved oxygen
   (b) toxic chemicals
   (c) organic wastes
   (d) sediment

_____ 5. pH is
   (a) a measure of phosphorous concentration
   (b) a measure of hydrogen ion concentration
   (c) measured by counting the number of different kinds of aquatic insects
   (d) measured on a scale from 1 to 100

_____ 6. Dissolved oxygen content in stream water is affected by
   (a) organic matter in the water
   (b) overgrowth of algae
   (c) water nutrient levels
   (d) all of the above

_____ 7. Which pollutant would most likely be the source of coliform bacteria in a stream?
   (a) fertilizers from a golf course
   (b) discharge from a factory
   (c) forestry activities
   (d) sewage or animal wastes

_____ 8. Most of the aquatic macroinvertebrates in a stream are
   (a) clams
   (b) insect larvae
   (c) worms
   (d) arthropods such as crayfish

_____ 9. Group 1 aquatic macroinvertebrates
   (a) require high-quality, unpolluted water
   (b) require polluted water
   (c) are tolerant of a wide range of water quality conditions
   (d) include leeches

_____ 10. nonpoint source pollutants include
   (a) fertilizers from lawns
   (b) pesticides from farm fields
   (c) soil erosion from construction sites
   (d) all of the above

EXERCISE **21** **Human Impact on the Environment**

POST-LAB QUESTIONS

21.1 Impact of Land Use Changes

1. A wooded area is converted to a housing development and strip mall. Predict and briefly describe three ways this change in land use could affect atmospheric carbon dioxide levels.

2. How can plants reduce the temperature of their surroundings? Would this effect be as great during winter in northern temperature regions as during the summer?

3. Describe how changing the ways that land surfaces are used by humans causes changes in water resources.

21.2 Evaluation of Water Quality of a Stream

4. Why is it important for sewage treatment plants to remove nitrates and phosphates from sewage before discharging it into a river?

5. Trout are fish that require relatively low-temperature water. Suppose a shaded trout stream has its overhanging streamside trees removed. Explain the possible effects of this action on the trout population over the course of a summer.

21.2.A.1. Physical and Chemical Sampling

6. Below are the results of physical and chemical sampling of a stream. What do these results tell you about the suitability of this stream as fish habitat? What are likely causes for each of these values?

Parameter	Measurement
pH	5.2
Phosphate	0.15 mg/L
DO	3 mg/L

21.2.B. Laboratory Analysis—Evaluating Biological Tests and Samples

7. Below are the data from one stream's aquatic macroinvertebrate sample. Calculate the SQI value of this stream. What does this sample indicate about the water quality?

Taxa Present		
Group 1 (Pollution Intolerant)	**Group 2 (Somewhat Pollution Tolerant)**	**Group 3 (Pollution Tolerant)**
Caddisfly larvae	Crane fly larvae	Midge fly larvae
Mayfly larvae	Dragonfly larvae	Blackfly larvae
Riffle beetles	Crayfish	
Dobsonfly larvae		

8. Referring to the stream in question 7, suppose you also know there are very few individuals of the group 1 taxa present, but midge fly larvae are extremely abundant. Does this information alter your assessment of the water quality of the stream? Why or why not?

10. A stream flows through marshes, farmland, and suburban housing developments before emptying into the ocean. Describe three ways that the water quality of the stream is affected by the different land areas along its course.

FOOD FOR THOUGHT

9. Describe two ways that stream water quality that is healthy for trout differs from unhealthy water quality.

Measurement Conversions

Metric to American Standard	American Standard to Metric
Length	
1 mm = 0.039 inch	1 inch = 2.54 cm
1 cm = 0.394 inch	1 foot = 0.305 m
1 m = 3.28 feet	1 yard = 0.914 m
1 m = 1.09 yards	1 mile = 1.61 km
1 km = 0.622 mile	
Volume	
1 mL = 0.0338 fl oz	1 fluid ounce = 29.6 mL
1 L = 4.23 cups	1 cup = 237 mL
1 L = 2.11 pints	1 pint = 0.474 L
1 L = 1.06 qt	1 quart = 0.947 L
1 L = 0.264 gal	1 gallon = 3.79 L
Mass	
1 mg = 0.0000353 oz	1 ounce = 28.3 g
1 gm = 0.0353 oz	1 pound = 0.454 kg
1 kg = 2.21 lb	

Terms of Orientation in and Around the Animal Body

A. Body Shapes

1. **Symmetry.** The body can be divided into almost identical halves.

2. **Asymmetry.** The body cannot be divided into almost identical halves (e.g., many sponges).

3. **Radial symmetry.** The body is shaped like a cylinder (e.g., sea anemones) or a wheel (e.g., sea stars).

4. **Bilateral symmetry.** The body is shaped like ours in that it can be divided into halves by only one symmetrical plane (midsagittal).

B. Directions in the Body

1. **Dorsal.** At or toward the back surface of the body.

2. **Ventral.** At or toward the belly surface of the body.

3. **Anterior.** At or toward the head of the body—ventral surface of humans.

4. **Posterior.** At or toward the tail or rear end of the body—dorsal surface of humans.

5. **Medial.** At or near the midline of a body. The prefix *mid-* is often used in combination with other terms (e.g., *midventral*).

6. **Lateral.** Away from the midline of a body.

7. **Superior.** Over or placed above some point of reference—toward the head of humans.

8. **Inferior.** Under or placed below some point of reference—away from the head of humans.

9. **Proximal.** Close to some point of reference or close to a point of attachment of an appendage to the trunk of the body.

10. **Distal.** Away from some point of reference or away from a point of attachment of an appendage to the trunk of the body.

11. **Longitudinal.** Parallel to the midline of a body.

12. **Axis.** An imaginary line around which a body or structure can rotate. The midline of longitudinal axis is the central axis of a symmetrical body or structure.

13. **Axial.** Placed at or along an axis.

14. **Radial.** Arranged symmetrically around an axis similar to the spokes of a wheel.

C. Planes of the Body

1. **Sagittal.** Passes vertically to the ground and divides the body into right and left sides. The *midsagittal* or *median plane* passes through the longitudinal axis and divides the body into right and left halves.

2. **Frontal.** Passes at right angles to the sagittal plane and divides the body into dorsal and ventral parts.

3. **Transverse.** Passes from side to side at right angles to both the sagittal and frontal planes and divides the body into anterior and posterior parts—superior and inferior parts of humans. This plane of section is often referred to as a cross-section.